Methods of
Experimental Physics

VOLUME 2

ELECTRONIC METHODS

SECOND EDITION

PART B

METHODS OF
EXPERIMENTAL PHYSICS:

L. Marton, *Editor-in-Chief*

Claire Marton, *Assistant Editor*

Volume 2

Electronic Methods

Second Edition

PART B

Edited by

E. BLEULER

Department of Physics
The Pennsylvania State University
University Park, Pennsylvania

R. O. HAXBY

Department of Physics
Iowa State University
Ames, Iowa

1975

ACADEMIC PRESS · New York San Francisco London

A Subsidiary of Harcourt Brace Jovanovich, Publishers

ACADEMIC PRESS, INC.
111 Fifth Avenue, New York, New York 10003

United Kingdom Edition published by
ACADEMIC PRESS, INC. (LONDON) LTD.
24/28 Oval Road, London NW1

Library of Congress Cataloging in Publication Data

Bleuler, Ernst, (date)   ed.
    Electronic methods.

    (Methods of experimental physics, v. 2)
    Includes bibliographical references.
    1.   Electronic measurements.   2.   Physics—Research.
I.   Haxby, Robert Ozias, (date)   joint ed.   II.   Title.
III.   Series.
QC535.B63 1975   621.381   74-13656
ISBN 0-12-476042-2 (pt.B)

CONTENTS OF VOLUME 2, PART B

CONTRIBUTORS TO VOLUME 2, PART B

Numbers in parentheses indicate the pages on which the authors' contributions begin.

RICHARD BARNES, *Department of Physics, Iowa State University, Ames, Iowa* (133)

E. BLEULER, *Department of Physics, The Pennsylvania State University, University Park, Pennsylvania* (1)

E. R. CHENETTE, *Department of Electrical Engineering, University of Florida, Gainesville, Florida* (461)

MARVIN CHODOROW, *Microwave Laboratory, Stanford University, Stanford, California* (264)

RALPH W. ENGSTROM, *RCA Corporation, Lancaster, Pennsylvania* (301)

F. J. FRIEDLAENDER, *School of Electrical Engineering, Purdue University, West Lafayette, Indiana* (359)

EDWIN A. GOLDBERG, *RCA Astro Electronics Division, Princeton, New Jersey* (15)

PAUL D. HUSTON,* *RCA Electronic Components, Lancaster, Pennsylvania* (345)

FERDO IVANEK, *Fairchild Microwave and Optoelectronics Division, Mountain View, California* (264)

DANIEL MAEDER, *Nuclear and Corpuscular Physics Department, Ecole de Physique, Université de Genève, Geneva, Switzerland* (77, 170)

MYRON H. NICHOLS, *Consultant, Del Mar, California* (220)

JAMES K. ROBERGE, *Department of Electrical Engineering, Massachusetts Institute of Technology, Cambridge, Massachusetts* (375)

* Deceased.

PAUL E. RUSSELL, *College of Engineering Sciences, Arizona State University, Tempe, Arizona* (152)

ROBERT P. STONE, *Design Laboratory, RCA Corporation, Lancaster, Pennsylvania* (335, 336, 355)

J. E. TOFFLER, *Hughes Aircraft Company, Fullerton, California* (38)

R. S. TURGEL, *Electricity Division, IBS, National Bureau of Standards, Washington, D.C.* (48)

K. M. VAN VLIET, *Centre de recherches mathématiques, Université de Montréal, Montreal, Canada* (461)

ROBERT D. WANSELOW, *Radio Frequency Laboratory, TRW SYSTEMS Group, Redondo Beach, California* (251, 283, 287)

CHARLES W. WILLIAMS, *ORTEC, Oak Ridge, Tennessee* (29)

P. N. WINTERS, *Hughes Aircraft Company, Fullerton, California* (38)

FOREWORD

When the first edition of this volume was prepared somewhat over ten years ago, both vacuum-tube and solid-state electronic devices were in general use, and many advanced concepts were based on vacuum-tube circuitry. We have since seen a rapid shift from vacuum tubes to solid-state devices, and to a large degree from the use of discrete components to integrated circuits. A comprehensive revision of the volume became imperative and only a few sections remain unchanged.

Editorship of the first edition was carried out by one of us (Ernst Bleuler) jointly with Professor R. O. Haxby. Midway in the preparation of this revised edition, we suffered a grievous loss with the passing away of Professor Haxby. We not only lost an outstanding physicist but also a very good friend. This volume stands as a memorial to him and his achievements.

Professor Haxby's unexpected death put a much greater burden on E.B. The second signatory (L.M.) of this foreword wishes to use this opportunity to express his appreciation for all the effort and skill displayed in the preparation of this revised edition. Both of us would like to thank the contributors to this volume for their collaboration.

ERNST BLEULER

L. MARTON

CONTENTS OF VOLUME 2, PART A

CONTRIBUTORS TO VOLUME 2, PART A

ALLAN I. BENNETT, *Westinghouse Research Laboratory, Pittsburgh, Pennsylvania*

E. F. BUCKLEY, *Emerson and Cuming, Inc., Canton, Massachusetts*

EDWARD J. CRAIG, *Department of Electrical Engineering, Union College, Schenectady, New York*

ROBERT P. FEATHERSTONE, *Central Engineering Company, Minneapolis, Minnesota*

WILLIAM J. KEARNS, *Duplicon Corporation, Irvine, California*

I. A. LESK,* *Motorola Semiconductor Products Division, Phoenix, Arizona*

R. J. McFADYEN, *Electronics Laboratory, General Electric Company, Syracuse, New York*

R. M. SCARLETT, *Vicom Division, Vidar Corporation, Mountain View, California*

F. H. SCHLERETH, *Electronics Laboratory, General Electric Company, Syracuse, New York*

T. A. SMAY, *Department of Electrical Engineering, Iowa State University, Ames, Iowa*

* Present address: New Ventures, Motorola, Inc., Scottsdale, Arizona.

9. MEASUREMENTS

9.1. Counting*

9.1.1. Introduction

A system for counting periodic or random events can be separated into the following parts:

(1) A transducer, with associated amplifier, transforms the physical event into an analog signal with an amplitude of the order of 1 to 10 V. This component will not be discussed in this chapter.

(2) A shaping circuit generates a logic signal for each analog signal that meets some selected criteria.

(3) A memory with associated timing control stores the logic signals. Although the distinction is somewhat blurred, one may discern two types of measurements: (a) The quantity of interest is the total number of counts accumulated under certain conditions, e.g., from one of a number of samples; in this case, accumulation typically proceeds for a sizeable interval of time, say, in excess of 5 sec. (b) The quantity of interest is the near-instantaneous rate of events; accumulation in counting-rate meters typically lasts less than a few seconds and the memory is either reset or made to forget earlier information gradually. Digital or analog storage is possible for either measurement class, though an analog memory is rarely used for total-count accumulation.

(4) The result of the accumulation must be displayed and/or recorded.

Counting systems are characterized by a pulse-pair resolution (e.g., 10 nsec) or by a maximum frequency (e.g., 100 MHz). The measurement of periodic events (frequency determination) is described in Section 9.2.3.3 which also discusses the timing control. In that case, the counter will give the correct information if the event rate is below its maximum frequency and will give no or erratic results for higher frequencies. For

---

* Chapter 9.1 is by E. Bleuler.

1

random events, on the other hand, a counter with a finite pulse-pair resolution will always lose events ("dead-time losses," see Section 9.1.6).

9.1.2. Shaping Circuit

In most counters, a Schmitt trigger is used to produce a square wave from the analog signal (see Section 8.4.1, Fig. 1). The trigger level and hysteresis chosen will depend on the amplitude and shape ("wiggliness") of the input signal. The width of the square wave generated varies with the signal. In some cases, the Schmitt trigger is replaced or followed by a monostable multivibrator (Section 8.4.2) in order to produce a square wave of fixed width and area. This is required in some types of simple counting-rate meters; also, if the width of the one-shot is larger than the possible paralysis times of the transducer, it provides an approximation to a fixed dead time for the system.

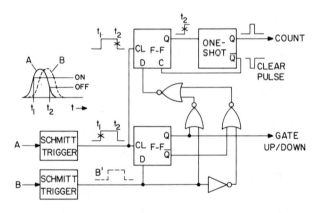

Fig. 1. Gating scheme for up/down counting. A is the counting input, B the control input. Invalid counts are prevented by requiring that B' at time t_2 must be high (low) if it was low (high) at time t_1. The flip-flops and the one-shot must be edge triggered as shown by the asterisks.

If one wants to count interference fringes or moiré lines, as when digitizing coordinates, one must be able to distinguish between the two directions of motion. Two transducers are needed, with the time sequence of their signals indicating whether the event should be added to the memory or subtracted from it. Two Schmitt triggers then produce square waves from which a gate signal (e.g., 1 for "up," 0 for "down") and a clock pulse for the counter must be derived. Figure 1 shows a possible scheme.

9.1.3. Digital Counters[1-4]

The overwhelming majority of digital counters are based on cascaded flip-flops similar to the T and JK flip-flops of Section 8.4.3. They have practically eliminated earlier devices such as the glow-transfer dekatron[5,6] and various electron-beam switching tubes.[5,7,8] There is no need, however, to construct scalers from separate flip-flops, since a large variety of IC scalers, decoders, and display drivers are available. In assembling a counter system, the decision on the type of logic to be used may depend on speed requirements, cost, existing equipment to which the counter may be interfaced, and probably personal preference. In the following sections, a few basic circuits are treated, with positive (TTL) logic as in Section 8.4.3. JK flip-flops are shown for illustration, though they may not be implemented completely in any given IC package.

9.1.3.1. Binary Counter. Figure 2 shows a 4-bit "ripple counter" in which the input square wave is applied to the clock input of the first flip-flop, whereas the clock of each later binary is connected to the Q output of the preceding stage. It is assumed that the binaries change state at the trailing (dropping) edge of the clock pulse. Before starting to count, the binaries are cleared ($Q = 0$) by applying a Clear $= 0$ pulse (not shown). The four output lines A, B, C, and D give the accumulated number in binary code; e.g., after the 11th pulse, ABCD $= 1101$.

9.1.3.2. Decade Scaler. Unless the counter is interfaced directly with a computer, the information from a decade package is vastly preferable to that from the scale of sixteen of Fig. 2. It is thus necessary to add six counts before all four bits are reset to $Q = 0$. This is achieved by adding the 3-input NAND gate. For the output of the gate to go to zero, it is necessary that there be a 1 at A, D, and the input. The waveforms

[1] R. L. Morris and J. R. Miller, "Designing with TTL Integrated Circuits," Chapter 10. McGraw-Hill, New York, 1971.

[2] D. E. Lancaster, "RTL Cookbook," Chapters 6, 7. Howard Sams, Indianapolis, Indiana, 1969.

[3] H. V. Malmstadt and C. G. Enke, "Digital Electronics for Scientists," Chapter 6. Benjamin, New York, 1969.

[4] E. Kowalski, "Nuclear Electronics," pp. 299–308. Springer, New York, 1970.

[5] J. Millman and H. Taub, "Pulse, Digital, and Switching Waveforms," pp. 698–706. McGraw-Hill, New York, 1965.

[6] K. F. Bacon, Circuit for a reversible dekatron counter, *Electron. Eng.* **31**, 172 (1959).

[7] S. Kuchinsky, *IRE Nat. Conv. Rec. Part 6*, 43 (1953).

[8] H. F. Stoddart, *Nucleonics* **17**, No. 6, 78 (1959).

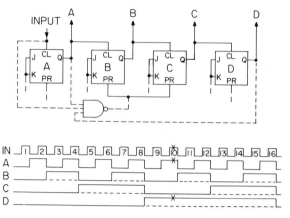

FIG. 2. Binary and decade counter. If the dashed connections are omitted and the preset inputs of B and C are kept at 1, the four flip-flops form a scale of sixteen with the waveforms shown. If the NAND gate is added, B and C are reset to 1 at the time indicated by asterisks, and the counter resets to zero at the trailing edge of input 10 [see R. L. Morris and J. R. Miller, "Designing with TTL Integrated Circuits," Figs. 10.1 and 10.5. McGraw-Hill, New York, 1971].

show that this will be the case when the tenth input pulse goes to 1. At that moment, then, B and C are preset to 1; at the trailing edge of the input, all binaries flip to $A = B = C = D = 0$. The information available at the lines ABCD is the same as before, with 1's meaning 1, 2, 4, and 8, respectively, but it will only go up to 9 (1001). This is known as BCD output (binary coded decimal, 1-2-4-8). It is the best-known code, but there are several others.[9,10]

9.1.3.3. Synchronous Counter. Since the information in the ripple counter travels from one flip-flop to the next, it may arrive with considerable delay at the binary giving the most significant bit. For example, for an 8-digit number, there will be 32 flip-flops. If the propagation delay per flip-flop is 50 nsec, the total delay will be about 1.6 μsec. This would not be of any consequence in a simple counting system, but problems arise when the information is to be strobed out to a computer or when one has to count in both directions. In the latter case, the gate must be present for the full interval of time during which the signal propagates. The desire to reduce the delay led to the design of synchronous counters where the input clocks simultaneously all flip-flops of the unit.

[9] Malmstadt and Enke,[3] pp. 253–254.
[10] Morris and Miller,[1] Chapter 8.

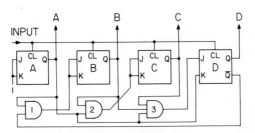

FIG. 3. Synchronous decade counter [from R. L. Morris and J. R. Miller, "Designing with TTL Integrated Circuits," Fig. 10.12. McGraw-Hill, New York, 1971].

Figure 3 shows the principle of a synchronous decade counter. The operation is as follows: Initially all outputs are at 0, thus the three AND gates are off and the JK controls of flip-flops B, C, and D are 0. The first clock pulse does not change these binaries. After the first pulse, however, the output of gate 1 and JK of B go to 1. The second clock pulse, thus, will switch A back to 0 and B to 1, etc. Table I shows the sequence of events. The output lines again show the BCD information. Synchronous decades are cascaded as ripple counters, with output D providing the clock pulse for the next decade. Many of the available counter packages are provided with an up/down gate input; details of the logic are given in Morris and Miller[1] and Lancaster.[2]

Another type of synchronous counter is the ring counter which is a

TABLE I. States of Synchronous Decade Counter after nth Input

Input	Output lines				J-K inputs				Truth Table		
n	A	B	C	D	JK(B)	JK(C)	J(D)	K(D)			
0	0	0	0	0	0	0	0	0	J	K	Q_{n+1}
1	1	0	0	0	1	0	0	1			
2	0	1	0	0	0	0	0	0	0	0	Q_n
3	1	1	0	0	1	1	0	1	0	1	0
4	0	0	1	0	0	0	0	0	1	0	1
5	1	0	1	0	1	0	0	1	1	1	$\overline{Q_n}$
6	0	1	1	0	0	0	0	0			
7	1	1	1	0	1	1	1	1			
8	0	0	0	1	0	0	0	0			
9	1	0	0	1	0	0	0	1			
10	0	0	0	0	0	0	0	0			

shift register closing on itself.[11-13] The ring principle was applied for some early high-speed counters using tunnel diodes,[14,15] but presently the same performance of 500 MHz direct counting can be obtained with MECL III decades.[16]

9.1.3.4. Divide-by-N Counters. By suitable feedback connections applied to a 4-bit binary counter, the input event rate can be divided by any integer N up to 16, the decade scalers of Figs. 2 and 3 being only the most widely used examples.[17,18]

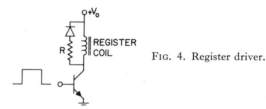

FIG. 4. Register driver.

9.1.3.5. Mechanical Counters. If the event rate is small, a mechanical counter, possibly with a few prescaling decades, is still useful as one of the less expensive high-capacity memories. Figure 4 shows a drive circuit. The width of the square wave and the supply voltage are specified by the manufacturer. The reverse diode and the damping resistor R prevent ringing and damage to the transistor when it is shut off.

9.1.4. Digital Displays and Recording

9.1.4.1. Types of Visual Displays. The discussion will, in general, be limited to the display of a decimal digit, i.e., the numbers 0 to 9. There are a variety of options, depending on requirements concerning power consumption, space, visibility, reliability, and cost.[19-21]

[11] Morris and Miller,[1] pp. 292–298.
[12] Lancaster,[2] pp. 116–120, 160–166.
[13] Malmstadt and Enke,[3] pp. 271–273.
[14] E. Baldinger and A. Simmen, *Nucl. Instrum. Methods* **57**, 141 (1967).
[15] Z. C. Tan, *Nucl. Instrum. Methods* **74**, 297 (1969).
[16] An MSI 500 MHz Frequency Counter Using MECL and MTTL, Appl. Note AN-581. Motorola, Inc., Phoenix, Arizona.
[17] Morris and Miller,[1] pp. 271–283.
[18] Lancaster,[2] pp. 146–158.
[19] L. Berringer, *Electron. Products* **15**, No. 2, 106 (1972).
[20] M. J. Riezenman, *Electronics* **46**, No. 8, 91 (1973).
[21] A. Sobel, *Sci. Amer.* **228**, No. 6, 64 (1973).

(a) If plenty of space is available, one of ten incandescent or gas-discharge bulbs can be lit, illuminating a ground glass showing the proper number. Usually, the ten numbers are arranged in a vertical line for each digit.

(b) The Nixie tube contains all ten numerals in the form of bent cathode wires in a single glass envelope. A neon gas discharge goes to whichever cathode is connected to ground through a turned-on transistor, and the cathode glow sheet outlines the desired number.

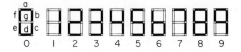

FIG. 5. Seven-segment displays.

The other devices to be described are all based on the seven-segment scheme shown in Fig. 5. Nine-segment and dot-matrix displays are used less frequently. Alphanumeric characters made up from straight-line segments are used extensively in cathode-ray-tube computer data displays,[22] but will not be discussed here.

(c) The brightest display is obtained with incandescent filament segments.

(d) A flat gas-discharge display, with several digits in the same envelope, consists of a transparent front plate with a transparent anode layer for each digit, and seven cathode segments behind it. As in the Nixie tube, the gas discharge will go to those cathodes that are grounded through a transistor switch.

(e) The gas space of (d) can be replaced by a thin sheet of fluorescent material which emits light when an ac voltage is applied between the electrodes.

(f) The segments may be fluorescent strips used as anodes in a vacuum tube, attracting electrons from a filament heated to a temperature sufficient for thermionic emission, but low enough so that its glow does not interfere with the fluorescent display.

(g) Perhaps the most widely used segments are small pieces of plastic diffusing the red light from a GaAsP light-emitting diode (LED, see Section 2.2.2). They are not very bright; since the light output efficiency

[22] S. Davis, "Computer Data Displays." Prentice-Hall, Englewood Cliffs, New Jersey, 1969.

increases with power, these displays are often used in a pulsed mode (with high power), multiplexing a single decoder-driver to the various digits (see Section 9.1.4.2).

(h) Liquid-crystal displays are in a state of rapid development.[20,21,23] The crystals do not emit light, but the transmission or scattering of light is changed when an electric field is applied. The display system consists of a thin layer (of the order of 100 μm) of a liquid crystal between two glass plates, the front one coated with transparent segment electrodes, the back one with a single electrode for each digit. Depending on the mode used (reflection or transmission, change in scattering or change in polarization), the digits will appear bright on a dark background or dark on a dull white background. These devices consume little power and can be made large (several centimeters) and bright.

9.1.4.2. Driving Circuits. The display devices (a) and (b) require ten driver lines per digit. Decoder/driver units are available in the 16-lead package (4 BCD inputs, 10 decimal outputs, 2 power leads). The decoder/drivers for the 7-segment displays are available in both 14-lead and 16-lead packages. The decoders are just an array of inverters and gates,[10] the drivers are output transistors with floating collectors. When the output is "on," the collectors provide a ground connection for the external device, similar to the register driver of Fig. 4.

In the case of the flat gas discharge (d) and the liquid-crystal (h) devices, and optionally in other devices, the leads to the seven segments are not brought out individually for each digit, but corresponding segments are connected internally. An 8-digit unit would have 15 inputs: 7 lines (a to g) for the character segments (say, cathodes) and 8 lines for the opposing digit electrodes (anodes). Only one decoder/driver will be needed, but the BCD input lines to it are taken in turn from the counter decades 1 to 8, while simultaneously the anodes 1 to 8 are energized in turn. Since there already exists a timing clock in most counting systems, to control accumulation and display time, the expense added by the multiplexing system is not large, and one saves, in this example, seven decoder/drivers.

9.1.4.3. Recording. The timing control which starts and stops the accumulation may also control the recording. A variety of devices are available.

---

[23] T. Kallard (ed.), "Liquid Crystals and Their Applications," pp. 31–39, 143-150. Optosonic Press, New York, 1970.

(a) *Printing register.* This is the cheapest method after paper and pencil, but it is applicable only if the count rate is so low that the prescaling factor is small and the content of the prescaler can be neglected.

(b) *Digital printer.* The information from each decade is used to control one printing wheel. All information must exist in parallel before the printing stroke. These instruments will print up to 20 lines per second, with 10 to 20 digits per line.

(c) *Paper punch or teletype.* In this case, the digits must be presented in series, necessarily starting with the most significant for the teletype. BCD inputs are accepted directly, and the paper tape is normally punched in that code. Typical speeds are about 50 digits per second for paper tape, 10 for teletype.

(d) *Magnetic tape.* This is, of course, the fastest and most expensive system. The questions to be decided are (1) whether one can transfer the information from the counting system to a general-purpose computer which controls the tape or whether one has to design a specific interface between the counter system and the tape drive; (2) whether to use a slewing or an incremental tape drive. The latter is somewhat slower, but less expensive and may be particularly suitable for a direct interface with the counting system.

9.1.5. Counting-Rate Meters

The primary use of counting-rate meters is in survey and monitoring instruments where one wants to know the "instantaneous" rate of events. The meter really shows the average counting rate during a preceding time interval whose length is chosen according to the statistical accuracy desired. The design of counting-rate meters varies from simple analog circuits to highly accurate and complex digital instruments.

9.1.5.1. *Linear Analog Meter.* The principle of the simple diode-pump meter is shown in Fig. 6a. The input is the square wave from a Schmitt trigger or one-shot multivibrator, with low source impedance. The small charge transferred at the rise of the pulse to the memory capacitor C_0 is $C_1(E - e_0)$. At a steady input rate n, the current balance is

$$nC_1(E - e_0) = e_0/R + C_0 \, de_0/dt, \tag{9.1.1}$$

with the solution, for $e_0(0) = 0$,

$$e_0(t) = \frac{nC_1RE}{1 + nC_1R} \left\{ 1 - \exp\left[- \frac{t(1 + nC_1R)}{RC_0} \right] \right\}. \tag{9.1.2}$$

The equilibrium reading,

$$e_0(\infty) = nC_1RE/(1 + nC_1R),\qquad(9.1.3)$$

is not linear, but approaches a saturation value E at high counting rates. Equilibrium is approached with a rate-dependent time constant $RC_0/(1 + nC_1R)$.

The meter can be linearized as shown in Fig. 6b. In this case, the charge transferred for each count is C_1E, since the inverting input of the operational amplifier is kept at zero potential. Thus

$$e_0(t) = nC_1RE[1 - \exp(-t/RC_0)]\qquad(9.1.4)$$

and the equilibrium output increases linearly with n. For a simple version of Fig. 6b, see Zadicario et al.[24]

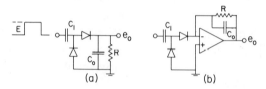

Fig. 6. (a) Simple diode-pump counting-rate meter. (b) Linearized counting-rate meter.

9.1.5.2. Logarithmic Counting-Rate Meters. Logarithmic response is desirable when a meter has to cover a large range of counting rates. If the feedback resistor of Fig. 6b is replaced by a logarithmic diode with a characteristic $e = V \ln(i/I)$, where V and I are constants, the equilibrium reading of the meter will be: $e_0 = V \ln(nC_1E/I)$. The long-term stability of this simple circuit is poor because of the temperature dependence of the diode characteristic.[†]

The Cooke-Yarborough-type meter[25,26] shown in Fig. 7 does not suffer from this temperature dependence. It approximates the loga-

[24] J. Zadicario, J. Grünberg, and U. Sold, *Nucl. Instrum. Methods* **33**, 238 (1965).
[25] E. H. Cooke-Yarborough and E. W. Pulsford, *Proc. IEE (London)* **98**, part II, 196 (1951).
[26] V. N. Kostić and B. J. Kovač, *Nucl. Electron.* **2**, 445 (1962).

[†] A temperature-compensated instrument is described by J. M. Rochelle and E. J. Kennedy: Miniaturized Logarithmic Count-Rate Circuit, *Rev. Sci. Instrum.* **44**, 1638 (1973).

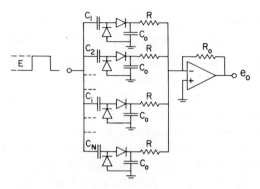

FIG. 7. Cooke–Yarborough-type logarithmic counting-rate meter. $C_{i+1} = aC_i$, $a \gg 1$.

rithmic response by a number of parallel diode pumps. According to Eq. (9.1.3), the equilibrium current flowing through the ith diode pump is $nC_iE/(1 + nC_iR)$. The output voltage is thus

$$e_0(\infty) = \frac{ER_0}{R} \sum_{i=1}^{N} \frac{nC_iR}{1 + nC_iR}. \tag{9.1.5}$$

The input capacitors are chosen to increase in size geometrically, with $C_{i+1} = aC_i$ ($a = 10$ in the original circuit). For a given counting rate, all pumps up to a certain index contribute essentially zero to the sum ($nC_iR \ll 1$), all pumps above, essentially unity ($nC_iR \gg 1$). An increase by a factor a in counting rate will add 1 to the sum—which is the essence of logarithmic response.

9.1.5.3. Statistics of Analog Counting-Rate Meters. The linear analog counting-rate meter (Fig. 6b) is a statistically poor device, being wasteful of information: For constant rate, the equilibrium reading is approached to within 1% only after $t = 4.6RC_0$, and the standard error for a single reading of the equilibrium output is $\sigma(e_0) = e_0/\sqrt{2nRC_0}$.[27] The relative error of the counting rate, $\sigma(n)/n = \sigma(e_0)/e_0$, is thus the same as that obtained by direct accumulation for a time of only $2RC_0$. The analysis of the accuracy of logarithmic meters is more complicated.[28]

Analog counting-rate meters of the type shown in Figs. 6 and 7 are rather inexpensive and convenient, since the output may be a direct meter reading or recorder tracing. But when maximum use of information is needed, e.g., in the case of fairly rapidly changing event rates,

[27] R. D. Evans, "The Atomic Nucleus," pp. 803–809. McGraw-Hill, New York, 1955.
[28] C. H. Vincent, Nucl. Instrum. Methods **47**, 157 (1967).

periodic accumulation and recording of counts without memory loss is indicated, either in analog form[29] or, preferably, in digital form.[30]

9.1.5.4. Digital Counting-Rate Meters. The frequency counters described in Section 9.2.3.3 are, of course, digital counting-rate meters which accumulate the input events during a selectable fixed time interval and then display the result for a short time, reset the counter memory to zero, and accumulate again. If a buffer register is used for the display, accumulation can proceed without loss of events except during the very short time interval needed for transfer and resetting.

Instruments specifically described as digital ratemeters may deviate in different ways from the frequency counter. For example, the time base may be adjusted automatically depending on the event rate and the desired (preset) accuracy.[31] Or the memory loss of the linear analog ratemeter may be simulated by subtracting periodically a fixed fraction of the memory content. [32,33] Finally, a digital feedback loop able to follow rapid variations in event rate has been described.[34]

9.1.6. Counting Losses

If the time intervals between events to be counted have a lower limit, no counting losses will occur provided the pulse-pair resolving time is below this limit. In the important special case of random events, however, the distribution of intervals is given by

$$p(t) \, dt = ne^{-nt} \, dt \tag{9.1.6}$$

where n is the average event rate. Because events can be arbitrarily close, a certain counting loss is unavoidable. The basic considerations have been given quite early[35,36] and have been redescribed repeatedly,[37,38] including the losses in slow recording equipment. For considerations of losses in pulse-height analysis systems, see Section 9.6.3.

[29] C. M. Kukla, *Rev. Sci. Instrum.* **38**, 804 (1967).

[30] B. Kawin and F. V. Huston, *Nucleonics* **22**, No. 7, 86 (1964).

[31] R. W. Tolmie and Q. Bristow, *IEEE Trans. Nucl. Sci.* **NS-14**, No. 1, 158 (1967).

[32] C. H. Vincent and J. B. Rowles, *Nucl. Instrum. Methods* **22**, 201 (1963).

[33] M. Werner, *Nucl. Instrum. Methods* **34**, 103 (1965).

[34] V. Polivka, *IEEE Trans. Nucl. Sci.* **NS-19**, No. 1, 545 (1972).

[35] L. J. Rainwater and C. S. Wu, *Nucleonics* **1**, No. 2, 60 (1947).

[36] W. C. Elmore, *Nucleonics* **6**, No. 1, 26 (1950).

[37] Evans,[27] pp. 785–790.

[38] I. De Lotto, P. F. Manfredi, and P. Principi, *Nucl. Instrum. Methods* **30**, 351 (1964).

9.1.6.1. Input-Stage Counting Losses. With the availability of fast scalers, it should be sufficient to consider the dead time due to transducer, amplifier, input shaper, and first scaling stage only. A simple treatment is possible for two idealized cases.

(a) *Nonextended dead time.* The system will be paralyzed for a fixed time τ after each event *counted*. The losses are equal to the number of events expected during these paralysis times; $l = m \cdot n\tau$, where m is the observed counting rate. Thus, $n = m + mn\tau$ and

$$m = n/(1 + n\tau). \tag{9.1.7}$$

(b) *Extended dead time.* The system is dead for a fixed time τ after each event *received*, even if the event occurred when the system was already paralyzed. Only the events preceded by an eventless interval $\geq \tau$ will be counted. The probability for this, from Eq. (9.1.6), is $e^{-n\tau}$ and thus

$$m = ne^{-n\tau}. \tag{9.1.8}$$

Most systems are closer to (a) than (b).

9.1.6.2. Determination of Counting Loss. It is tempting and often sufficient[39] to measure the pulse-pair resolution by simulating the analog inputs with a pulser which produces double pulses of adjustable separation; usually Eq. (9.1.7) is assumed to hold for correcting observed random event rates.

Direct methods which test the whole system, including transducers, have been developed in nuclear radiation physics where random events are prevalent.

(a) *Two-source method.*[37] Two sources A and B of approximately the same intensity, emitting a single gamma ray, are used. They are counted singly (m_A, m_B) and together (m_{AB}). If the losses are small, Eqs. (9.1.7) and (9.1.8) are the same and the difference between $m_A + m_B$ and m_{AB} allows the calculation of the dead time τ and of the loss at the observed rate m_{AB}. By moving the sources closer so that a single rate is close to m_{AB} and repeating the procedure, the losses at higher event rates can be determined.

(b) *Radioactive decay method.*[40] If a source decaying with a single, convenient, well-known half-life is available, the deviation of the measured

[39] C. B. Nelson, J. M. Hardin, and G. I. Coats, *Nucl. Instrum. Methods* **97**, 309 (1971).
[40] G. R. Martin, *Nucl. Instrum. Methods* **13**, 263 (1961).

time dependence of the radiation intensity from the expected exponential decay can be used to determine the counting losses as a function
of event rate.

(c) *Inverse square-law method.* This method should not be used since
the event rate at the detector rarely depends by an inverse square law on
the source-detector distance because of scattering and absorption.

It should be emphasized that the problem of counting losses is important for precision counting and that the losses may depend on phenomena
outside electronics, such as the physics of transducers or detectors.

9.2. Frequency Measurements*[1]

9.2.1. Wavemeters

Wavemeters are often used for frequency measurement if extreme accuracy is not required. Two different varieties of wavemeters will be described in this section. The first one is generally used if the power output of the source of unknown frequency is large enough (order of magnitude of one watt or more) and will be referred to as an *LC wavemeter*. The second one is capable of measuring the frequency, or frequencies, to which a passive circuit is tuned, and will be referred to as a *grid-dip* meter. Both types of instruments are frequently used even though more elaborate and accurate instruments are available.

Fig. 1. *LC* wavemeter circuit.

9.2.1.1. *LC* Wavemeters. The *LC* wavemeter consists simply of an inductance, a capacitance, a detector, and means for varying the value of either the inductance or the capacitance. For fixed coupling to a source of electrical oscillations, maximum energy is absorbed when the oscillation frequency and the wavemeter resonant frequency are identical. The latter is either read directly from the dial, or obtained indirectly through use of a calibration curve correlated with the dial reading. Figure 1 is a schematic diagram of an *LC* wavemeter which uses a variable capacitor for changing the resonant frequency, and a thermocouple milliammeter

[1] F. E. Terman and J. M. Petit, "Electronic Measurement," Chapter 5. McGraw-Hill, New York, 1952; C. G. Montgomery, "Technique of Microwave Measurements." McGraw-Hill, New York, 1947. See also Vol. 1, Section 8.3.4.

* Chapter 9.2 is by **Edwin A. Goldberg**.

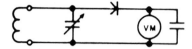

FIG. 2. *LC* wavemeter circuit with diode detector.

as a detector. In the diagram of Fig. 2, the current detector has been replaced by a voltage detector consisting of a diode and a dc voltmeter. The power required by the wavemeter is a function of the detector sensitivity. For some applications, a small incandescent lamp may be used in lieu of the thermocouple meter shown in Fig. 1.

The minimum coupling necessary to produce a satisfactory detector indication at resonance should be used to prevent detector overload and variation of the frequency of the source through excessive loading by the wavemeter.

Accuracy depends on a number of factors. The Q of the wavemeter should be as high as possible in order that the tuning be "sharp," and the instrument should be well built in order that calibration be maintained. The method is generally suitable if the required accuracy is not better than about 1%.

9.2.1.2. Transmission Line Wavemeters. A variable length transmission line (Lecher wire) may be used in lieu of a lumped tuned circuit consisting of an inductance and a capacitor (LC) as the tuned element for an absorption-type wavemeter. This technique is described in Section 9.2.3.4.

9.2.1.3. Grid-Dip Meters.[2] The grid-dip meter consists of a calibrated self-excited oscillator (see Vol. 2A, Chapter 7.2) and a micro- or milliammeter which measures the oscillator grid current or that of the equivalent transistor element. It is used primarily for determining the resonant frequency of an unexcited tuned circuit. The tank coil of the grid-dip meter oscillator is coupled to the circuit, and the oscillator frequency is varied. When the oscillator frequency coincides with the resonant frequency of the circuit being measured, the latter absorbs power from the oscillator and causes the rectified grid current of the oscillator to decrease or "dip" below its normal value. Figure 3 is a basic schematic diagram of a transistorized grid-dip meter.[3]

[2] W. M. Scherer, Applications of the Grid Dip Oscillator, CQ (magazine) Jan. 1949; The Radio Amateur's Handbook, 35th ed., p. 520. Amer. Radio Relay League, Hartford, Connecticut, 1958.

[3] *RCA Transistor Manual*, Tech. Ser. SC-13, p. 509. RCA Electron. Components and Devices, Harrison, New Jersey.

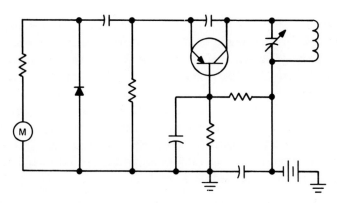

FIG. 3. Grip-dip meter circuit.

This device may also be used as a signal generator for rough alignment of radio receivers. It is simple, inexpensive, and quite handy to use. The accuracy of resonant frequency measurements of circuits is a function of the Q of the circuit being measured, and of the calibration accuracy of the oscillator, and is generally in the neighborhood of 1 or 2%.

9.2.2. Constant-Frequency Sources[4]

An unknown frequency may be measured by comparison with a known frequency, and once a single accurately known frequency source is established, it is possible to obtain a wide range of known frequencies through frequency multiplication, division, addition, and subtraction. The source of known frequency may be a highly stable crystal oscillator, a broadcast radio station, or transmissions from stations operated by the NBS. The details of transmissions from NBS stations may be obtained from NBS Special Publication 236, 1970 Edition.[5]

9.2.2.1. Calibration of Laboratory Oscillator. A typical standard-frequency laboratory oscillator might have a fundamental frequency of 100 kHz. This oscillator may be checked and adjusted to a very high degree of accuracy by coupling its output to a receiver which is tuned

[4] K. Henney (ed.), "Radio Engineering Handbook," 4th ed., Frequency Stabilization, p. 419. McGraw-Hill, New York, 1950.

[5] NBS Frequency and Time Broadcast Services Radio Stations WWV, WWVH, WWVB, and WWVL (P. P. Viezbicke, Ed.). Nat. Bur. Std. U.S. Spec. Publ. 236 CODEN: XNBSA Order by SD Catalog No. C 13.11: 236 from Superintendent of Documents, U.S. Govt. Printing Office, Washington, D.C. 20402. Price 25 cents.

to a standard-frequency broadcast, for example, a transmission on 5000 kHz. The fiftieth harmonic of the laboratory oscillator should coincide with the carrier frequency of the standard broadcast, and any deviation would be detected by the radio receiver, the difference between the two being the same as the audio-frequency output of the receiver. An error of one part per million would result in a 5-Hz detector output which could be observed with an oscilloscope, or an electronic frequency meter such as the Hewlett Packard Model 500B. Adjustment of the laboratory oscillator for as near a zero beat frequency as possible should be made, and the adjustment checked occasionally.

9.2.2.2. Frequency Division. Frequencies which are accurate submultiples of a standard frequency may be generated by several techniques.[6] Submultiple frequencies are used for establishing marker frequencies between the harmonically related frequencies of the standard oscillator, and frequencies lower than the fundamental of the standard.

9.2.2.3. Multivibrators. Multivibrators are oscillators which are normally free running,[†] but which may be caused to oscillate at some submultiple frequency of another signal, injected usually at one of the bases. The free running frequency, of course, must be adjusted to be close to the proper submultiple frequency of the input. The output is generally of rectangular waveform and thus very rich in harmonic content. When properly designed and driven from a suitable signal source, the operation is quite stable. In practice, it is necessary to provide means for monitoring the operation to make sure that division by the proper integer is being made. Improved stability may be obtained through incorporation of an LC tank circuit in one collector load, tuned to the desired subharmonic frequency.

9.2.2.4. Regenerative Dividers.[7] Regenerative frequency dividers are more complex than multivibrators, but have several operational advantages. Properly designed, they will either divide by the proper integer, or will not work at all. Figure 4 is a block diagram which illustrates the

---

[6] C. G. Montgomery, "Techniques of Microwave Measurements," Chapter 6–7, p. 354. McGraw-Hill, New York, 1947; J. Millman and H. Taub, "Pulse and Digital Circuits," Chapter 12. McGraw-Hill, New York, 1956.

[7] F. E. Terman, "Radio and Electronic Engineering," 4th ed., Sect. 18–16, p. 663. McGraw-Hill, New York, 1956.

[†] See Vol. 2A, Section 7.3.3.

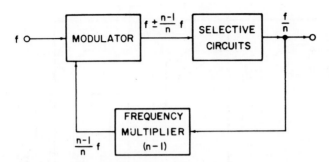

Fig. 4. Regenerative divider system diagram.

principle of operation. Assume that the output frequency is the input frequency f divided by the integer n. This frequency f/n is multiplied in a frequency multiplier (see Section 9.2.2.6) by $n - 1$. The output frequency of the multiplier $(n - 1)(f/n)$ is then in turn mixed with the input frequency f in the modulator and the frequency difference f/n selected by the filter.

9.2.2.5. Counters. Electronic counters make very satisfactory frequency dividers since they will operate over a very wide frequency range without affecting the dividing integer. A number of counter circuits exist, but only the application of one type, the bi-stable multivibrator or flip-flop, will be discussed here. When driven from a source of suitable waveform, the output frequency will always be one-half the input frequency. Combinations of flip-flops may be used for dividing by any predetermined integer. If several flip-flops are cascaded as in Fig. 5, the output frequency will be $f/2^n$, where n is the number of flip-flops. Division by other inte-

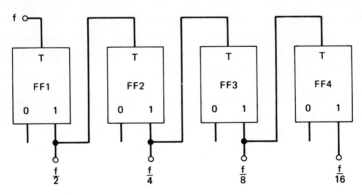

Fig. 5. Cascaded flip-flop stages connected as 16-to-1 frequency divider.

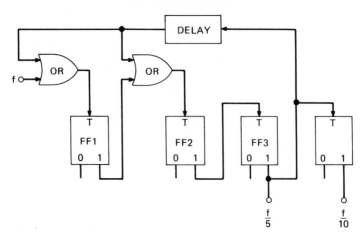

Fig. 6. Binary counter stages connected as 10-to-1 frequency divider.

gers is readily accomplished through feedback connections which will reset the states of the counter stages to some predetermined initial count other than zero after a particular counter stage flips over. A 10-to-1 frequency divider using this principle may be arranged as shown in Fig. 6.

A quinary divider is constructed using three cascaded binary stages. Feedback is so arranged that every time the last stage changes from state B to state A, the first two stages are reset to state B which corresponds to the count of three. Thus, the third stage changes from state B to A once every fifth cycle of the input rather than once every eighth cycle as would be the case if there were no feedback. The fourth stage changes state every time the third stage changes from B to A, but not when the change is from A to B; thus its fundamental output frequency is one-half that of the third stage or one-tenth that of the input signal. The harmonic content is high, but is predominately odd since the waveform is symmetrical. Other combinations of feedback may be used to produce division by any other integer, and arranged so that the output waveform will contain both even and odd harmonics.

9.2.2.6. Harmonic Generation. Harmonic frequencies, or integer multiples of a standard or known frequency, are very useful for frequency measurement purposes. Generally the output of oscillators inherently contain a high enough harmonic content for normal frequency measurement applications. The outputs of multivibrators and counters are especially rich in harmonic content. In cases where greater than normal harmonic content is desired, it may be obtained by amplifying the signal in a nonlinear device. Saturable reactors and overdriven vacuum tubes or transistors are excellent harmonic generators. A high-Q tuned circuit in the plate or collector circuit may be used to select the desired harmonic.

9.2.2.7. Frequency Synthesizers. It is possible to generate signals bearing well-defined frequencies referred to a standard frequency source. Devices for doing this are often referred to as *frequency synthesizers*. Frequency synthesizers suitable for laboratory use are commercially available from a number of instrument firms. Frequency range, method of frequency selection, and the increments between available frequencies depend upon the particular instrument chosen; however, some typical synthesizers allow for the selection of any frequency from 0 to 10^7 Hz in steps of 0.01 Hz. Selection of frequency is done digitally through setting of decade dials.

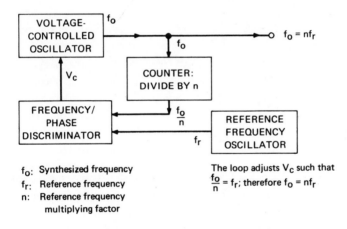

f_o: Synthesized frequency
f_r: Reference frequency
n: Reference frequency
  multiplying factor

The loop adjusts V_c such that
$\frac{f_o}{n} = f_r$; therefore $f_o = nf_r$

FIG. 7. Simple frequency synthesizer diagram.

Synthesizers may incorporate digital counters, modulators and filters, and one or more frequency reference sources. A simplified block diagram of a synthesizer employing a reference oscillator, a counter, and a voltage-controlled oscillator (VCO)—an oscillator whose frequency is a function of a control voltage—is illustrated in Fig. 7. The feedback loop will force the VCO control voltage to a value which, in turn, forces the frequency f_o/n to equal f_r at the inputs to the frequency/phase discriminator. For example, if the frequency source f_r is 1000 Hz, and the counter divides by 27, the value of f_o, once the loop has stabilized, will be 27,000 Hz. The system design of wide-range synthesizers is considerably more complex than that shown in the preceding illustration, however, and the reader is referred to representative articles in the literature for further information.[8-11]

9.2.3. Measurement Techniques

The particular method of making a frequency measurement depends on the nature of the equipment available and the degree of accuracy required. Means for using such devices as the *LC* wavemeter and the grid-dip meter have been described in Section 9.2.1. Frequency measurements through the use of calibrated oscillators, and frequency dividers using the "heterodyne" principle, will be discussed in this section. The accuracy of this type of measurement is primarily a function of the accuracy of calibration of the measurement equipment which can be in error by only a very small fraction of a percent.

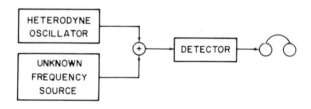

FIG. 8. Frequency measurement setup using calibrated heterodyne oscillator.

9.2.3.1. Heterodyne Oscillator. A heterodyne oscillator is simply a calibrated variable-frequency oscillator. It may be used in conjunction with a detector circuit (see Chapter 6.10) and a pair of headphones for measuring the frequency of another source. The oscillator output and source to be measured are both coupled to the detector, which in turn is connected to the headphones as in Fig. 8. The oscillator frequency is adjusted for an audible tone in the headphones, and further adjusted until the frequency of this tone becomes a minimum (zero, ideally). The oscillator

[8] A. Noyes, Jr., Coherent Decade Frequency Synthesizers. General Radio Experimenter, September 1964.

[9] G. H. Lohrer, Remote Programming for G. R. Synthesizers. General Radio Experimenter, May 1965.

[10] Applications for Coherent Decade Frequency Synthesizers. General Radio Company, October 1968.

[11] W. F. Byers, Synthesizing at Higher Frequencies. General Radio Experimenter, January/February 1970.

frequency, or a harmonic or subharmonic of the oscillator frequency, is then equal to the frequency of the other source. Determination of the correct harmonic of the oscillator which corresponds to the unknown frequency may be made as follows:

The oscillator is tuned to a frequency which results in a zero beat. This frequency will be designated f_1 and the unknown source frequency will be nf_1, where n is an integer to be determined. The frequency of the oscillator is then reduced, and the next lower oscillator frequency for which a zero beat occurs, f_2, is determined. The unknown source frequency is $(n + 1)f_2$. Let f_s designate the unknown source frequency. Then

$$f_s = nf_1$$

and

$$f_s = (n + 1)f_2$$
$$f_s = f_1 f_2 / (f_1 - f_2).$$

9.2.3.2. A High-Precision Technique. One method for making an extremely precise measurement of frequency will be described here. The equipment used will consist of a frequency standard, consisting of a 100-kHz oscillator and frequency dividers with fundamental output frequencies of 10 kHz and 1 kHz, a calibrated heterodyne oscillator, a calibrated audio-frequency oscillator, a tunable detector, and an oscilloscope. The 100-kHz frequency-standard oscillator is assumed to have been accurately calibrated, possibly through comparison of the frequency of the fiftieth harmonic with the 5000-kHz carrier frequency of the WWV transmission.

1. Couple the unknown frequency source to the detector and tune to this frequency. (If a regenerative detector is used, the detector should be oscillating, and should be tuned for a "zero beat" in the headphones. Regeneration is then reduced until the detector ceases to oscillate.)

2. Couple the heterodyne oscillator to the detector and adjust to the unknown frequency by tuning for a zero beat tone in headphones.

3. Leave the detector tuning fixed, couple the output of the 10-kHz frequency divider stage of the frequency standard and the heterodyne frequency oscillator to the detector, and then decrease the frequency of the heterodyne frequency meter until the first zero beat is reached as detected by the detector. The heterodyne oscillator is then exactly tuned to some integer multiple of 10 kHz, and the proper integer should be evident from the calibration on the dial.

4. Couple the heterodyne oscillator and unknown frequency source to the detector, and measure the beat frequency by comparing the beat frequency output of the detector with the output frequency of the calibrated audio oscillator using the oscilloscope. If horizontal deflection is provided by the detector output, and vertical deflection by the audio oscillator, the audio oscillator will be tuned to produce a Lissajous figure indicating equal horizontal and vertical frequency. The unknown frequency is then equal to the accurately determined heterodyne oscillator frequency plus the audio oscillator frequency.

It was assumed that the audio oscillator had been accurately calibrated. The 1-kHz frequency divider on the frequency standard may be used for checking the audio oscillator calibration at the frequency of 1 kHz and harmonics thereof using the oscilloscope as a frequency comparator. Audio oscillator output would be connected to the horizontal deflection plates, and the 1-kHz divider output would be connected to the vertical plates. Stationary Lissajous patterns would indicate when the audio oscillator was tuned to a harmonic of 1 kHz, and the number of the harmonic.

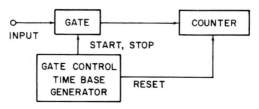

FIG. 9. Frequency measurement with a gated electronic counter—determination of cycles per predetermined interval of time.

9.2.3.3. Frequency Measurement Using a Time Interval Gated Electronic Counter.[12] Gated electronic counters provide a convenient means of frequency measurement, and the method is applicable for frequencies from a fraction of a hertz up to approximately 100 MHz. The frequency being measured is directly displayed by the counter register. Two methods of operation may be used, the first being best for frequencies above perhaps 500 Hz, and the second for the measurement of lower frequencies.

Figure 9 is a block diagram which indicates the principle of operation of the first method. The time base generator is a device which puts out two signals, one a reset signal, which sets the counter to zero, and the

[12] J. Millman and H. Taub, "Pulse and Digital Circuits," p. 345. McGraw-Hill, New York, 1956.

other a gate control signal which immediately follows the reset signal and whose duration is accurately known. The duration should be $10^{\pm n}$ seconds, where n is an integer chosen on the basis of the frequency being measured. Cycles of the input signal are counted by the counter for the time interval of the gate control signal, and the result displayed on the counter register. The process can be made repetitive in order to provide periodic measurements of frequency. Proper placement of the decimal point provides a direct indication of frequency. For instance, if the gate is opened for a period of 10^{-2} sec, the frequency of the input signal will be 10^2 times the number of cycles counted during this interval.

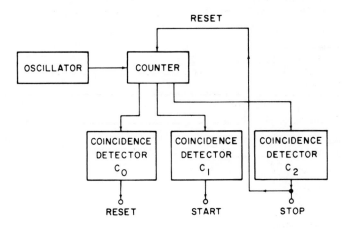

FIG. 10. Generation of pulses spaced by a predetermined time interval using an oscillator of accurately known frequency and an electronic counter.

The gate control signals are usually generated by means of the system illustrated in the block diagram of Fig. 10. Each coincidence unit is connected to the counter stages in such a manner that it will produce a sharp pulse when a prescribed count is reached.[13] The C_1 coincidence circuit is connected to produce a "start" pulse when the count reaches C_1. The C_2 coincidence circuit is connected to produce a "stop" pulse when the count reaches C_2 and also resets the counter to zero. A third coincidence circuit connected to produce a pulse at the count of C_0 may be incorporated to provide an external reset pulse. This arrangement produces a repetitive sequence of control pulses. It can, of course, be used in other applications. A very wide range of timing intervals is possible through

[13] Millman and Taub,[12] Sect. 16–10, p. 508.

variation of the oscillator frequency and of the counts (C_0, C_1, and C_2) which actuate the coincidence circuits.

The second method, used for low frequencies, consists in measuring the period of one cycle or of some predetermined number of cycles, usually 10^n (see Section 9.3.4).

9.2.3.4. Ultra-High-Frequency Measurements. In general, frequencies within the range of 300 to 3000 MHz are defined as *ultra-high* frequencies. Basic principles for frequency measurements which have already been described in this chapter are in general applicable to measurements in this frequency range; however, the detailed techniques employed are somewhat different from those used for lower frequency measurements.

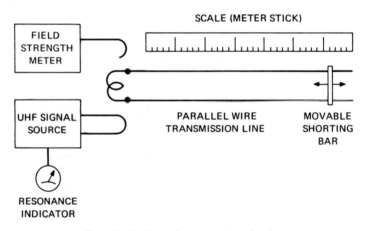

FIG. 11. Lecher wire wavemeter circuit.

A very simple technique which has been widely used for many years employs *lecher wires*.[14] The lecher wire method is analogous to the *LC* wavemeter (Section 9.2.1.1) in that it is a device which absorbs energy from a source whose frequency is the same as the resonant frequency of the device. It differs in that a distributed parameter circuit (a transmission line) is employed rather than a lumped parameter circuit (a coil condenser tuned circuit).

The lecher wire (Fig. 11) consists of a transmission line of parallel wires, a means for moving a shorting bar along the line, a scale for determining the distance the shorting bar has been moved, means for coupling the lecher wires to the frequency source being measured, and an

[14] The Radio Amateur's VHF Manual, pp. 289–291. Amer. Radio Relay League.

indicator to indicate when the line is resonant. The indicator could be a grid current meter, or a plate current meter connected to the oscillator whose frequency is being determined, or a field strength meter. The shorting bar on the lecher wires is moved, and the distances between successive locations of the shorting bar which cause the lecher wires to absorb energy as indicated by the chosen indicator is noted. This distance corresponds to one-half the wavelength of the energy source being measured. Therefore, the frequency, f, may be computed as follows:

$$f = 150/d$$

where f = frequency in MHz and d = distance in meters between successive locations of the shorting bar which results in energy absorption by the lecher wire line. The precision of frequency measurement obtainable with this method is of course a function of the care used in constructing the lecher wire transmission line, and the care used in making the measurements. Generally, it is possible to determine frequency to a precision in the order 0.1 to 1.0% with this method.

If a higher degree of precision is desired than is possible with a lecher wire system, a heterodyne technique may be employed, the specific technique being a function of the precision required, and the test equipment being utilized. The heterodyne method requires that a known frequency source be available, and also a means for determining the difference frequency between the known frequency source and the frequency being measured. A known frequency near that being measured may be obtained from a frequency multiplier whose input is a standard known frequency. For example, assume that the frequency of a source whose frequency is estimated to be between 380 and 400 MHz is to be determined. A frequency of 380 MHz could be derived by selecting the nineteenth harmonic of a 20-MHz frequency standard from the output of a frequency multiplier. This 380-MHz reference signal would be mixed with the signal whose frequency is to be determined in a modulator. The output of the modulator is passed through a low-pass filter, and this difference frequency is measured by a conventional method such as a direct reading digital frequency meter. The frequency of the unknown signal is the sum (or difference) of the 380-MHz reference signal, and the frequency at the output of the low-pass filter. Figure 12 is a block diagram of this measuring system.

A second measurement made with a 400-MHz reference (twentieth harmonic of the 20-MHz basic standard) or a 360-MHz reference (eigh-

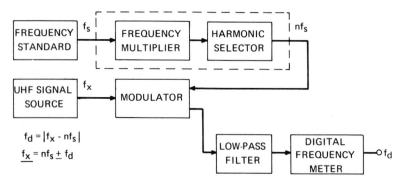

FIG. 12. A heterodyne method for measuring frequencies in the UHF range.

teenth harmonic of the 20-MHz standard) may be made to resolve the sum or difference ambiguity which exists with only one measurement.

A number of electronic instrument manufacturers produce frequency measuring equipment, and the reader is referred to their catalogs and other publications for further information.[15-17]

[15] Beckman Electronic Instrumentation, 1967–68, pp. 101–167.
[16] Frequency and Time Standard, Application Note 52. Hewlett-Packard Co.
[17] General Radio, Catalog U, pp. 188–201.

9.3. Time Measurement*

Time measurement as a means of performing experimentation in physics has advanced markedly over the past few years. The reasons are largely the availability of equipment and the increasing ease with which the experiments may be instrumented. The limitations which are encountered in performing time measurements are quite different for "long" time intervals versus "short" time intervals. Long time interval measurements are classified here as those significantly greater than one second, whereas short time interval measurements are classified as those significantly less than one second.

9.3.1. Errors

There are three distinctly different types of errors involved in time measurements which are general enough to be applied to any time measurement. These are drift, walk, and jitter. In long time interval measurements drift is the dominant source of error, whereas for short time interval measurements drift errors are compounded by the errors of turn-on–turn-off and indeed for very short time interval measurements ($<10^{-10}$ sec); the turn-on–turn-off errors often become the limiting error.

9.3.1.1. Drift. The error identified as drift is that error which may be associated with variations in the time measurement due to variations in temperature, or effects of time, which cause aging or wear on components involved in the time measurement clock.

9.3.1.2. Walk. Walk (sometimes called time slewing) may be created by two distinctly different effects. These two sources of walk may be extracted from Fig. 1. First, consider an ideal discriminator with a threshold, V_t, operating on signals of finite risetime. Although the two signals in Fig. 1 were caused by events occurring at the same time, t_0, the larger signal (V_1) crosses the discriminator threshold before the smaller signal (V_2). This shift in the crossing time causes the output pulse of the discriminator to "walk" along the time axis as a function

---

* Chapter 9.3. is by Charles W. Williams.

of pulse amplitude. The "walk" is most noticeable for pulse heights near the discriminator threshold.

The second walk effect is due to the charge sensitivity of real time discriminators.[1,2] Even though the discriminator's threshold has been exceeded, a finite amount of charge Q is required to trigger the discriminator. In Fig. 1 the times to collect a fixed charge Q are indicated by the shaded triangles. Clearly, the delay time introduced by this effect is longer for smaller pulse heights $(T_2 > T_1)$; in fact, for a flat top pulse of infinite duration the time to collect a charge Q approaches infinity as the pulse height approaches the discriminator threshold. In practical cases the walk, due to charge sensitivity, is limited by the width of the pulse above the discriminator threshold.

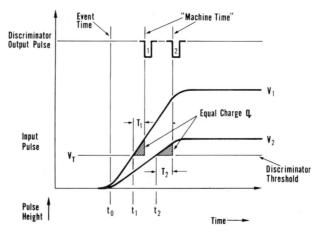

FIG. 1. Demonstration of walk caused by the amplitude effect and the charge sensitivity of the discriminator.

The two sources of walk add to determine the time at which the discriminator produces an output. This output time is frequently called *machine time*, and is used in the measuring device (clock) for performing the time analysis.

9.3.1.3. Jitter. Independent of the walk effect is another source of timing error which is statistical in its nature. This source of error is due to an intrinsic time jitter of the signal source as seen by the time pickoff

[1] R. Nutt, *IEEE Trans. Nucl. Sci.* **NS-14**, No. 1, 110 (1967).

[2] P. D. Compton, Jr., and W. A. Johnson, *IEEE Trans. Nucl. Sci.* **NS-14**, No. 1, 116 (1967).

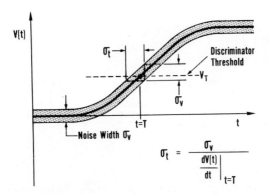

FIG. 2. The source of time jitter due to noise.

device. There are two main sources of jitter. These are noise and charge collection statistics.

Figure 2 shows the effect of system noise on the timing accuracy. This noise may be generated in the detector itself, in the electronics required to amplify the signal, or may simply be the equivalent input noise of the timing discriminator. If the steady-state noise has an amplitude width σ_v, then the width of the timing uncertainty in crossing the threshold V_t is given to a rather good accuracy by the triangle rule:

$$\sigma_t = \sigma_v \Big/ \frac{dv(t)}{dt}\bigg|_{t=T}. \tag{9.3.1}$$

The width of the timing uncertainty is σ_t, and $dv(t)/dt\,|_{t=T}$ is the slope of the average signal shape as it crosses the discriminator threshold. For validity, the noise and slope should be constant in the region of measurement. This source of jitter is a dominant one for many types of signals, e.g., semiconductor detector signals and normal signals derived from electronic sources, but is usually not important with detectors of the scintillator–photomultiplier type.

For scintillator–photomultiplier combinations, the number of elementary charge carriers released (photoelectrons) is small, typically between 10 and 1000. Consequently, it is the statistics of the emission and collection of the photoelectrons which dominate the timing properties. The net result is observed as fluctuations in the shape of the photomultiplier output pulse.[3,4] Figure 3 demonstrates this effect. The outer two lines define the rms envelope of the shape fluctuations $\sigma_s(t)$ about the mean

[3] M. Bertolaccini et al., Nucl. Instrum. Methods 51, 325 (1967).
[4] L. G. Hyman et al., Rev. Sci. Instrum. 35, 393 (1964).

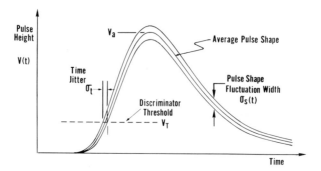

Fig. 3. Time jitter due to statistical pulse shape fluctuations with scintillator–photomultiplier detectors. The pulse shape is typical of the current pulse obtained at the anode (except for polarity) for a fast phototube and a fast scintillator such as Naton 136.

pulse shape. In contrast to the case of noise discussed above, the amplitude fluctuation width $\sigma_s(t)$ is a function of the time and the initial number of photoelectrons. Again one can see that the timing jitter σ_t can be derived by a triangle law,

$$\sigma_t = \sigma_s(t) \bigg/ \frac{dv(t)}{dt} \bigg|_{t=T} \tag{9.3.2}$$

where now both the numerator and denominator must be evaluated at the time T, when the pulse crosses the discriminator threshold, and subject to the restrictions given for Eq. (9.3.1).

9.3.2. Time Derivation Methods

The choice of the appropriate time derivation method depends upon the desired time resolution, counting efficiency, and the required dynamic range of pulse heights. There are two basic methods in use which will be elaborated upon. These are: (1) constant amplitude discriminators, normally referred to as leading edge discriminators; and (2) constant fraction discriminators which include all "crossover"-type discriminators.

The constant amplitude or leading edge methods are subject to both jitter and walk but have an advantage in simplicity. Until recently this has been the most widely used method, especially where narrow dynamic range of pulse heights are involved.[5-7] This method will continue to be

[5] A. Schwarzschild, *Nucl. Instrum. Methods* **21**, 1 (1963).
[6] J. A. Miehe *et al.*, *IEEE Trans. Nucl. Sci.* **NS-13**, No. 3, 127 (1966).
[7] R. E. Bell, *Nucl. Instrum. Methods* **42**, 211 (1966).

used in those applications where pulse height variation is small, and the walk effect is of little consequence. When dealing with detectors, such as scintillation detectors, the leading edge discriminator exhibits excessive jitter even for narrow dynamic ranges because of the statistical nature of the pulse formation.[3,8]

If a system derives its time properties from a constant fraction of the signal amplitude,[8,9] i.e., $V_t(t) = fV_{max}$, there will be no "walk" in the system provided that the input signal does not exhibit a variation in bandwidth versus signal amplitude. An example of a system of this type is what is sometimes referred to as slow or conventional crossover timing. Here the fraction of pulse height used for timing represents 50% of the total charge collection when this type of system is used with a nuclear radiation detector. The time resolution for this method with a slow scintillation detector such as NaI(Tl) is poorer than leading edge timing for a narrow dynamic range.[7] The reason for this is that the 50% collection point represents a poorer statistical point on the charge collection waveform from which the time measurement is derived.

One of the more recent developments, the constant fraction of pulse height trigger, or constant fraction discriminator, is shown in the block diagram in Fig. 4. With this type of system, one can both obtain the optimum triggering point in terms of jitter and minimize the effect of walk so as to obtain a true optimum timing system.[3,4] With the correct choice of a combination of delay and triggering fraction, a system comprising this circuit may also be used to greatly reduce the effect of charge collection variation in various detectors.[10]

9.3.3. Short Time Interval Measurements

Short time interval measurements may normally involve either analog processing methods or digital processing methods (see Section 9.3.4) or in some instances combinations of the two.[11,12] When the two methods are combined, the major time interval measurement is a digital process and an interpolation is done by an analog process for increased accuracy. Short time interval measurements are often limited by the errors in

[8] D. A. Gedcke and W. J. McDonald, *Nucl. Instrum. Methods* **55**, 377 (1967).

[9] D. A. Gedcke and W. J. McDonald, *Nucl. Instrum. Methods* **58**, 253 (1968).

[10] R. L. Chase, *Rev. Sci. Instrum.* **39**, 1318 (1968).

[11] I. De Lotto *et al.*, Automatic acquisition and reduction of nuclear data, p. 291. *Proc. EANDC Conf., Karlsruhe, Germany* (1964).

[12] R. Nutt, *Rev. Sci. Instrum.* **39**, 1342 (1968).

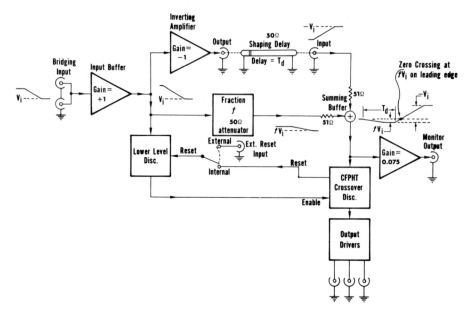

Fig. 4. Simplified block diagram of constant fraction of pulse height trigger.

starting and stopping the time measuring device, i.e., walk and jitter (see Section 9.3.1).

9.3.3.1. Analog Processing. Probably the most used method of short time interval measurement is the oscilloscope. Here the accuracy is quite poor but the measurement most simple. The method is to apply a start (trigger) pulse which starts a linear trace at a predetermined sweep speed on the oscilloscope screen. The second pulse at $t_0 + t$ is applied to the vertical deflection. This causes a deflection in the trace from the horizontal which allows the time difference measurement to be interpreted from the screen. Of course other measurements may be made simultaneously if the signal applied contains further information such as amplitude and shape.

This same principle is used in radar applications for ranging, except that the information is provided by intensity variation of the oscilloscope trace instead of deflection. The sweep is started at the rf burst transmission time and the trace is intensified when the reflected signal is received, indicating the elapsed time to measure the distance.

The most accurate method for ultra-short time measurements and one which is widely used is the *time analyzer* composed of a time-to-amplitude converter and a multichannel pulse height analyzer (see Chapter

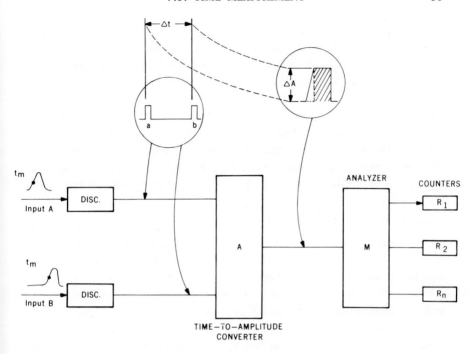

FIG. 5. Time analyzer block diagram.

9.6). This type of system,[13,14] shown in Fig. 5, can be capable of accuracies better than 0.01% and may have a fundamental inaccuracy limit of less than 1×10^{-11} sec. The instrument measures and classifies the time interval between pairs of events. To classify pairs of events into a finite number of channels, each channel is characterized by a width δt and a position t. By measuring a series of events the instrument gives a histogram of the differential, or time difference spectrum df/dt of the intervals measured. At completion of measurement the number of events collected in a channel is equal to $(df/dt)\ \delta t$. This process is further explained by Fig. 5, where the time-to-amplitude converter changes the time interval $\varDelta t$ into a pulse height interval $\varDelta A$, which is then measured by the multichannel pulse height analyzer M, and assigned to the nth channel, corresponding to the output n which drives the counter R_n.

There are two characterized types of time-to-amplitude converters. These are designated the start-stop and the overlap types. The overlap-type converter has one fundamental advantage, that of high input count

[13] E. Gatti and V. Svelto, *Nucleonics* **23**, No. 7, 62 (1965).
[14] M. Bonitz, *Nucl. Instrum. Methods* **22**, 238 (1963).

rate tolerance. It is, however, a narrow time range device with relatively poor stability and linearity characteristics. The start-stop-type converter, on the other hand, accepts slightly lower input count rates but has a wide-range capability and offers excellent stability and linearity.

The overlap time-to-amplitude principle is very similar to, or can be considered a special application of the overlap coincidence concept. In essence the time-to-amplitude converter measures the amount of pulse time overlap between two signals, whereas an overlap coincidence device provides a logic output signal indicating only that the pulses do indeed overlap in time. Bonitz[14] should be consulted for further treatment of these concepts.

9.3.3.2. Digital Processing. The primary method of digital measurement is the same for short or long time measurements (see Section 9.3.4); however, a limit on resolution is imposed by the basic counting rate of the digital clock. Practical limits are around 200 MHz which limits the inherent resolution to about 5×10^{-9} sec. This limit has been improved in a few applications by the use of the "vernier-chronotron" principle,[15] which has demonstrated a resolution value of $<1 \times 10^{-11}$ sec. This process has been essentially abandoned in favor of analog or combined analog–digital processes because of stability problems.

9.3.4. Long Time Interval Measurement

For long time intervals the electronic methods used normally comprise a combination of stable oscillator and electronic counter. Figure 6 shows a block diagram of such a system. The counter control consists of inputs such as start, stop, and reset. These may be either manual or electronic. In many cases when the counter is started, the logic gate is opened and the counter is allowed to count the oscillator frequency. Simple clocks of this type have an inherent error of ± 1 cycle per time interval measured; however, some designs incorporate a function which allows the measurement to be performed at the same phase point on the oscillator input signal for both start and stop, thereby reducing this error to $\pm \frac{1}{2}$ cycle. Preset clocks (timers) of this type provide a measured time interval that is essentially free of error.

With the elimination of the ± 1 cycle ambiguity the primary error remaining is simply the instability of the oscillator. Commercial clocks use as a time base a variation from the least expensive, the ac power line

---

[15] H. W. Lefevre and J. T. Russell, *Rev. Sci. Instrum.* **30**, 159 (1959).

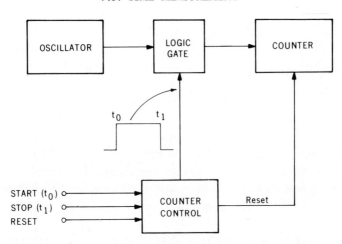

FIG. 6. Simplified block diagram of oscillator–counter time measurement system.

frequency, to crystal oscillators enclosed in ovens where the temperature is very carefully regulated. The aging rate of very good crystal controlled clocks is better than 3×10^{-9} per day.

9.3.4.1. Atomic Clocks. An item worthy of special notation is the atomic clock. This is a device incorporating an oscillator and counter technique. So accurate is the cesium beam clock designed by the National Bureau of Standards, that by international agreement the "second" has been defined in terms of this atomic frequency. The atomic process here involves a transition in the precession axis (hence electromagnetic field) of the outer electron of the cesium atom. The required radio frequency is 9,192,631,770 Hz. The cesium beam tube resonator is normally used to stabilize the output frequency of a high quality quartz oscillator and gives an overall accuracy of $\pm 1 \times 10^{-11}$ for the life of the tube.

9.4. Phase Measurements*

In dealing with variables of a sinusoidal nature, the experimenter is frequently faced with the task of measuring the phase relationships, or time delays, among them. For steady-state voltage sine waves of the same frequency, phase measurements consist of meter or counter readings, interpretation of Lissajous figures, or similar techniques which provide steady indications of phase difference.

Phase-sensitive detectors are widely used as system components. In many applications, such as phase-locked loops or signal detection, the inputs are complex waveforms of different frequencies. The phase shift signal is processed electronically. Because of its rapid variations, it cannot be measured by the above steady-state methods.

Phase shifting networks and devices are included in this chapter because of their use in phase measurement and related applications.

9.4.1. Oscilloscope Methods

The versatile cathode-ray oscilloscope provides several methods of measuring steady-state phase shift over a wide frequency range. Before making a measurement, it is necessary to check the phase shift introduced by the oscilloscope or other portions of the test setup and subtract this value from the reading obtained. For measuring very low frequencies or maintaining permanent records, an oscilloscope camera is a useful accessory.

9.4.1.1. Dual Trace and Calibrated Time Base. This basic method may be considered a prerequisite for more accurate methods. In addition to providing a phase measurement accurate to a few degrees, it reveals the presence of harmonic distortion, amplitude limiting (clipping), phase jitter, spurious oscillations, dc offset, or other conditions which may require correction.

A dual-channel oscilloscope is used, with one voltage applied to each vertical input. The reference voltage is used to synchronize the horizontal

---

* Chapter 9.4 is by J. E. Toffler and P. N. Winters.

sweep. The phase difference is determined from the time interval between zero crossings and the common period of the applied signals, by the following formula:

$$\text{Phase difference in degrees} = \frac{360 \times \text{time interval}}{\text{period of either signal}}.$$

9.4.1.2. Lissajous Patterns.[1] When sine wave voltages of the same frequency are applied to the horizontal and vertical deflection plates of an oscilloscope, the resulting pattern is a measure of the phase difference between the two waves. If the two sine waves are of equal amplitude and in phase, the pattern appears as a straight line at a 45° angle with the horizontal. If the two waves are not in phase, the pattern is generally an ellipse. The sine of the phase angle difference may be calculated from the eccentricity of the ellipse. Some typical Lissajous figures are shown in Fig. 1.

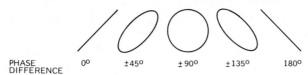

PHASE DIFFERENCE   0°   ±45°   ±90°   ±135°   180°

FIG. 1. Lissajous patterns.

If a dual-trace oscilloscope is available, this method usually has no advantage over the previous method, and has the distinct disadvantage that an ambiguity exists as to which wave is leading or lagging.

The straight-line patterns at 0 and 180° provide a means for detecting "phase null" conditions accurately and without ambiguity. A phase measurement technique based on this method of null detection is described in Section 9.4.4.3.

Lissajous patterns were first studied in the 19th century in connection with the paths traced by freely swinging pendulums.

9.4.2. Electronic Counter Method

Phase measurement with a counter is a special application of time interval measurement. The measurement is generally made between points where the signals pass through zero volts going in the same direction.

[1] A. Tiedemann, "Elements of Electrical Measurements," p. 19. Allyn and Bacon, Boston, Massachusetts, 1967.

The zero crossing is used because it is the easiest point to determine accurately with a counter, and it is generally in the region of maximum slope.

In addition to the time between zero crossings, the period is required. If the two signals are of unequal amplitude, the larger should be used for the period measurement.

Phase difference in degrees is computed from the formula of the preceding section.

9.4.3. Phase Shifting Networks and Devices

Variable phase shifting devices for sine waves fall into two categories: (1) resistance–reactance phase shift networks, which are limited to a finite number of degrees; and (2) continuous phase shift devices in which phase shift is proportional to shaft rotation for any number of turns.

9.4.3.1. Resistance–Reactance Networks. Two examples of this class of phase shifters are shown in Fig. 2. The circuit of Fig. 2a can provide a maximum phase shift of almost 180°, while the more complex circuit of Fig. 2b can provide almost 360°.[2] Both circuits are examples of *all-pass* networks, so-called because their input–output transfer functions have constant amplitude, independent of phase or frequency.

These circuits have the advantage that only a single input voltage is required, compared to the multiphase signals required by continuous phase shifters. In addition, they can often be built in the laboratory with common components. They have the disadvantages that available phase shift is limited, and is not proportional to shaft rotation of the potentiometers. The circuits should be driven from a low-impedance source and operate into a high-impedance load.

9.4.3.2. Continous Phase Shifters. Electromechanical devices such as synchro transformers or multitapped potentiometers may be used to generate a continuous phase shift proportional to shaft rotation. The phase advances linearly as the shaft is rotated any number of turns. There is no discontinuity or interruption as the shaft position passes through 360°. As mentioned above, these devices usually have the disadvantage of requiring two or more input phases (such as sine and cosine).[3]

For extremely accurate phase shifts (within 0.05° of dial reading) the

[2] W. D. Cannon, *Western Un. Tech. Rev.* **10**, 63 (April 1956).

[3] B. Chance, V. Hughes, E. F. MacNichol, D. Sayre, and F. C. Williams (eds.), "Waveforms," p. 491. McGraw-Hill, New York, 1949.

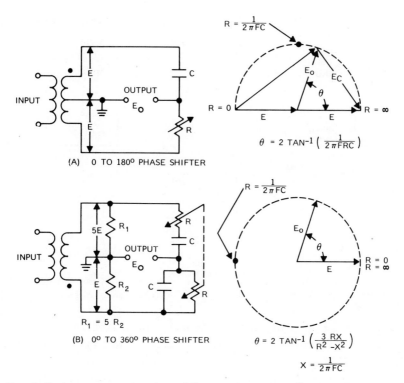

FIG. 2. Resistance–reactance phase shift networks. $E_0 = Ee^{j\theta}$, where E is the voltage at the transformer end with the dot.

phase shifter is designed to operate at a single frequency supplied by an internal frequency standard within the equipment.

9.4.4. Commercial Phase Meters

Phase measurements at frequencies from a fraction of a hertz to approximately 10 GHz can be made with commercially available test equipment. Due to the many different types of instruments, the user must be careful to select a type which meets his particular requirements. Basic principles of the more common types are described below.

9.4.4.1. Time-Delay Method. A favorite scheme for implementing a direct-reading phase meter is to convert both input signals to square waves and measure the relative time delay between them by digital techniques.[4] A block diagram is shown in Fig. 3. The signal and reference

[4] I. F. Kinnard, "Applied Electrical Measurements," p. 216. Wiley, New York, 1956.

9. MEASUREMENTS

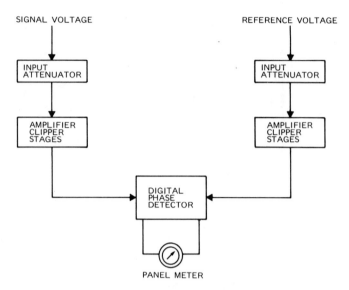

Fig. 3. Block diagram of typical commercial phase meter.

voltages are applied to separate channels and passed through several stages of amplification and symmetrical clipping to form square waves. Circuits which perform this useful function are known by a variety of names, such as clippers, slicers, limiters, or zero-crossing detectors.

The two square wave pulse trains which appear at the outputs of the clipper stages can be used in various ways to provide an average output voltage linearly proportional to phase. Digital phase detectors, designed to operate with square wave inputs, are described in detail in the following section. The output pulses from the phase detector are applied to a panel meter. Meter deflection is proportional to the average pulse amplitude. The meter scale is linear and calibrated to read in degrees. Meters based on these principles are accurate to within one or two degrees of the correct reading.

9.4.4.2. Subtraction Method. A simple phase meter can be built on the principle of subtracting two signals of equal amplitude, either sine waves or square waves, and rectifying the output. If the two signals are in phase, the subtraction yields zero. If they are 180° out of phase, the subtraction yields a sine wave (assuming sine wave inputs) twice the amplitude of either input signal. The average magnitude of the rectified output is a linear function of phase difference for angles of 10° or less. The method is particularly useful for measuring very small angles, such as a fraction of a degree. The subtraction and rectification can often be done by connecting a rectifier-type ac voltmeter between the two voltages, thereby measuring their difference.

9.4.4.3. Calibrated Phase Shifter. Phase measurements based on the use of a precision phase shifter and a phase null detector can be made over a frequency range from audio to microwave. The principle is illustrated in Fig. 4, which shows an oscilloscope used as the null detector. With the switch set to position 1, the phase shifter is adjusted until the Lissajous pattern (Section 9.4.1.2) becomes a straight line, indicating phase null. The dial reading of the phase shifter is read and recorded. With the switch set to position 2, the phase shifter is again adjusted for a straight line on the oscilloscope, and the dial reading is again noted. The difference between the two readings is the desired phase angle.

The phase shifter and the null detector are often packaged together as one equipment. These devices must be suitable for the desired frequency range. For example, at audio frequencies the phase shifter might consist of one of the types discussed in Section 9.4.3.2, whereas at microwave frequencies it would consist of a continuously variable coaxial delay line.

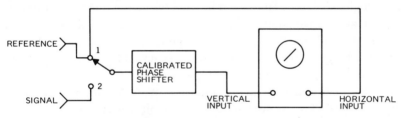

FIG. 4. Phase measurement with a calibrated phase shifter.

9.4.4.4. Phase Meters with Digital Readout. If the front panel meter on an "analog" instrument is replaced by an averaging network (low-pass filter) and a digital voltmeter, a digital readout is obtained. The digital voltmeter responds to the average voltage out of the phase detector and may be calibrated to read in degrees, radians, or percentage of cycle. Some instruments obtain a digital readout by means which are essentially equivalent to the electronic counter method of Section 9.4.2. Any phase meter with digital readout is called a *digital phase meter*.

9.4.5. Basic Phase Detector Circuits

Phase detectors take a wide variety of forms. Regardless of form, the function of the detector is to provide an output voltage which is a func-

tion of the phase difference between two input voltages. Besides their
use in phase meters, the circuits are useful system components for purposes such as error sensing in feedback control systems and coherent
detection in communication systems.

9.4.5.1. Diode Phase Detectors. Many types of diode phase detectors
are described in the literature.[5] They are often classified on the basis
of the number of diodes. While these comparisons are of theoretical interest, practical application requires consideration of over-all circuit complexity, rather than just the number of diodes. For most applications,
four-diode, double-balanced circuits, such as those described herein, offer superior rejection of undesired harmonics and intermodulation products, compared to one-diode and two-diode circuits. Figure 5a is a form
of the well-known four-diode "ring" circuit. In this form, the circuit
produces two outputs with respect to ground, equal in magnitude and
opposite in sign. The relation of the average dc output to the signal and
reference inputs depends on the relative amplitudes of the inputs as well
as their phase difference.[6] If the reference is much larger than the signal,
the output will be proportional to signal amplitude and the cosine of the
phase angle. Thus, the circuit can be used for determining the "in-phase"
component of an unknown signal. A similar circuit with the reference
shifted 90° yields the "quadrature" component. If the reference (assumed
to be constant) is much smaller than the signal, the output depends only
on phase angle and not on signal amplitude.

If the signal and reference frequencies are different, the average dc
output is zero. This feature makes the circuit a useful *coherent detector*
for extracting signals embedded in noise. Another useful feature of the
circuit is a *double-balanced* output, which means that neither input frequency appears in the output. For this reason, the circuit is useful for
applications where the phase difference (sometimes called error signal)
frequency may approach the frequency of one of the two inputs.

Circuit operation is most readily understood if it is assumed that one
of the two input voltages, either the signal or the reference, is much
larger than the other. The larger voltage, together with the diodes, performs a switching function which reverses the polarity of the smaller
voltage whenever the larger voltage passes through zero. A square wave
may be used as the switching voltage. Useful output is obtained on both

[5] E. Pappenfus, W. Breune, and E. Schoenike, "Single Sideband Principles and
Circuits," p. 127. McGraw-Hill, New York, 1964.
[6] S. Krishnan, *Electron. Radio Eng.* **36**, 45 (1959).

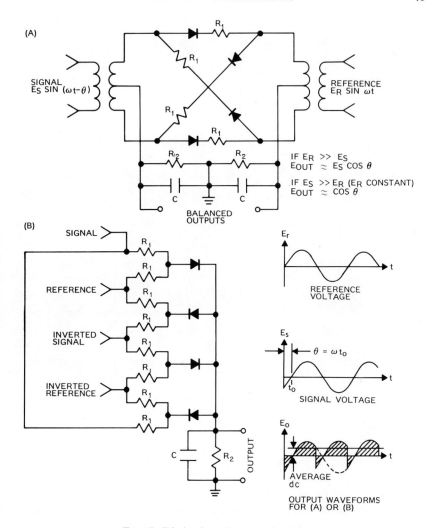

FIG. 5. Diode phase detector circuits.

the positive and negative half-cycles of the switching voltage—an important advantage over simpler circuits which have an output only on *alternate* half-cycles.

A similar circuit used by the authors is shown in Fig. 5b. This circuit has the advantage that the input signals are referenced to ground. Therefore, they may be obtained from transistor or IC *phase splitters*, which provide two push-pull signals 180° out of phase. No transformers are required, as they are in the previous circuit.

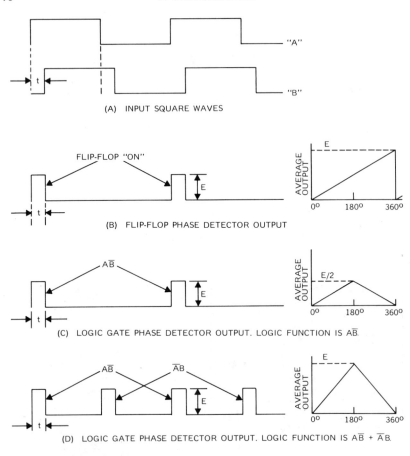

FIG. 6. Output waveforms of digital phase detectors.

9.4.5.2. Digital Phase Detectors. As indicated above, many commercial phase meters standardize the amplitudes of input sine wave signals by successive stages of amplification and limiting, until the signals become square waves. With only two levels, *high* and *low*, these waves are suitable for digital processing with discrete or IC logic elements such as flip-flops and gates.[7]

A phase detector may be built with a flip-flop having direct *set* and *reset* inputs. If the leading edge of a square wave turns the flip-flop *on*, and the leading edge of a succeeding square wave turns it *off*, the average output will be proportional to the fraction of a cycle of *on* time,

---

[7] "Reference Data for Radio Engineers," 5th ed., pp.32–41. Sams, Indianapolis, Indiana, 1968.

and hence to the phase difference. The circuit provides a linear output from 0 to 360°, but is unstable for phase angles near the lower or upper limit due to the abrupt transition in the output characteristic at this point. The difficulty can be avoided by inverting one of the applied signals. An inverter which can be switched in or out of the circuit has the added advantage of providing a convenient meter calibration check. Waveforms are shown in Fig. 6, a and b.

A phase detector may be built with logic elements by implementing functions such as $A\overline{B}$, or $A\overline{B} + \overline{A}B$.[†] Typical waveforms are shown in Fig. 6, c and d. The waveform of Fig. 6d has the advantage of two output pulses per cycle of input, thereby facilitating the succeeding operation of low-pass filtering to obtain the average dc output. The useful range is 0 to 180°. There are no abrupt transitions in the output characteristic, as in the case of the flip-flop phase detector.

[†] See "Reference Data for Radio Engineers,"[7] pp. 32–38 for Boolean algebra notation.

9.5. Voltage, Current, and Charge Measurements* †

9.5.1. Dc Measurements

The basic electrical quantity defined within the system of SI units[1,2] is the ampere, the unit of current. Other electrical units, such as the volt, are derived from the ampere using the well-known laws interrelating electrical and physical quantities. Thus the units used in the measurement of dc quantities are rigorously defined by the SI system.

Because the practical implementation of the defined units by means of so-called absolute measurements is a difficult and lengthy process, physical reference standards are maintained by various national laboratories and similar institutions from which calibrations of measuring instruments can be obtained. Usually such calibrations at the user level are derived through one or more intermediate steps of secondary standards, etc., depending on the accuracy requirements.

In experimental work probably the most common dc measurement is that of voltage. This is because voltage measurements, especially when made with instruments having high input impedance, are least likely to disturb or influence the circuit or device under test. Measurement of current usually requires that the circuit be broken to insert the measuring instrument. Instead, current is often determined by measuring the voltage drop across a known resistor. Similarly, charge can be computed from the voltage drop across a known capacitor.

The result of the measurement must be communicated to the observer and this can be effected by a continuously variable or analog indication,

---

[1] ASTM Metric Practice Guide, Appendix A2. U.S. Dept. of Commerce, Nat. Bur. Std. Handbook 102, 1967; The International System of Units (SI). U.S. Dept. of Commerce, Nat. Bur. Std. Spec. Publ. 330, 1971.

[2] W. L. Erickson and D. V. Wells, "Experimental Backgrounds for Electronic Instrumentation," Chapter 14. Laboratory Systems Research, Inc. Boulder, Colorado, 1968.

† Not prepared as a part of the author's official duties as Physicist at the National Bureau of Standards.

* Chapter 9.5 is by R. S. Turgel.

or by discrete stepwise or digital indication. These methods will be discussed in the following sections.

9.5.1.1. Analog Indicators. The technology of electrical measurements was developed around analog indicators that translated electrical signals into mechanical deflections. A wide variety of devices exist that use the electromagnetic, electrostatic or electrothermal effect for the energy conversion into motion.[3,4] In all cases the output is a function of the electrical input signal and is continuously variable over the measurement range. Generally a pointer or light-reflecting mirror or a recording pen or stylus is used. Of the various mechanisms that produce the deflection, the permanent magnet-moving coil (D'Arsonval) instrument is the most prevalent.

Compared to digital instruments, analog indicators have the advantage of simplicity and low cost. They are superior for producing graphical records or when the rate of change of the signal is of interest. However, their resolution and accuracy are limited to 0.2% of full scale at best and in this respect they fall short of the capability of digital readouts.

9.5.1.1.1. D'ARSONVAL INSTRUMENTS. The operating principle of the D'Arsonval movement is based on the torque produced by the electromagnetic interaction of a current carrying coil suspended in a field of a permanent magnet. The coil is deflected through an angle approximately proportional to the current through the coil. The restoring torque is provided by the suspension. While this movement is basically a current measuring device, it can be used to measure voltage by adding an appropriate resistor so that the voltage drop across the instrument at a given current corresponds to the voltage to be measured.

The same principle of the D'Arsonval movement is used in instruments ranging from laboratory-type wall galvanometers to small rugged portable panel meters. The difference is mainly in the type of suspension, sensitivity, and response time.[3-5]

Portable and panel instruments use taut band suspensions, or pivot and jewel movements with hair springs, to provide restoring torque. In a practical design, because structural considerations limit the torque-to-weight ratio of the moving coil, a compromise has to be made between

[3] F. K. Harris, "Electrical Measurements." Wiley, New York, 1952.

[4] D. V. Drysdale and A. C. Jolley, "Electrical Measuring Instruments," Part I, 2nd ed. revised by G. F. Tagg. Wiley, New York, 1952.

[5] D. H. Gallagher and A. E. Paschkis, Moving Coil Galvanometers. L and N Tech. Publ. A12, 1110/1965. Leeds and Northrup, Philadelphia, Pennsylvania, 1965.

sensitivity, response time, and internal resistance. A typical value is 1000 Ω for a 50 μA full scale movement with a response time of less than one second. Most panel meters use pointers sweeping over an arc of 90 or 100°, while galvanometers generally use small mirrors attached to the coil frame to deflect a light beam. The higher the sensitivity of the movement, the greater its susceptibility to shock and vibration. For this reason, high-sensitivity galvanometers usually require fixed installations in a laboratory environment. Similarly, the limit of sensitivity for rugged portable instruments is of the order of 50 μA full scale.

9.5.1.1.2. AMPLIFIERS TO INCREASE SENSITIVITY. In many applications the deflection sensitivity of a D'Arsonval-type meter by itself is not sufficient. Its usefulness can be extended, however, by means of input amplifiers to provide greater voltage and current sensitivity. Commonly, a high input impedance "buffer" amplifier precedes the voltmeter so that the current required to deflect the meter movement is furnished by the amplifier rather than the voltage source to be measured. The meter amplifier combination thus "loads" or disturbs the circuit under test only to a negligible extent. The gain available from the amplifier can also increase the voltage sensitivity of the instrument and thus extend the measurement capability to lower voltage ranges.

Stable calibration is assured with operational amplifiers (see Chapter 6.5). Their over-all response is primarily determined by a passive feedback network, and variations in amplifier gain have only second order effects. A typical diagram for a voltmeter amplifier is shown in Fig. 1.

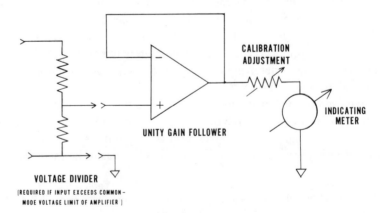

FIG. 1. Dc voltmeter with high-impedance buffer amplifier (zero adjustment circuit omitted).

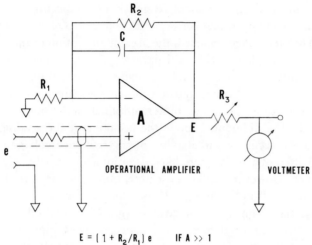

$$E = (1 + R_2/R_1) e \qquad \text{IF } A \gg 1$$

FIG. 2. Simplified circuit of dc voltmeter amplifier with voltage gain (zero adjustment circuit omitted).

If voltage gain is required, the circuit can be modified as shown in Fig. 2.

Very high impedances at room temperature ($10^{11} \Omega$, 5 pF) can be obtained with FET transistors in the input stage of the operational amplifier. However, the FET gate current doubles approximately every 7°C of temperature rise, so that bipolar transistors with high current gain become preferable at increased temperatures. The drift characteristics of commercial operational amplifiers, both FET and bipolar, are satisfactory for the majority of applications. It is only when long-term stabilities of better than a microvolt are required that chopper-stabilized amplifiers are necessary. Several commercial amplifiers capable of measuring nanovolts are available.

At the microvolt and nanovolt level particular attention must be paid to thermal emfs (see Section 9.5.5) and careful design of the physical layout is required to reduce their effect. At these low voltage levels the Johnson noise in the source resistance is a limiting factor in the attainable resolution.

9.5.1.1.3. DIFFERENTIAL VOLTMETERS. The resolution and accuracy of analog voltmeters can be effectively increased by using the meter to measure a small voltage difference between the unknown and an adjustable voltage standard instead of the total voltage. For instance, with a four-decade dial voltage standard and a 1% indicating meter having a full scale sensitivity which corresponds to one step of the lowest decade (1 in 10^4) of the voltage standard, the useable resolution of the instrument becomes $10^{-4} \times 10^{-2} = 10^{-6} = 1$ ppm.

A differential voltmeter generally contains an internal reference voltage, a precision voltage divider, and a voltmeter amplifier combination. In addition, sometimes an input buffer amplifier and range switches are included. In the differential mode the accuracy of the reading depends mainly on the accuracy and stability of the reference voltage and of the divider. Differential voltmeters like potentiometers can be adjusted to a null balance. However, unlike most potentiometers they are designed to give calibrated off-null readings. Those equipped with a buffer amplifier have a constant high input impedance regardless of whether the instrument has been adjusted to a null balance. Others approach "infinite" resistance at the null balance, providing no input voltage dividers are used. Such dividers are not necessary if the internal reference voltage is greater than the applied voltage. To cover a wide range, some instruments have an 1100 V internal reference combined with a resolution of better than 1 ppm. A simplified schematic is shown in Fig. 3.

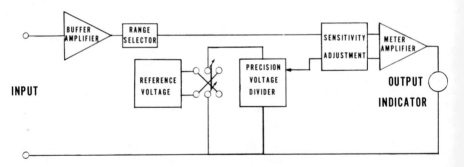

Fig. 3. Simplified diagram of differential voltmeter.

9.5.1.2. Digital Indicators. Digital indicators present the result of a measurement in the form of discrete numerals. They have the advantage of easy readability and are capable of high resolution. They are unambiguous in that they do not require interpolation between scale marks as do analog instruments, and for this reason they are potentially less prone to reading errors. They do, however, have an inherent quantizing uncertainty of $\pm \frac{1}{2}$ of the least significant digit.

An important property of most digital indicators is their ability to store the reading without any degradation for an indefinite period. Because of this storage capability and because the speed of response of the indicator and the associated electronic circuitry are not tied to a relatively slow mechanical system as in the analog case, measurements of very rapidly occurring events can be made and the results retained for readout at a convenient later time. Digital voltmeters, for example, can sample the input voltage and make a measurement with a resolution of one part in 10^4 or more in a fraction of a second. Yet the results of such a measurement can be displayed long enough for an observer to record them.

Digital indicators are by their nature easily adapted to an automatic data acquisition system so that measurements can not only be recorded automatically, but also processed further by a computer.

9.5.1.2.1. CONVERSION METHODS. Conversion of the analog input signal to digital output is accomplished in digital voltmeters by a number of methods.[6] Three of the more common ones are mentioned below.

(a) *Successive approximations.*[7,8] This method produces accurate and fast conversions but does not discriminate against noise. In fact, if the noise component is greater than the resolution, no final value will be obtained. Filtering to reduce noise can be provided at the cost of speed of response.

The successive approximation converter operates on the potentiometric principle. At each step of the conversion the input is compared to a feedback voltage derived from an internal reference source by means of (semiconductor) switches and resistance ladder networks. The increments in feedback voltage are scaled so that the switch positions correspond to the digitized value of the input when the conversion is completed. The increments are made successively smaller so that the first step corresponds to the most significant digit or bit and the last to the least significant. Conversion to an accuracy of the order of 0.01% can be made in less than a microsecond per bit.

(b) *Voltage-to-frequency converter.* Voltage (or current)-to-frequency conversion[9] is an integrating type of process and therefore provides a

[6] B. Kay and L. Harmon, *Hewlett-Packard J.* **20**, No. 7 (1969).

[7] H. Schmid, "Electronic Analog/Digital Conversion." Van Nostrand Reinhold, Princeton, New Jersey, 1970.

[8] A. K. Susskind (ed.), "Notes on Analog-to-Digital Conversion Techniques." MIT Press, Cambridge, Massachusetts, 1957.

[9] D. J. Taylor, *Rev. Sci. Instrum.* **40**, 559 (1969).

degree of noise cancellation. In theory, any symmetric noise frequency component that is harmonically related to the integrating period will cancel and will therefore be totally rejected.

The input voltage is applied to an integrating operational amplifier (see Chapter 6.5 and Section 9.5.2). When the amplifier reaches a predetermined level, a pulse generator produces a pulse of constant charge that is fed back to partially discharge the integrating capacitor. At the same time an output pulse is produced. If both positive and negative input voltages are to be measured, a polarity sensing circuit and bipolar pulse generators must be provided.[10] Alternatively, a bias voltage can be applied that shifts the input signal into the positive region for all values. Zero input voltage then corresponds to the center frequency of a symmetric frequency band.

The output pulses are counted for a predetermined period which can be chosen to make the instrument direct reading. The average output pulse rate is proportional to the average input voltage during the measurement interval.[†] For high resolution the pulse generating circuit must have a frequency range extending from 0 to typically 100 kHz at full scale. The accuracy of the output pulse rate depends on the calibration and stability of the converter feedback pulse generating circuit.

While the resolution can be increased by extending the measurement interval, in practice the over-all linearity of the converter sets the useful limit. Typically, for a converter with 0.01% linearity and a 0–100 kHz frequency range the measurement time is 100 msec. If the output is to be displayed in "engineering units" (e.g., measurement of temperature), such change (as long as it is linear) can be achieved easily by appropriate adjustment of the measurement time and thereby the total number of pulses accumulated in the counter. In other applications it may be desirable to select a measurement interval that is a multiple of an interfering ac signal and make use of the "noise" cancellation property of the integrating system.[11] If necessary, the output pulses from several integrating periods can be accumulated in order to increase the total number of counts.

[10] J. A. Rose, *EDN* **11**, No. 7, 52 (1966).
[11] K. J. Jochim and R. Schmidhauser, *H-P J.* **21**, No. 8, 7 (1970).

[†] If the input voltage fluctuates around the zero level and an offset bias is not used, a bidirectional (up-down) counter is necessary for proper integration.

(c) *Dual slope integration*. Dual slope integration[12-15] is the most popular conversion method for digital voltmeters and panel meters. It is an integrating method and therefore has inherent noise rejection capability. It is outstanding in that it can achieve high accuracy without using either precision resistor networks or a high-stability oscillator, but requires merely an accurate reference voltage.

The input signal is connected to an operational amplifier in an integrating configuration. The integrating capacitor is charged from a threshold voltage V_t to a voltage V_0 at a rate proportional to the input voltage E_i for a time interval T_i. The input of the amplifier is then connected to the reference voltage E_r which is of the opposite polarity to the input voltage. The capacitor is then discharged during the time interval T_r at a rate proportional to E_r until the amplifier output again reaches the threshold level.

If the frequency of the internal clock pulse oscillator is f, and T_i is chosen by counting N_0 pulses, then

$$T_i = N_0/f. \qquad (9.5.1)$$

Similarly,

$$T_r = N/f \qquad (9.5.2)$$

where N is the number of counts accumulated during period T_r. If the equivalent time constant of the integrator is RC, then

$$V_0 - V_t = \Delta V = E_i T_i/RC = E_i N_0/RCf^{\dagger} \qquad (9.5.3)$$

and for the discharge interval

$$V_t - V_0 = -\Delta V = -E_r T_r/RC = -E_r N/RCf. \qquad (9.5.4)$$

Combining these equations,

$$T_r = \frac{\Delta V RC}{E_r} = \frac{E_i N_0}{RCf} \cdot \frac{RC}{E_r} = \frac{E_i}{E_r} \cdot \frac{N_0}{f} = \frac{N}{f}. \qquad (9.5.5)$$

[12] S. K. Ammann, *Electronics* **37**, No. 29, 92 (1964).
[13] L. L. Schick, *IEEE Trans. Instrum. Meas.* **IM-17**, 186 (1968).
[14] H. B. Aasnaes and T. J. Harrison, *Electronics* **41**, No. 9, 69 (1968).
[15] S. K. Ammann, U.S. Patent No. 3,316,547; R. W. Gilbert, U.S. Patent No. 3,051,939.

---

† Within the limits of linearity of the operational amplifier.

Therefore

$$N = E_\mathrm{i} N_0 / E_\mathrm{r} .$$

By choosing E_r equal to the full-scale voltage and N_0 equal to the full-scale indication, the voltmeter will be direct reading and $N = E_\mathrm{i}$.

It will be seen that neither time nor clock frequency nor the integrating time constant enters into the final equation, and as long as the internal clock frequency remains constant during the period of measurement (less than 100 msec) its exact value is of no importance. By contrast, in the voltage-to-frequency converter the circuit time constant and the period of measurement directly affect the result and their long-term stability is therefore critical.

9.5.1.2.2. DIGITAL VOLTMETER PERFORMANCE. The outstanding features of digital voltmeters (DVMs) are their high resolution and an accuracy better than that of analog instruments. The resolution and repeatability of readings often exceeds the accuracy, however, which must be taken into consideration when interpreting the measurement results. Accuracy specifications[16] are frequently of the form: percent of reading + percent of full scale.[17†] Table I illustrates the actual reading uncertainties for four different methods of specifying the accuracy of a hypothetical 4-digit DVM. Evidently method C permits better accuracy in midrange than the others. Note the rather sharp increase in percent uncertainty as the reading decreases from 20% of range to 10% of range. This shows the desirability of 100% "over-ranging," that is, extending the range beyond the nominal 100 to 200% (to a reading of 19999 in the example). Range changing can then be done at the 20% point where the reading uncertainty is appreciably lower than at 10% of range.

In practice, reading uncertainties can be many times larger than those specified, if the effects of unwanted normal-mode and common-mode voltages are not considered (see Section 9.5.5). Overloading the input amplifier with a superimposed ac signal can lead to incorrect readings that are not always obvious. Such undesired voltages can arise from

[16] D. F. Schulz, *WESCON Tech. Papers* **9**, Part 6 (Instrum. Meas.), No. 8.5 (1965).

[17] American National Standard Requirements for Automatic Digital Voltmeters and Ratio Meters. Amer. Nat. Std. Inst., Inc., New York, C39.6 (1969).

---

† If the resolution is 1 μV, an additional uncertainty of a number of microvolts is often specified.

TABLE I. Reading Uncertainty of Hypothetical 4-Digit DVM for Different Accuracy Specifications

Nominal reading		Uncertainty of reading (%)				Uncertainty in last digit			
		A	B	C	D	A	B	C	D
100%	9999	0.03	0.03	0.03	0.1	3	3	3	10
50%	5000	0.06	0.05	0.04	0.1	3	2.5	2	5
30%	3000	0.1	0.08	0.05	0.1	3	2.3	1.6	3
20%	2000	0.2	0.11	0.07	0.1	3	2.2	1.4	2
10%	1000	0.3	0.21	0.12	0.1	3	2.1	1.2	1
1%	100	3.0	2.0	1.0	0.5[a]	3	2	1	0.5

Accuracy specifications: A   0.03% of range
B   0.01% of reading + 0.02% of range
C   0.02% of reading + 0.01% of range
D   0.1% of reading

[a] Taking into account quantizing effect of digital resolution.

ground loops or common-mode voltages that may be present when the source of voltage and the measuring instrument are connected to different ac grounds. To reduce this type of interference, high-accuracy digital voltmeters should be constructed with a guard circuit and effective power supply isolation in addition to the usual electrostatic shield (see also Section 9.5.5.1).[18]

Rejection of noise and other unwanted superimposed ac signals is improved with low-pass filters, integrating amplifiers, or a combination of both.[19] These methods necessarily increase the time required to make a measurement. Nevertheless, with normal- and common-mode rejection ratios of typically 50 and 120 dB (near 60 Hz), most digital voltmeters complete a measurement in a fraction of a second. This relatively high speed can be important for automatic data acquisition, and it permits a

[18] C. Walter, H. MacJuneau, and L. Thompson, *Hewlett-Packard J.* **22**, No. 5, 4 (1971).

[19] S. E. Tavares, *IEEE Trans. Instrum. Meas.* **IM-15**, 33 (1966).

different approach to experimental design than is possible with purely analog instrumentation.[20,21]

9.5.1.3. Indirect Measurements. When voltage and current measurements are made with analog or digital indicating meters, a result is directly obtained from the meter readout. There is a class of measuring instruments, however, where the value of the measured quantity must be inferred from circuit parameters. Usually one or more circuit elements are adjustable and a null detector is used in the adjustment process. This method of measurement is preferred when the highest precision and accuracy are required because the critical circuit components can be passive elements which are generally more stable, more easily calibrated and have lower noise than active elements. They therefore contribute less to the uncertainty in the result of the measurement process. Such indirect measurements require more skill and time than similar measurements using direct indicating instruments and are not as easily adapted to automatic data acquisition or automatic control.

One of the best known examples of such instruments is the laboratory potentiometer.

9.5.1.3.1. POTENTIOMETERS. Potentiometers[22-27] are the basic voltage measuring device used in the laboratory. The unknown voltage is compared to that of a standard cell or other reference using calibrated resistance ratios and a null detector. Because resistance ratios can be made to be both highly accurate and very stable, the potentiometer is well suited for precision measurements. Moreover, the null methods employed draw little current from either the unknown or the reference so that corrections due to loading of the voltage source are not necessary.

The majority of potentiometers use the constant current method. Various designs differ in details of how problems such as switch contact resistance and parasitic emfs have been handled. The basic principle of

[20] J. W. Frazer, *Sci. Technol.* No. 79, 41 (1968).

[21] W. M. Bullis *et al.*, Use of Time Shared Computer Systems to Control Hall Effect Experiment. U.S. Dept. of Commerce, Nat. Bur. Std. Tech. Note 510 (1969); F. B. Seeley and W. J. Barron, *IEEE Trans. Instrum. Meas.* **IM-19**, 245 (1970).

[22] F. K. Harris, "Electrical Measurements," Chapter 6. Wiley, New York, 1952.

[23] T. M. Dauphinee, *Instrum. Contr. Syst.* **35**, No. 7, 149 (1962).

[24] F. L. Hermach, J. E. Griffin, and E. S. Williams, *IEEE Trans. Instrum. Meas.* **IM-14**, 215 (1965).

[25] A. Abramowitz, *Rev. Sci. Instrum.* **38**, 898 (1967).

[26] E. E. Geraci, *ISA Trans.* **8**, 163 (1969).

[27] R. L. Bishop and R. C. Barber, *Rev. Sci. Instrum.* **41**, 327 (1970).

operation, however, is the same. The unknown emf is compared to the drop across part of a calibrated voltage divider which is adjusted until a null balance is obtained. Before the measurement, the current through the divider is "standardized" by comparing the voltage drop across a designated section of the divider with the standard cell or reference voltage. By suitable choice of resistance ratios, the instrument can be made direct reading in terms of voltage.

The constant resistance method used in more specialized types of potentiometers is particularly suited to measurements at the microvolt and nanovolt level. The comparison voltage is developed across a fixed resistance and there are no switch or slide wire contacts in the measuring circuit which could give rise to spurious emfs. In operation the comparison voltage is adjusted by varying the current through the resistor. The resolution and accuracy of the instrument are therefore limited by the ability to measure the potentiometer current. In some designs both potentiometer methods are combined in a single instrument.

Potentiometers are available with divider accuracies better than 1 ppm. The range of voltages that can be measured extends from a few volts down to 0.01 μV, though not in any one instrument. Higher dc voltages are measured with a voltbox which consists of a multitap fixed-ratio resistive divider with an output voltage compatible with the potentiometer range.

Although theoretically a potentiometer has infinite input impedance at the null balance point, in practice the input current is not zero but varies within the limits that correspond to the threshold of sensitivity of the null detector. The input impedance is therefore finite and variable. As a consequence, the internal resistance of a voltage source cannot always be neglected. A detector with a threshold of 1 nA together with a 1000-Ω voltage source will give rise to an uncertainty of 1 μV.

It should be noted that voltboxes have a constant input impedance, typically of the order of 300 to 1000 Ω/V, and that this impedance remains at the same value even when the potentiometer is balanced.

High-precision instruments are normally provided with a guard circuit to reduce both dc leakage currents and electrostatic pickup. This is particularly important if electronic null detectors are used, because ac pickup can completely mask the dc unbalance signal.

9.5.1.3.2. NULL DETECTORS. Many precision measurements require the detection of zero voltage difference or a zero current condition. For this purpose the detector does not have to be calibrated, but merely needs adequate sensitivity, stability, and low internal noise so that the measuring circuit can be adjusted to the null balance point. Because the insertion point of a null detector is usually not near the circuit ground, its common-mode rejection ratio is an important factor.

Galvanometers are widely used as null detectors. They have high sensitivity, excellent common-mode rejection, adequate stability, and because of their relative long time constants their bandwidth is small so that their noise rejection is good. They do require, however, mechanically stable and vibration-free mountings, and they are susceptible to damage from overload. Their recovery even from nondamaging overloads is usually slow.

Electronic null detectors are electrically and mechanically more rugged. They permit a wider choice of input impedance and can also provide an output for a chart recorder. Low-drift, low-noise amplifiers are available that have adequate sensitivity and fast response. However, common-mode and normal-mode interference present a design problem. Many dc bridges and potentiometers are not adequately shielded, and induced ac signals at the null detector terminals represent normal-mode interference which can be orders of magnitude larger than the desired dc signal. If the measuring circuit is not guarded, a large common-mode voltage may also be present at the input to the null detector. Both of these can cause improper operation of the detector and give rise to false balance indication.

Spurious dc signals can also give false indications. These can arise from thermal emfs in the detector circuit or sometimes from rf pickup which is rectified at switch contacts or from some other functions in the circuit. Under such conditions it is best to insert a high quality thermofree reversing switch into the detector circuit. By reversing the detector, thermal and other spurious emfs in the detector are effectively cancelled and the sensitivity of the detector is doubled. A null balance is reached when reversing the detector produces no change in the output indication.

The highest null detector sensitivity can be obtained with a galvanometer amplifier. This consists of a light beam galvanometer and a pair of balanced photocells connected in a negative feedback circuit.

A fraction of the output current of the photocells is fed back to the galvanometer and helps to stabilize the operating point and to keep the reflected light beam on the photocells. The gain of the system is high, because a very small deflection of the light beam is sufficient to produce a significant change in the net output current of the photocells. In one commercial version the mechanical stability of the galvanometer is enhanced by special fluid damping.

It should be noted that because galvanometer amplifiers do not require external power supplies (other than for the light source which is not electrically connected to the rest of the circuit), they avoid the problems associated with ground loops and thus retain the large common-mode rejection typical of simple galvanometers.

9.5.2. Charge Measurement

Electric charge is a quantity that is not easily measured directly. However, since it is equivalent to the time integral of a current, it can be measured with an analog or digital current integrator. An analog current integrator in its simplest form is a capacitor with a high leakage resistance. The amount of charge is then proportional to the voltage drop across the capacitor which can be measured with an electrometer amplifier (see below). This simple integrator has limitations and will

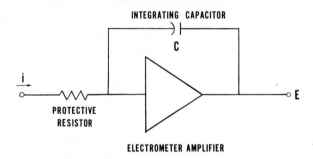

FIG. 4. Integrating charge amplifier. $E = (1/C) \int i\, dt = Q/C + \text{const.}$

introduce an error if the current source has a finite resistance or is voltage-limited. The current flow will then depend on the voltage across the capacitor and will give rise to the well-known exponential charging characteristic. Also, if the current flow is intermittent, the capacitor can discharge through the source.

The method using an operational amplifier[28-30] in an integrating configuration is therefore preferred (see Fig. 4). Since the amplifier summing

[28] F. M. Glass et al., IEEE Trans. Nucl. Sci. **NS-14**, No. 1, 143 (1967).

[29] V. Radeka, IEEE Trans. Nucl. Sci. **NS-17**, No. 1, 269 (1970); **NS-17**, No. 3, 433 (1970).

[30] F. S. Goulding et al., IEEE Trans. Nucl. Sci. **NS-17**, No. 1, 218 (1970).

node is at virtual ground the current flow will be independent of the charge already on the capacitor, which also cannot discharge through the source resistance. The accuracy of the measurement will depend not only on the leakage current of the integrating capacitor but also on the drift of the integrating amplifier. The range of measurements is dependent on the output voltage capability of the amplifier. When the average rate of charge accumulation has to be measured as, for instance, in some pulsed beam experiments, the integrating circuit accumulates charge over a given period which is then averaged. Integrating amplifiers require manual or automatic reset or charge leak circuitry to remove some or all of the charge from the capacitor.

For long integration periods better stability is obtained with a current-to-frequency converter and digital counter.[9] (See Section 9.5.1.2.1.)

9.5.2.1. Electrometer Amplifiers (see also Section 6.3.4). Electrometers are dc electronic voltmeters specifically designed to work at very high impedance levels,[31] as high as $10^{16} \Omega$ on some commercial models. They can be used for the measurement of voltage, current, or charge depending on the circuit configuration. The lower limits for current and charge measurements are typically 10^{-15} A and 10^{-14} C. Currents are measured with high-resistance shunts (of the order of $10^{12} \Omega$ for the lowest range) connected preferably in the feedback circuit rather than across the input to reduce the input voltage drop and the response time of the circuit. For charge measurements the feedback resistor is replaced by an integrating capacitor.

The high impedance of the amplifier is obtained with MOSFET transistors,[32,33] electrometer tubes, or a vibrating capacitor in the input stage.[34,35] Negative feedback further increases the inherently high input resistance of these devices. The ultimate limit is the offset or bias current which is of the order of 10^{-15} A for the MOSFET and 10^{-14} A for the electrometer tube (Raytheon CK9886). Insulation leakage currents set the limit for the vibrating capacitor.

At these high-impedance levels particular attention must be paid to the insulating materials used and the absence of surface contamination

[31] V. Radeka, *Proc. Int. Symp. Nucl. Electron., Versailles,* 1969, **1**, 46-1 (1969). See also V. Radeka, State of the Art of Low Noise Amplifiers for Semiconductor Radiation Detectors, Rep. BNL-12798, Brookhaven Nat. Lab., Upton, New York (1968).

[32] V. C. Negro *et al., IEEE Trans. Nucl. Sci.* **NS-14**, No. 1, 135 (1967).

[33] E. J. Kennedy, *IEEE Trans. Nucl. Sci.* **NS-17**, No. 1, 326 (1970).

[34] J. Dimeff and J. W. Lane, *Rev. Sci. Instrum.* **35**, 666 (1964).

[35] J. H. Fermor and A. Kjekshus, *Rev. Sci. Instrum.* **38**, 771 (1967).

such as solder flux or dust. Sapphire and quartz have the highest resistivity; Teflon, polyethylene, or polystyrene are satisfactory for less critical applications up to about $10^{14}\ \Omega$. The usual precautions with regard to electrostatic shielding and guarding (see Section 9.5.5.1) should be observed. Because of the high impedance, parasitic voltages and currents such as those generated by flexing coaxial cables will be noticeable and will have to be considered in the design of electrometers.

9.5.3. Ac Measurements

The fundamental electrical units are defined only for direct current. Equivalent alternating current units can be derived from the unit of power, since in a self-consistent system of units (such as the SI[1]) the unit of power must be independent of the way in which the power is generated. We can therefore define the effective value of alternating current or voltage as the quantity that produces the same power dissipation in a pure resistance as a numerically equal direct current or voltage. The effective value is the root-mean-square (rms) value of the amplitudes of the sinusoidal components of the alternating current or voltage.

Although most ac measurements are expressed in terms of rms values, many instruments do not have rms response. Their output indications, however, are generally calibrated in rms values with the implied assumption that the quantity measured is a pure sine wave. This condition is not fulfilled in the majority of measurements, and the presence of harmonics in some cases can lead to appreciable errors (see Section 9.5.3.2).

9.5.3.1. Rms Measurements. Only instruments that have true rms response (and sufficient bandwidth) can make ac measurements that are independent of waveform. To measure current or voltage in the lower audio and power frequency range, electrodynamic and moving iron meters can be used. At higher frequencies thermocouple and bolometer instruments are available for rms measurements.[36] Analog electronic voltmeters usually convert the ac input to dc by one of the methods mentioned in the following sections and then use a dc indicating instrument (D'Arsonval movement).

9.5.3.1.1. THERMOELEMENTS AND TRANSFER STANDARDS. Measurements of the highest accuracy are made with transfer standards by means of which ac quantities can be related physically to the equivalent dc quantities for which the electrical units are defined. Transfer standards must

[36] F. K. Harris, "Electrical Measurements," Chapter 10. Wiley, New York, 1952.

therefore respond equally to direct current and to alternating current over the frequency range in which they are used. There is no method for absolute calibration of ac–dc transfer standards, but they can be intercompared with great precision. For certain specially constructed transfer standards, maximum limits of ac–dc errors can be assigned based on knowledge of the details of construction and the electrical and thermal properties of the material used.[37,38] Routine calibrations of other transfer standards can then be made to accuracies of the order of 50–100 ppm for the audio-frequency range up to 50 kHz and accuracies of 0.2% up to 30 MHz.[39]

Practically all high-accuracy transfer elements for voltage and current measurements are thermoelements. These consist of a heater wire to which one or more thermocouples are attached to measure the temperature rise caused by the current through the heater. For voltage measurements special resistors with negligible differences in their resistance values for alternating and direct currents are connected in series with a thermoelement, usually one with a 5 or 10 mA current range. The ac–dc difference in the response of thermoelements is stable for periods of years even though their input–output characteristics often exhibit drift and remain stable within a few parts per million only for periods of the order of minutes. To overcome the drift problems during measurements, a number of procedures and circuit configurations have been devised.[40–42]

Another type of transfer standard is the electrodynamic ammeter or voltmeter.[36] This is a deflection instrument with a moving and a stationary coil. When the coils are connected in series, the interaction of their magnetic fields produces a torque proportional to the square of the current. Because of the inertia of the system, the instantaneous values of the torque are averaged and result in a deflection proportional to the mean square current (neglecting instrument errors).

Electrodynamic instruments are usually constructed with accuracies of a fraction of a percent at power frequencies. Their frequency response is

[37] F. L. Hermach and E. S. Williams, *IEEE Trans. Instrum. Meas.* **IM-15**, 260 (1966).

[38] F. J. Wilkins, T. A. Deacon, and R. S. Becker, *Proc. IEE* **112**, 794 (1965).

[39] F. L. Hermach and E. S. Williams, *Trans. IEEE* (*Commun. Electron.*) **70**, Part I, 200 (1960).

[40] J. E. Griffin and F. L. Hermach, *AIEE Trans.* **81**, Part 1 (Commun. Electron.), 329 (1962).

[41] E. S. Williams, U.S. Dept. of Commerce, Nat. Bur. Std., Tech. Note 257 (1965).

[42] F. L. Hermach, J. E. Griffin, and E. S. Williams, *IEEE Trans. Instrum. Meas.* **IM-14**, 215 (1965).

limited by the inductance of the coils to at most a few kilohertz. They are rugged and can withstand reasonable overloads, but they consume appreciable power which limits their usefulness as voltmeters in high-impedance circuits and as ammeters in low-voltage circuits.

9.5.3.1.2. RMS TO DC CONVERTERS. In electronic instrumentation it is often desirable to convert an ac input to an equivalent dc output which can then be measured by conventional dc techniques. If the ac input is other than a pure sine wave, correct results can only be obtained if the circuit has true rms response. Such converters have been developed using thermoelements, thermistors,[43] electronic multipliers, or other devices with characteristics approximating square-law response.

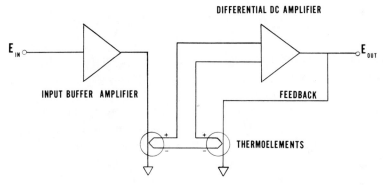

FIG. 5. Thermocouple rms to dc converter.

In a typical application two thermoelements are used in a comparison circuit (see Fig. 5) in which one of them is connected to the ac input through a buffer amplifier and the other to a dc voltage source.[44] The outputs of the thermoelements are fed into a differential amplifier which controls the dc voltage source such that if the thermoelement outputs are equal the dc voltage is the rms equivalent of the ac input. The accuracy of this circuit is limited by the characteristics of the thermoelements. Even closely matched thermoelements track only over a limited part of their operating range. The higher the desired accuracy, the more restricted that range must be. The circuit can be modified to work as a differential voltmeter with the thermoelements operating at a fixed level. In this case the gain of the ac input amplifier is varied by changing the

[43] P. Richman, *IEEE Convention Record* Part 10 (1966).
[44] K. Jessen, *EEE* **18**, 50 (1970).

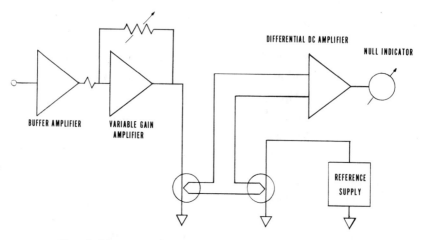

FIG. 6. Thermocouple converter with stabilized operating point.

value of the feedback resistors so that the signal fed to the thermo-element remains constant (see Fig. 6). Accuracies of 0.05% can be obtained by this method.

If both inputs to an electronic multiplier[45-48] are connected together, the output will be proportional to the square of the input voltage and therefore capable of making rms measurements. A number of such rel-atively inexpensive electronic multipliers are available with linearities in the range of fractions of a percent. They can be combined with input amplifiers and attenuators as rms-to-dc converter as shown in Fig. 7.

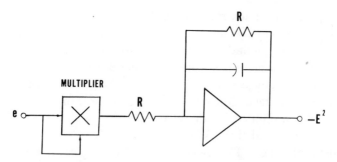

FIG. 7. "Mean square" circuit. $E^2 = (1/T) \int_0^T e^2 dt$.

[45] B. Gilbert, *IEEE J. Solid State Circuits* **SC-3**, 365 (1968).
[46] E. L. Renschler, *EEE* **17**, No. 5, 60 (1969).
[47] T. Cate and H. Handler, *EEE* **17**, No. 5, 68 (1969).
[48] R. C. Gerdes, *EDN* **15**, No. 20, 27 (1970).

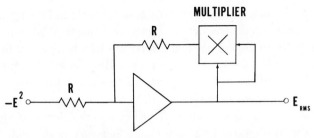

FIG. 8. Square root circuit. $E_{rms} = \sqrt{E^2}$.

The output from the multiplier is the square of the instantaneous input voltage, suitably scaled. This must be averaged in an integrating circuit if the dc signal proportional to the mean square of the input is desired. With another multiplier unit in a feedback configuration the square root can be obtained to make the circuit output proportional to the rms of the input as shown in block diagram Fig. 8. Packaged units of such rms converters are commercially available.

9.5.3.1.3. APPROXIMATE METHODS. Square-law characteristics can be approximated with selected semiconductors in their nonlinear operating region which are sold as proprietary items by a number of manufacturers. Alternatively, an approximation to square-law response can be obtained with a segmented approach using a number of biased diodes to obtain the linear segments bounded by a parabolic curve.[49,50] With these methods, obtainable accuracies are of the order of 1%. The segmented approach does permit higher accuracies at the cost of considerable complications in circuitry.[51]

9.5.3.2. Average and Peak Measurements. Measurement of ac voltage or current by the average or peak reading methods involves rectification of the input signal. Rectification can be full-wave or half-wave and if the output is an indicating meter, the scale is usually calibrated in terms of rms values. The meter scale will read correctly for pure sine waves, but in the presence of harmonics will be in error (see Section 9.5.3.2.3). Average reading meters are widely used in electronic voltmeters both digital and analog, mainly because of their simplicity and low cost.

[49] J. W. Sauber, *Electronics* **28**, No. 11, 170 (1955).

[50] T. Pavlidis, Fixed error piecewise linear uniform approximation of functions. Tech. Rep. No. 128, Informat. Sci. and Syst. Lab., Princeton Univ., Princeton, New Jersey (1969).

[51] G. Ochs and P. Richman, *Electronics* **42**, No. 20, 98 (1969).

9.5.3.2.1. SIMPLE RECTIFIER INSTRUMENTS. A dc indicating meter can measure average audio-frequency voltage or current in conjunction with a rectifier.[36] Normally copper oxide rectifiers in a bridge circuit are used because of their low forward voltage drop and low cost. The finite voltage drop of the rectifiers causes linearity distortion at low-voltage ranges, which is compensated by adjusting the scale markings. At higher audio frequencies the capacitance at the rectifier junction leads to appreciable errors in the average reading meter and therefore peak reading instruments are used for measurements at frequencies up to the megahertz range. In a peak reading instrument a capacitor is charged through a diode and the voltage drop across the capacitor is measured with a dc voltmeter. The diode and capacitor are often contained in the probe unit, thus reducing the input capacitance of the ac section of the instrument. With special diodes peak voltmeters can operate at frequencies up to the UHF range.

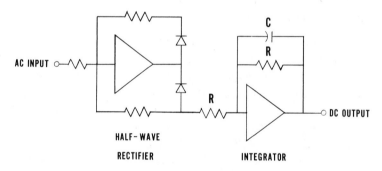

FIG. 9. Operational rectifier.

9.5.3.2.2. OPERATIONAL RECTIFIERS. Some of the limitations in making average ac measurements with simple rectifiers can be overcome by placing the rectifiers into the feedback circuit of an (operational) amplifier.[52,53] Because of the open loop gain of the amplifier, the nonlinearity at low voltages is considerably reduced in this circuit. For the same reason a low forward voltage drop is no longer a requirement, so that the copper oxide meter rectifiers can be replaced by silicon diodes which have higher reverse impedance and smaller junction capacitance. The upper frequency limits are therefore dictated by the amplifier rather than the rectifier. A typical circuit is given in Fig. 9. Ac average to dc

[52] P. L. Richman, U.S. Patent No. 3,311,835 (March 28, 1967).
[53] L. A. Marzetta, *J. Res. Nat. Bur. Std.* **73C**, 47 (1969).

voltage conversion can be obtained to an accuracy of better than 100 ppm over the audio-frequency range up to 50 kHz.[53]

9.5.3.2.3. ERRORS CAUSED BY HARMONICS. Nearly all ac meters are calibrated in terms of rms values though they may not have true rms response. Rms calibrations for average and peak reading meters give correct indications only for pure sine waves. They respond differently in the presence of harmonics and the indicated values are dependent not only on the percentage of harmonics present, but also on their order and relative phase. For odd harmonics of order n and relative amplitude k percent, the largest error which is equal to k/n percent occurs when the harmonic is in phase with the fundamental. In general, the characteristics of the harmonics, particularly their relative phase angles, are not known and therefore specific corrections cannot be determined. Only limits of error can be estimated. These have been calculated for a few special cases.[54] As an example, using values derived from the cited reference, for a total of 10% harmonic content an average reading meter would indicate values falling within the limits shown in Table II.

TABLE II. Deviations of Average Reading Ac Meter Indication from Rms Value due to Harmonic Content

Harmonic content (%)		Deviations (%)	
Second	Third	Upper limit	Lower limit
0	10	+3	−3.5
$3\frac{1}{3}$	$6\frac{2}{3}$	+1.5	−4.5
5	5	+2	−2
$6\frac{2}{3}$	$3\frac{1}{3}$	+1	−2.5
10	0	0	−1

As the percentage of harmonics increases, the errors become larger and reach a maximum. For instance, a 50% third harmonic content results in readings that may be anywhere in a range from 104 to 80% of the true rms reading.

[54] M. R. Negrete, WESCON Tech. Papers 9, Part 6 (Instrum. Meas.), No. 8.7 (1965).

A peak reading meter will indicate the maximum amplitude of the applied signal and will therefore read high if the peaks of the fundamental and harmonic are in phase since the amplitudes are added. Under these conditions, a 10% harmonic content will result in a reading that is 10% higher than the reading without harmonics irrespective of the order of the harmonics. A minimum reading will be obtained when the peak of the fundamental coincides with a peak of the harmonic of the opposite polarity. In this case the meter indication will depend not only on the amplitude of the harmonic but also on the order.

It is evident that an average reading meter will give better accuracy than a peak reading meter in the presence of harmonics and will be adequate for moderate accuracy requirements (of the order of a few percent), particularly if the total harmonic content is less than 10%. If higher accuracy is required, as for instance with a digital voltmeter having more than two digits, the voltage to be measured must have very low harmonic content for the result to be meaningful. Usually in such applications an rms responding instrument should be used instead of an average or peak reading meter.

9.5.3.3. Current Comparators.

Current comparators are designed for high-precision measurement of alternating current ratios to an accuracy of the order of 1 ppm.[55] They operate on the principle of zero magnetic flux balance in a specially constructed toroidal core. In the basic design[56] the primary and secondary windings are connected so that the magnetic flux produced by one opposes that of the other. In addition, a detector winding and a "deviation" winding are provided. If the primary to secondary current ratio differs from the value given by the reciprocal of the turns ratio, a net flux results in the core and a current through the deviation winding is required to balance it. The amplitude and phase of this deviation current is a measure of the departure of the actual current ratio from the nominal.

The use of current comparators has been extended to dc measurements by a modification in the core construction and the addition of a modulator. This provides an alternate method of direct current ratio determination.[57]

[55] N. L. Kusters, *IEEE Trans. Instrum. Meas.* **IM-13**, 197 (1964).

[56] N. L. Kusters and W. J. M. Moore, *AIEE Trans.* (*Power Apparatus Syst.*) **80**, Part 3, 94 (1961).

[57] N. L. Kusters, W. J. M. Moore, and P. M. Miljanic, *IEEE Trans.* (*Commun. Electron.*) **83**, 22 (1964).

9.5.4. Oscilloscopes[†]

Cathode ray oscilloscopes are perhaps the most versatile instruments in circuit analysis. In addition to providing a means of amplitude measurement, they permit the shape of the waveform and its repetition frequency to be determined at the same time. Moreover, with dual beam oscilloscopes or a time division multiplexing circuit more than one signal can be studied and the time correlation between signals can be observed.

Oscilloscopes permit the measurement of signal amplitude at any point of a waveform which cannot easily be done by other methods. Accuracy of measurement is generally of the order of 3% over a frequency range from dc to several megahertz. Sampling oscilloscopes extend the upper frequency limit into the gigahertz range. They achieve this by sampling repetitive high-frequency input signals and reconstructing the waveform from the amplitude and phase information obtained from the samples, which is then displayed at a slower speed. Often only one sample is taken from each cycle of the input, so that if 1000 samples are used to reconstruct the waveform, the repetition frequency of the reconstructed signal has been reduced by a factor of 1000. This technique eliminates difficulties associated with fast deflection circuits and high writing speeds that would be necessary to display the high-frequency input signal directly.

At the lower end of the frequency spectrum, signals below 20 Hz down to fractional hertz can be measured with storage oscilloscopes or by photographing the oscilloscope trace. Storage oscilloscopes use special cathode ray tubes that retain the trace for a limited time or until erased.

The storage feature is particularly useful for the analysis of servo and similar circuits with time constants of the order of a second. This response speed is too fast for most pen recorders and too slow for observation by ordinary oscilloscopes.

9.5.5. Measurement in the Presence of Electrical Noise

All electrical measurements are made in the presence of "noise" which ultimately limits the resolution of the measurement.[58] Noise is here understood in its broadest sense as any disturbing signal superimposed on the desired signal. There are various types of electrical noise that can

[58] P. G. Baird, *WESCON Tech. Papers* **9**, Part 6 (Instrum. Meas.), No. 8.6 (1965).

[†] See also Chapter 11.2.

be classified according to their origin (see also Part 13). Some can be reduced or eliminated by proper design; others are fundamental properties of materials or components and can be influenced only by external conditions.

Johnson noise is a function of absolute temperature, resistance, and bandwidth. The rms Johnson noise voltage is expressed as

$$e_N = \sqrt{4kTR\,\Delta f} \quad (V)$$

where

$$k = \text{Boltzmann's constant (J/K)}$$

$$T = \text{absolute temperature (K)}$$

$$R = \text{resistance } (\Omega)$$

$$\Delta f = \text{effective bandwidth (Hz).}$$

It is usually impractical to reduce the temperature or the resistance sufficiently to appreciably affect the noise; however, the bandwidth can often be decreased by filtering, by lock-in amplifiers (see Section 9.5.5.2) or, for dc signals, by integration of the output. In normal laboratory measurements the Johnson noise is liable to be a limiting factor only at the nanovolt level or for electrometer measurements with extremely high source resistance. In most cases Johnson noise will be negligible compared to noise from other sources.

Every amplifier, whether tube or transistor, generates internal noise which appears at the output in addition to the noise contained in the input signal. Unlike Johnson noise which is inherent in the resistor, amplifier-generated noise can be reduced by good design and careful selection of components. In feedback amplifiers the noise contribution of the input stage usually predominates, as the contribution of the following stages is reduced by the gain in the loop. Commercial broadband operational amplifiers can be obtained with an equivalent input noise of 5 μV rms, while some microvoltmeters are manufactured with noise levels in the nanovolt range.

Unwanted signals can be introduced into a circuit by electrostatic and electromagnetic pickup. Shielding (including the leads to the source) will reduce electrostatic pickup and the use of twisted or coaxial leads will tend to cancel electromagnetically induced noise signals (see Section 9.5.5.1).

Parasitic emfs are a form of "noise" that is particularly troublesome in low-level measurements where they sometimes are of the same order of magnitude as the signal to be measured. In dc circuits they can be thermal emfs that are present whenever a thermal gradient exists across a junction of dissimilar metals. They can be minimized by selecting metals having low relative thermoelectric force using, for instance, all (annealed) copper wiring, special low thermal solder, or copper-to-copper crimped connections and gold-plated copper switch contacts. Another approach is to minimize thermal gradients by isothermal enclosures, or if that is not possible, by providing thermal symmetry along critical paths, so that thermal emfs generated will tend to cancel.

When coaxial cables are being flexed, small parasitic voltages are generated by electrostatic charges. The effect is reduced in special low-noise cables with a conducting layer coating the insulation where it touches the shield.

In locations where there are strong rf fields from nearby transmitters or induction furnaces, parasitic voltages in low-level dc measuring circuits are sometimes generated by the rectifying action of nonlinear circuit elements. The rectification efficiency does not have to be high to create this kind of disturbance. Slight oxidation on the surface of switch contacts may be sufficient. In such cases it is best to surround the entire experimental setup with a shielded enclosure.

9.5.5.1. Shielding and Guarding. Noise introduced into the signal circuit, examples of which have been given in the previous section, is often referred to as *normal-mode* noise. Of the noise sources mentioned, electrostatic pickup can be substantially reduced by shielding. A numerical estimate of the effectiveness of shielded wire lines is given by Mohr.[59] To be effective, however, shielding cannot be done haphazardly. In a complex system each signal environment (for instance, transducer, amplifier, power supply) is best shielded separately. Each shield should be at a uniform potential and connected to only one ground or common point to avoid ground loops. If there is more than one ground in the system, it must be assumed that these "grounds" can be at different potentials which will cause currents to flow between ground points. It is important to keep these circulating currents out of the signal line. Because stray capacitances exist between all parts of the circuit, all possible ground current paths must be analyzed before deciding on how shields are to be connected in the system.[58,60]

[59] R. J. Mohr, *IEEE Trans. Electromagn. Compatibility* **EMC-9**, 34 (1967).

[60] R. Morrison, "Grounding and Shielding in Instrumentation." Wiley, New York, 1967.

When the low terminal of the signal source is at a different potential than the ground or common point of the measuring circuit, the difference is referred to as *common-mode voltage*. Such voltages can arise when there are several grounding points in a system, or if the source is connected to a power supply as are bridge circuits, strain gages, etc. Some differential amplifiers for such applications are designed to have a high common-mode rejection ratio (CMRR). The CMRR is highest at dc and decreases with frequency and may also be affected by gain settings. Many manufacturers fail to specify this frequency dependence. High common-mode rejection can also be achieved by "guarding." Guards (see Fig. 10) completely surround the measuring circuit and are electrically connected to the common-mode potential of the source.

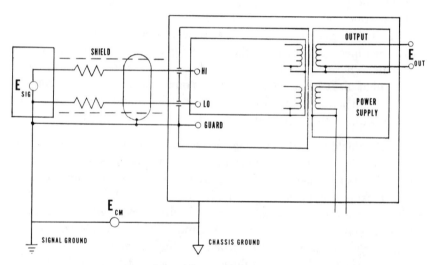

FIG. 10. Guard circuit.

The potential difference between the measuring circuit and the guard circuit will therefore be small relative to the voltage between the measuring circuit and ground so that currents from the measuring circuit through leakage resistances and stray capacitances will be much reduced. The guard thus effectively intercepts leakage currents from ground points to the measuring circuit, and currents due to the common-mode voltage flow along the guard circuit without disturbing the measurement.

When providing a guard, it is important to isolate every path into and out of the circuit it encloses. Both the power supply and the output connections must be coupled through transformers having preferably triple internal shields connected to the measuring circuit common, guard, power line, or output ground respectively. Most high-accuracy measuring devices such as digital voltmeters use guarded construction with typical common-mode rejection ratios of 120 dB.

For ac measurements, unwanted ground loop currents can be attenuated using coaxial chokes.[61] These consist of small high permeability toroidal cores through which several turns of a coaxial cable have been threaded. If the currents through the outer shield and inner conductor of the cable are not equal and opposite (because currents other than the desired signal are present), the net difference current will be attenuated by the choke. A typical value of inductance is 10–40 mH resulting in an attenuation of the order of 10^4 at 1 kHz for leads with 0.01 Ω resistance. These chokes offer a convenient way of eliminating the disturbing effects of ground loops without having to remove the loops themselves. Because the impedance of the chokes is frequency-dependent, they are not effective at low frequencies.

9.5.5.2. Phase-Lock and Signal Averaging. Periodic signals can be measured even though they are buried in "noise" and could therefore not be detected with ordinary wideband detectors.[62] The specialized methods sometimes referred to as *lock-in amplifiers* and *signal averaging* improve the signal-to-noise ratio by restricting the effective bandwidth or by taking repeated samples at a particular point of the signal waveform. The first method relies on the uniformity of the noise distribution with frequency. Consequently, the measured noise will be directly proportional to the bandwidth of the detector so that in theory at least it can be made arbitrarily small. The signal averaging method relies on the random distribution of noise amplitudes whose average will tend to zero if a sufficient number of measurements is made.

The basis of the phase-lock method is a phase-sensitive detector[63] (synchronous rectifier) driven by a reference signal with the same frequency as the input. Sometimes the reference signal is used to modulate (chop) a noisy slowly varying input.[64] A modification of the method uses the input signal itself to generate the reference.[65] The phase-sensitive

[61] D. N. Homan, *J. Res. Nat. Bur. Std.* **72C**, 161 (1968).

[62] C. A. Nittrouer, Modern Signal Processing Techniques for Overcoming Noise. Princeton Appl. Res. Corp. Tech. Note T-200.

[63] F. M. Gardner, "Phaselock Techniques," Chapter 5. Wiley, New York, 1966.

[64] R. Brower, *Electronics* **41**, 80 (1968).

[65] R. C. Hanson, *H-P J.* **18**, No. 9, 11 (1967).

detector followed by a low-pass filter is a means of limiting the bandwidth to a value centered about the reference frequency and commensurate with the rate of variation of the input information.

In the signal averaging methods,[66] the instantaneous amplitudes at any one point of the signal waveform are stored and averaged. Storage can be either analog or digital. In the analog system, high quality capacitors are used in circuits having the desired integrating time constants for the averaging process. During each sampling period charge is added to the capacitors until equilibrium is established. The amount of charge added at each step is proportional to the difference between the amplitude of the signal and the voltage across the storage capacitor. To measure slowly varying signals the capacitor is discharged at a predetermined rate. Usually more than one point on the waveform is averaged simultaneously and the result of the measurements is often displayed on a cathode ray tube.

In the digital method, the input signal amplitude is digitized with an analog-to-digital converter and the digital value is stored in a memory module after arithmetic processing with a suitable averaging algorithm. A commercial version samples 1000 points with a resolution of up to 9 bits (2^9) at each point.[67]

[66] C. R. Trimble, *H-P J.* **19**, No. 8, 2 (1968).
[67] J. E. Deardorff and C. R. Trimble, *H-P J.* **19**, No. 8, 8 (1968).

9.6. Pulse Amplitude Measurements* †

Experimental data are frequently obtained in the form of "events" which call for individual measurement although the statistical distribution of a certain parameter may be all that one is looking for. Events of simple structure, characterized by a *small, fixed number of parameters*, are called *pulses*, whereas the term *burst* denotes events of high complexity (such as a chain of pulses). Periodically occurring pulses of fixed shape and amplitude are easily and accurately measured on a calibrated cathode ray tube display, as described in Section 9.9.4. The present chapter, however, deals with the more elaborate methods required for measuring the distribution of pulses having *variable amplitude* and/or *shape*, and

---

† References are given in full at the bottom of the pages where they first occur, except for the following conference proceedings which are frequently referred to, and therefore will be quoted in abridged form (P–1, P–2, etc.) in the text:

P–1. *Proc. Gatlinburg Conf. Multichannel Pulse Height Analyzers, Sept. 26-28, 1956* (H. W. Koch and R. W. Johnston, eds.), Nucl. Sci. Ser. Rep. No. 20; Nat. Acad. Sci.—Nat. Res. Council, Publ. No. 467, Washington, D.C. (1957).

P–2. *Proc. Grossinger Conf. Utilization Multiparameter Analyzers Nucl. Phys., Nov. 12-15, 1962.* Organized by Columbia Univ., New York, under the auspices of the USAEC (L. J. Lidofsky, ed.). CU(PNPL)-227, NYO 10595, UC32, UC34 (available from Dept. of Commerce, Washington, D.C.).

P–3. *Proc. ISPRA Nucl. Electron. Symp.*, organized by EURATOM Ispra, 6-9 May 1969 (L. Stanchi, ed.); Publ. EUR 4289e, Center for Informat. and Documentation—CID, Brussels (June 1969).

P–4. *Proc. Monterey Conf. Instrum. Tech. Nucl. Pulse Anal., April 20-May 3, 1963.* Nucl. Sci. Ser. Rep. No. 40; Nat. Acad. Sci.—Nat. Res. Council, Publ. No. 1184, Washington, D.C. (1964).

P–5. *Proc. Int. Symp. Nucl. Electron.*, organized by S.F.E.R. (Soc. Fr. des Electron. et des Radioélect.), Versailles (Sept. 10-13, 1968).

P–6. *Proc. Int. Symp. Nucl. Electron.* organized by S.F.E.R. (Soc. Fr. des Electron. et des Radioélect.), Paris, 25-27 Nov. 1963. Publ. by Eur. Nucl. Energy Agency of the OECD (R. P. Perret, ed.), Paris (1964).

P–7. *Proc. EANDC Conf. Automatic Acquisition Reduction Nucl. Data, Kernforschungszentrum Karlsruhe, July 13-16, 1964* organized by Eur.–Amer. Nucl. Data Committee in collaboration with Gesellschaft für Kernforschung m.b.H.

* Chapter 9.6 is by Daniel Maeder.

occurring in *irregular sequence*. Such handicaps are typical of the signals from nuclear radiation detectors, where the probability distribution of some pulse variable is usually referred to as a *spectrum*.

9.6.1. Basic Problems of Pulse Spectrometry

9.6.1.1. Pulse-Shape Terminology. The physically significant information is generally contained in the early parts of a nuclear radiation detector signal: typically, the peak amplitude (A_p in Fig. 1) is a measure of either particle energy or ionization density, while the delay and risetimes ($t_{1/2}$ and T_r in Fig. 1) are of interest for particle identification based on time-of-flight and range. In Fig. 1, $t = 0$ indicates the initial nuclear event responsible for a pulse in one or several detectors, each of which may be evaluated for its timing information.

In practice, amplitude information is never derived from the pulse peak (A_p) alone since all measuring circuits require a finite sampling time. The measured amplitude A_m is therefore a mean value over the sampling time (T_s in Fig. 1); for accurate measurements, a nearly *flat top* having *low droop* (typically, $\Delta A/A_m < 1\%$) is desirable.

The later parts of the signal usually contain no additional information about the nuclear event. They are essentially determined by amplifier circuitry, designed with the purpose of achieving some compromise

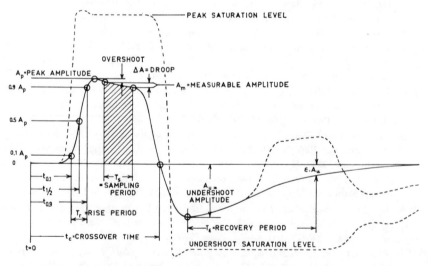

FIG. 1. Characteristics of pulses from a radiation detector. Broken lines show pulses that overload the amplifier chain, and should be rejected from amplitude measurement. $t = 0$ indicates the timing of the nuclear event.

between contradictory requirements such as:

— flat pulse top
— fast recovery (short-term tail)
— negligible base line shift (long-term tail)
— efficient noise filtering.

The *short-term recovery*, as sketched in Fig. 1, may be specified by the time after which the tail of a single pulse becomes smaller than a certain fraction ε of either the main peak or the undershoot peak. For example, an amplifier chain with a *single differentiating RC network* would produce a simpler pulse shape than the one shown in Fig. 1 (namely, a sharp peak followed by a smooth decay, for example, to $\varepsilon = 0.7\%$ within $T_\varepsilon \simeq 5\ RC$). Such a pulse shape is undesirable because of its low aspect ratio (i.e., $T_s \ll RC \ll T_\varepsilon$); *delay-line clipping* (see Section 9.6.4.5) is more favorable in that it permits a finite sampling period to be followed by a rapid decay, so that T_ε is not very much larger than T_s.

Long-term pulse tails, usually quite small for a single pulse (therefore not explicitly shown in Fig. 1), can become bothersome when a large number of them overlap, or when an overloading pulse causes an excessively large tail. Slow tails are equivalent to a shift of the reference level (or *baseline*) for the amplitude measurement of subsequent pulses. Such shifts are caused by ac couplings between amplifier stages which may appear necessary—in addition to the *RC* or delay-line clipping network already mentioned—to assure stability of the time-averaged dc operating point in a high-gain amplifier chain; in reality, stability of the *effective baseline under varying pulse rates* is the essential condition for high-precision pulse spectrometry (see Section 9.6.3.6).

Application of dc restorer techniques (Section 8.3.4) is limited to strictly unipolar pulses. On the other hand, double or higher order delay-line clipping (see Section 9.6.4.5) may be employed to reduce baseline shifts; generally speaking, any bipolar waveform containing approximately equal areas above and below the baseline (such as the one sketched in Fig. 1) is favorable in this respect, although special precautions may be necessary to preserve the balance of $\pm$ areas under overload conditions.

Noise filtering essentially amounts to a slowing down of the steeply rising and falling portions of the pulses, and is therefore incompatible with the convenience of a rectangular pulse shape. The extent to which filtering is necessary depends on the detector characteristics, in particular on the level of pulse amplitudes at the first amplifier stage (see Section 9.6.4.2).

9.6.1.2. Range of Input Variables. The design of a universal pulse amplitude measuring instrument appears difficult, on account of wide variations in experimental conditions. Typical ranges covered by present-day instrumentation are:

Amplitude level at detector output—10^{-4} V (ionization chambers) up to 1 V or more (high-gain phototubes used as scintillation detectors).

Optimum sampling period—10^{-9} sec (Cerenkov counters) to 10^{-4} sec (ionization chambers).

Amplitude information equivalent—1 channel (simple particle discriminator) to 10^4 channels (high resolution γ-ray spectrometry) or more (multiparameter analysis).

Pulse rate—10^{-3}/sec to 10^8/sec.

Time distribution

— continuous (truly random)
— exponentially decaying (after a pulsed irradiation)
— in bursts (durations between 10^{-6} and 10^{-1} sec, determined by the operation of pulsed accelerators)
— in bunches (as short as 10^{-9} sec, usually at repetition rates $>10^6$ Hz).

Capacity—10^2 to 10^8 pulses in a total distribution, with ratios up to 10^4 between different channels of the same spectrum.

Selection—coincidence requirements may cut down the fraction of accepted input pulses to less than 10^{-3} of all input pulses. Parasitic effects of rejected pulses on the measurement of the selected ones must be eliminated by means of gating circuitry.

Variables other than pulse amplitude

— timing with respect to a given "start" pulse (analog conversion into a pulse amplitude allows $\sim 10^{-11}$ sec resolution; see Section 9.3.3)
— autocorrelation using successive pulses from the same source as "start" and "stop" signals in a time-to-amplitude conversion
— pulse amplitude ratio (logarithmic conversion from which a difference pulse is derived; application for position-sensitive detectors, etc.)
— rise or collection times (passing the pulse of amplitude A_p through a differentiating network produces a signal whose peak amplitude A_p' is a measure of the maximum slope of the original pulse; use amplitude ratio technique indicated above to obtain $T_r = A_p/A_p'$).

As illustrated by these examples, almost any parameter of the incoming events can be converted into a pulse amplitude and then measured as such. However, when a detailed evaluation of complex events is desired (with a measurement of several parameters per event), resort to raw-data recording methods (Sections 9.6.5.1 and 9.6.6.4) is almost inevitable.

9.6.1.3. Actual Performance Limits. High-resolution spectrometry is very demanding in *stability of amplifier gain and channel positioning*, not only against ambient temperature and power supply fluctuations, but particularly against variations dependent on pulse rate. With extreme care, 10^4 amplitude levels can be resolved.

To measure spectra with good statistics (say, $<1\%$ uncertainty in the majority of 10^3 channels) in a reasonable time, modern equipment allows fairly *high pulse rates* (say, 10^5/sec) without a noticeable effect on amplitude resolution.

Heavy *counting losses* at such high pulse rates are in general acceptable, provided they have *no influence on the spectrum shape*, and that they can be accurately evaluated (see Section 9.6.3.2). Deadtime may well be of the order of 10 μsec even in a "fast" pulse spectrometer.

Achieving an intensity resolution of better than 1% requires a correspondingly good *channel width stability and uniformity*—say, to a few parts in 10^3. Also, in order to be able to accumulate at least 10^4 counts in every channel, some channels need a much larger counting capacity; 10^6 channel counts are possible with the more advanced instruments.

The total *number of amplitude channels* required for a spectrometer in order to match a given detector follows from a rule-of-thumb which recommends a minimum of three channels per peak. Thus, if the FWHM resolution of a Ge(Li) γ-ray spectrometer is 300 eV at $E_\gamma = 10$ keV, the channels should be at most 100 eV wide: 10,000 channels would therefore seem necessary to cover the range from 10 keV to 1 MeV in one run. However, as the line width increases in proportion to $\sqrt{E_\gamma}$, a substantial economy can be gained from nonequidistant channel arrangements or a nonlinear amplitude transformation preceding the linear channel array (for the given example, 1800 channels would be sufficient).

9.6.1.4. Desirable Output Presentations. Historically, pulse amplitude spectra were constructed manually from *analog records* of individual events on photographic paper or film.[1] Early film-analyzing machines

---

[1] E. Rutherford, F. A. B. Ward, and C. E. Wynn-Williams, *Proc. Roy. Soc.* **A129**, 211 (1930).

designed to eliminate the time-consuming manual procedures[2,3] had several drawbacks (excessive total playback time, unsatisfactory channel width definition), so that direct *curve-plotting* instruments of a very simple conception came into use.[4]

At about the same time, progress in electronics began to open the road towards a new, computer-like conception of pulse spectrometers. Early digital multichannel instruments presented the results on a large array of mechanical registers,[5-7] without any automatic read-out or computing facilities. Modern commercial pulse spectrometers present analyzed results in one or several of the following ways:

(a) curve plotting by mechanical recorder,
(b) curve or contour pattern on a CRT display,
(c) digital information to be read from a CRT display, or from an array of indicator lamps,
(d) digital print-out on paper,
(e) digital punch-out on paper tape,
(f) digital record on magnetic tape.

As most experimenters prefer results in digital form, options (a) and (b) are now mainly used for monitoring purposes; (c) is practical only for the case of a small number of channels (say, 20); (e) and (f) are particularly useful because results on tape can be fed directly into a computer for further evaluation. With relatively little additional equipment, spectrometers producing outputs (e) or (f) can be operated as a special type of computer, e.g., for background subtraction or for dissecting complex spectra into components.[8]

An alternative conception of modern pulse spectrometry is the use of a general-purpose, fast-access computer for all memory and output functions, so that the specific pulse measuring equipment (apart from the amplifier chain) is reduced to an analog-to-digital converter (ADC) and a suitable interface. This type of solution is extremely flexible; it allows

[2] W. A. Hunt, W. Rhinehart, J. Weber, and D. J. Zaffarano, *Rev. Sci. Instrum.* **25**, 268 (1954).

[3] A. B. Van Rennes, Tech. Rep. No. 3, Servomechanisms Lab.; Pulse Amplitude Anal. Nucl. Res., Massachusetts Inst. Technol., Cambridge, Massachusetts, 1955.

[4] D. Maeder, *Helv. Phys. Acta* **20**, 139 (1947).

[5] D. H. Wilkinson, *Proc. Cambridge Phil. Soc.* **46**, 508 (1950).

[6] E. Gatti, *Nuovo Cimento* (9) **7**, 655 (1950).

[7] E. H. Cooke-Yarborough, Ref. P-1, p. 48.

[8] T. R. Folsom and R. A. Cramer, Rep. on project NR 087-056. Univ. of California, Scripps Inst. of Oceanogr., LaJolla, California (Sept. 1958).

an almost unlimited number of channels, and is therefore well-suited for multiparameter analysis (Sections 9.6.2.5 and 9.6.6.8).

9.6.1.5. General Functional Diagram of a Pulse Spectrometer. Figure 2 shows the basic building blocks of a complete measuring system. Some of the timing and gating circuitry may not be required for all applications: for example, baseline and pulse interval inspection circuits are a necessity only when energy resolution is to be pushed to its theoretical limit at

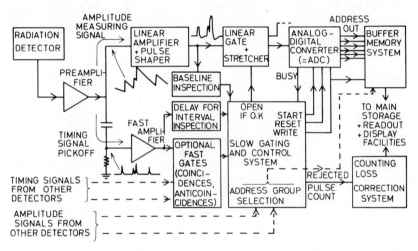

FIG. 2. Block diagram of a complete pulse measuring system. Dashed lines indicate connections required for multidimensional measurements.

high pulse rates (see Sections 9.6.3.5 and 9.6.3.6). No ADC will be found in individual-channel counters (see Section 9.6.2.3) nor in analog-type recording instruments (Section 9.6.5). Buffer memories are needed only when the main storage system does not allow sufficiently fast access; in some cases, the "main store" may simply consist of individual event records on paper punch or on magnetic tape, which must be run subsequently through a computer in order to obtain the spectra (see Section 9.6.6.4).

9.6.2. Comparison of Single Discriminator, Single Channel, and Multichannel Methods

9.6.2.1. Single Discriminator. The simplest equipment for pulse measurements consists of one discriminator (in most cases, a Schmitt trigger or a tunnel diode) with adjustable bias, and a counter circuit (Section

9.1.3) which registers the number of pulses (in a fixed period T_c) whose amplitude exceeded the chosen bias value. Such a small and simple instrument can be designed for extremely fast operation (e.g., $T_1 =$ deadtime $= 10$ nsec) and good stability (e.g., a few millivolts bias drift per day).

Even so, speed and stability are hardly satisfactory for determining details of the differential distribution from a series of single discriminator measurements: if the counts obtained at bias levels ΔU, $2\Delta U$, ... are denoted by N_1, N_2, ..., the difference $N_k - N_{k+1}$ is subject to the combined statistical error of both N_k and N_{k+1} (plus timing and bias drift errors which we shall neglect here). Dividing an approximately flat spectrum into many (say, K) intervals gives $N_{k+1} \approx N_1(1 - k/K)$, and the relative statistical error in the kth interval $(k = 1, 2, \dots, K)$

$$\varepsilon_k^{\text{(difference)}} = \frac{\sqrt{N_k + N_{k+1}}}{N_k - N_{k+1}} \approx \frac{\sqrt{2N_1(1 - k/K)}}{N_1/K} = K\sqrt{\frac{2(1 - k/K)}{N_1}}$$

increases linearly with K unless the counting periods T_c are increased in proportion to K^2 (total counting time proportional to K^3).

> *Example.* $\varepsilon_1 < 1\%$ with $K = 200$ channels requires $N_1 \gtrsim 2K^2/\varepsilon_1^2 \geq 8 \times 10^8$, and the time error in defining the counting periods δT_c has to be appreciably smaller than $T_c/\sqrt{N_1} \lesssim 0.35 \times 10^{-4}\ T_c$. Systematic corrections must be equally accurate; assuming that the deadtime T_1 is subject to a 10% uncertainty, total blocking and overlap times should be limited to $< 10^{-3}\ T_c$ so that not more than one out of every 1000 pulses is lost or falsified. The maximum input rate compatible with $T_1 = 10$ nsec is then $r_i < 10^5$/sec so that the counting period $T_c > 8 \times 10^3$ sec ≈ 2 h for $k = 1$. Assuming equal T_c at all 200 bias levels (only a factor 2 could otherwise be gained), the total counting time amounts to $KT_c > 400$ h or approximately two weeks, over which experimental conditions must be kept rigorously constant.

Even with a completely automatic system for the successive high-precision bias level adjustments (e.g., if $\Delta U = 10$ V/K $= 50$ mV, successive levels should be equally spaced within about $\pm \frac{1}{2}\varepsilon_k \cdot 50$ mV $= \pm 0.25$ mV), the single-discriminator method would be unsuitable for resolving fine details in a pulse spectrum.

9.6.2.2. Single Channel. Two discriminators combined with an anti-coincidence circuit (Fig. 3) permit direct counting of pulses in a chosen amplitude interval, called the *channel* (say, from $k \, \Delta U$ to $(k + 1) \, \Delta U$). Each channel count ΔN_k, being obtained in a single measurement, is affected only with its own statistical error, viz.,

$$\varepsilon_k^{(\text{channel})} = 1/\sqrt{\Delta N_k}$$

(in addition to systematic errors, as in Section 9.6.2.1). At a fixed total input rate, the counting periods must now be chosen proportional to K if again a flat spectrum is divided into K channels (total counting time increases in proportion to K^2).

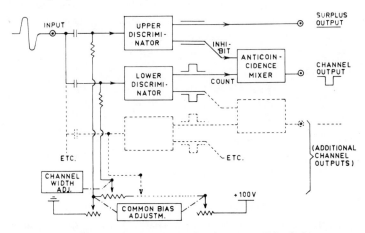

FIG. 3. Block diagram of a differential discriminator. Discriminator response is shown for the case when the input pulse amplitude falls within the channel limits, i.e., when the lower discriminator is triggered while the upper discriminator remains in its quiescent state. Dotted lines indicate connections for an additional channel. The anticoincidence boxes contain delaying and clipping circuits acting on the "count" pulses.

Example. $\varepsilon_k \leq 1\%$ requires $\Delta N_k \geq 10^4$. Blocking and overlap time corrections are to be based on the total input rate and the shape of the spectrum. At $r_i = 10^5/\text{sec}$ (see preceding example), $K = 200$ leaves an average of $r_i/K = 500$ counts/sec per channel, so that $T_c \gtrsim 20$ sec. The whole spectrum can be measured in $KT_c \gtrsim 4000$ sec ≈ 1 h.

The single channel method is thus seen to be feasible for such applications where there is no imperative need to minimize total run times, or where the number of channels is relatively small (note that, according to the above example, one run with 2000 channels would take four days). Actual measuring procedures are based mostly on maintaining a constant level difference between the two discriminators ("channel width", e.g., $\Delta U = 0.05$ V) while a common bias adjustment is swept through the whole amplitude range (say, 1 to 10 V), either in steps (when counts are recorded through a scaling circuit into an assembly of registers, or are printed, punched, or photographed after each step) or continuously (when a counting rate meter is used in connection with an analog curve plotter).

An alternative procedure, using a channel width that varies with channel position,[9] permits a reduction of counting rate variations between different portions of typical spectra. The special case of a constant fractional channel width is most easily realized by passing the input pulses through a calibrated attenuator whose transmission factor is logarithmically swept through a suitable range (say, 100 to 10%) while the discriminator channel settings are kept constant. This scheme is of particular interest with γ-ray spectrometers where it counterbalances the general trend of detector efficiencies to drop off with increasing γ-ray energy. For NaI scintillation detectors, for example, the apparent photopeak efficiency curves become almost flat over the range 0.1 to 1.5 MeV.[10]

For routine measurements of a peaked distribution with a single channel instrument, once the peak shape has been determined to be Gaussian with a full half-height width ω, it is convenient to set the channel width $\Delta U = \omega$ so that most (namely 76%) of the pulses belonging to the peak are contained in the channel when it is properly centered. It may be useful to know that a *centering error* of $\pm 0.1\omega$ would cause a loss of only 0.45%, while an equal *error in channel width* would change the counts by $\pm 6\%$ of the total peak.

9.6.2.3. Several Channels. Adding more discriminators in the manner indicated by the broken lines in Fig. 3 provides an assembly of adjacent amplitude channels, which yield in one simultaneous run the results obtained otherwise in a succession of single channel measurements. Since it would be impractical to build instruments of this type with a large number of channels, the measurement of an unknown pulse spectrum will, in general, still require a repetitive procedure.

[9] B. Aström, *Nucl. Instrum.* **1**, 143 (1957).
[10] D. Maeder, *Rev. Sci. Instrum.* **26**, 805 (1955).

Example. The spectrum considered in the preceding examples (Sections 9.6.2.1 and 9.6.2.2) can be measured at the same input rate, since blocking and overlap times of the individual discriminators are unchanged. Only 25 counting periods of $T_c \gtrsim 20$ sec are necessary if the 200 single channel steps are replaced by 25 runs of an 8-channel assembly.

Thus, total counting time for a complete spectrum would be considerably reduced. On the other hand, such an instrument is best suited for routine measurements of spectra having a limited number of peaks, whose positions (U_1, U_2, ...) and half-widths (ω_1, ω_2, ...) are known, so that each channel can be adjusted to fit one individual peak. Figure 4

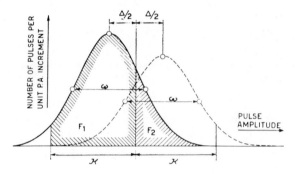

FIG. 4. Fitting two adjacent counting channels of width $\varkappa$ for the analysis of a double peak. The half-widths ω of the peaks are assumed to be equal [From D. Maeder, *Nucl. Instrum.* **2**, 299 (1958)].

refers to the basic problem of fitting two channels to two partially overlapping peaks of equal half-widths $\omega = \omega_1 = \omega_2$, at a given spacing $\varDelta = U_2 - U_1$. If the common discriminator level is set midway between peaks, the fraction of each peak covered by a channel of width $\varkappa$ can be read from Fig. 5, as illustrated by the following example.

Example. Assume that an unresolved group is interpreted (based on a detailed analysis of previous multichannel measurements) as two superimposed peaks spaced by $\varDelta = 1.5$ keV, in a spectral region where an undisturbed peak would be $\omega = 2.0$ keV wide. For a 2-channel fit, choose the channel widths rather large, say $\varkappa/\omega = 2.0$, if the shape of the spectrum permits; continuous background or neighboring peaks may in

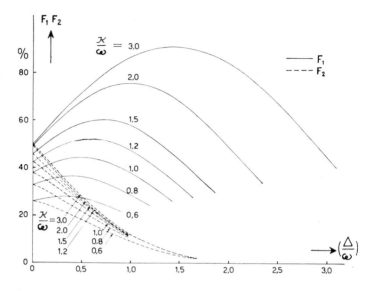

FIG. 5. Fractions of each peak corresponding to shaded areas in Fig. 4 [From D. Maeder, *Nucl. Instrum.* **2**, 299 (1958)].

other cases dictate a smaller $\varkappa/\omega$ value. From Fig. 5, at $\varDelta/\omega = 0.75$, the fractions $F_1 = 74\%$, $F_2 = 18\%$, which leads to the following conversion of actual channel counts c_1, c_2 into integrated peak counts:

$$N_1 = (c_1 F_1 - c_2 F_2)/(F_1{}^2 - F_2{}^2) = 1.44 c_1 - 0.35 c_2$$
$$N_2 = 1.44 c_2 - 0.35 c_1.$$

In more general cases, a matrix inversion procedure permits a satisfactory evaluation of a complex spectrum from measurements in a limited number of channels.[11]

9.6.2.4. Multichannel Measurements. Simultaneous counting in a large number of channels (say, $K \geq 100$) permits substantial savings in measuring time, even though instruments of this category (see Section 9.6.6) tend to have longer deadtimes—due to their increased complexity—than individual discriminators. They are, however, usually equipped with gating circuitry so that spectrum distortion is minimized and counting losses can be corrected for.

[11] H. M. Childers, *Rev. Sci. Instrum.* **30**, 810 (1959).

Example. The spectrum considered in the preceding sections can be measured at the same input rate ($r_i \lesssim 10^5$/sec) in a multichannel instrument. However, deadtime $T_1 = 64 \,\mu$sec now limits the effective acquisition rate to about $r_a \approx 1.4 \times 10^4$/sec (see Section 9.6.3.2). To accumulate an average of 10^4 counts in each of the 200 channels, a total run time of $2 \times 10^6/r_a \approx 140$ sec is sufficient.

9.6.2.5. Multidimensional Measurements.

In coincidence experiments, correlation of pulse amplitudes (or timing) from several (say D) detectors is to be analyzed. Each event then may be represented by a point in D-dimensional space; a complete classification will require $K_1 \times K_2 \times \ldots \times K_D$ channels if K_i channels are adequate for the ith detector.

Obviously, conventional ($D = 1$, or "$1D$"-type) multichannel spectrometers can be applied to multidimensional measurements by including, in the coincidence selection criteria, the requirement of an output (usually slow) from a single channel discriminator for each detector, except the one detector whose pulses are sorted in the spectrometer.[†] Single channel–multichannel combinations are, in general, unsatisfactory when a multidimensional distribution is to be measured in detail; their use is limited to cases where a few settings of each single channel yield sufficient information (in the sense explained in Section 9.6.2.3).

A more efficient investigation of "$2D$" spectra is possible with a suitably modified "$1D$" spectrometer having a large number of channels. For instance, splitting of an existing 4096 channel memory into 64×64 or 32×128 channels is a relatively simple matter[13]; with some additional modifications, existing spectrometers could even be adapted for a modest "$3D$" resolution (e.g., $64 \times 8 \times 8$). Excellent "$2D$" resolution (at least 100×100) is obtained by photographing dots from a cathode ray tube on a stationary film or by accumulating the corresponding electric charges on a storage screen for immediate display (Section 9.6.5.5). Still higher "$2D$" or "$3D$" resolution is possible by using a computer on-line with the experiment (Section 9.6.6.8).

---

[12] R. E. Bell, R. L. Graham, and H. E. Petch, *Can. J. Phys.* **30**, 35 (1952).
[13] R. L. Chase, *IRE Nat. Conv. Record*, Part 9, 196 (1959).

[†] The fast–slow coincidence technique, indicated in Fig. 2 by dashed lines marking the flow of information from additional detectors, is described in detail by Bell *et al.*[12]

The most general solution allowing an almost unlimited number of channels, as required for applications with $D \geq 3$, consists in the sequential recording of individual events (Section 9.6.6.4). Acquisition rates in multidimensional pulse measurements vary widely depending on limitations of the various analyzer systems: deadtimes of the order of 1 msec are typical, although advanced techniques allow reducing this to a few μsec. The first comprehensive review of these problems was provided in 1962 in the form of numerous papers presented at the Grossinger conference.[P-2] More recent references will be quoted in connection with a particular approach (see Section 9.6.6.7).

9.6.3. Systematic Errors and their Correction

9.6.3.1. Classification. *Deadtime* (or *busy time*, or *blocking time*) causes loss, or rejection, of pulses arriving shortly after another one. If there is no correlation between amplitudes of successive pulses, and if deadtime depends uniquely on the amplitude of the accepted pulse (but not on that of the rejected pulse), the shape of the spectrum is not distorted by these losses. The possibility of influences on spectrum shape would exist, for example,

— if every large pulse were preceded by a small one
— if the end of the blocking period were accompanied by a gradual recovery of the spectrometer threshold level.

Conversely, *spectrum shape* can have an *influence on deadtime*, and therefore on the relative counting loss at a given total input rate: the "busy" time of a typical AD-converter may vary from a few μsec (for small pulses) to $\sim$100 μsec (for large pulses). Correction schemes will be discussed in Sections 9.6.3.2–9.6.3.4.

Pile-up of successive pulses causes false amplitude measurements: the superposition of a close pair simulates a single pulse with increased amplitude ("*peak pile-up*" in Heath's terminology[14,15]). A distant pulse pair can be recorded as two separate pulses, of which the second one is usually shifted downward ("*tail pile-up*"). At high acquisition rate, the spectra therefore tend to shift and to smear out, or to show parasitic peaks (see Sections 9.6.3.5 and 9.6.3.6). Apart from these pile-up effects, pulse spectra may be distorted even at zero counting rate because of *unequal channel widths* (see Sections 9.6.3.7 and 9.6.3.8).

[14] R. L. Heath and L. O. Johnson, Ref. P–3, paper III-1, p. 141.
[15] L. O. Johnson and R. L. Heath, *IEEE Trans.* **NS-17**, No. 1, 276 (1970).

9.6.3.2. Counting Losses due to a Variable Blocking Time. In order for a pulse to be acceptable, it must not follow a previous pulse too closely: if the preceding pulse itself was rejected, there is a minimum waiting time determined by a pile-up inspector circuit, whereas the blocking time associated with an accepted pulse depends on the operation of classifying and recording circuitry. Therefore, of r_i pulses per second received at the input, those occurring during any one of the following periods will be lost:

T_0 after each rejected event, and

T_1 after each accepted event (usually $T_1 \gg T_0$).

T_1 need not be the same for all accepted pulses; in this case, an average value will be assumed whenever T_1 appears in the subsequent discussion. The acquisition rate now becomes

$$r_a = r_i/(1 + r_i T_{\text{eff}}) \tag{9.6.1}$$

where T_{eff}, the total average blocking time per accepted event, can be calculated from T_0 and T_1 as follows:

The probability δp for a T_1 period being extended to $T_1 + t$ (within δt) is the product of probabilities for

(1) a pulse at time $t - T_0$, within δt (time scale origin set at end of T_1 period),

(2) a previous extension, to any time between $t - T_0$ and t,

(3) no further pulse between $t - T_0$ and t.

Using the variable $x \equiv r_i t$ and the parameters $x_0 \equiv r_i T_0$, $x_1 \equiv r_i T_1$, these probabilities can be written:

$$(1)\ \delta x, \quad (2)\ \int_{\xi = x - x_0}^{x} p(\xi)\, d\xi, \quad (3)\ e^{-x_0}.$$

In the case $0 \leq x < x_0$, no previous extension is required, the integral is therefore replaced by unity, and it immediately follows that

(a) $\delta p(x) = e^{-x_0}\, \delta x,$ or $p(x) = e^{-x_0}, \qquad 0 \leq x < x_0.$

For $x_0 \leq x < 2x_0$, the integral is split into $\int_{x-x_0}^{x_0} \cdots + \int_{x_0}^{x} \cdots$, and the first portion is evaluated by substituting the result (a). Equating the sum to $e^{x_0} p(x)$ yields an integral equation whose solution is

(b) $p(x) = e^{-x_0}\{1 - (1 - x_0 e^{-x_0}) a^{(x - x_0)}\}, \qquad x_0 \leq x < 2x_0,$

using the abbreviation $a \equiv e^{e^{-x_0}}$. Similarly,

(c) $p(x) = e^{-x_0}\left\{1 - (1 - x_0 e^{-x_0})\left(1 + \dfrac{2x_0 - x}{a^x_0 e^{x_0}}\right)a^{x-x_0}\right\},$

$\qquad\qquad\qquad\qquad\qquad\qquad\qquad\qquad 2x_0 \leq x < 3x_0.$

The respective contributions to the average blocking time extension are

(a') $\dfrac{1}{r_i} \displaystyle\int_0^{x_0} x p(x)\, dx = \dfrac{e^{-x_0}}{r_i} \cdot \dfrac{x_0^2}{2} = \tfrac{1}{2} x_0 T_0 e^{-x_0}$

(b') $\dfrac{e^{-x_0}}{r_i} \{\tfrac{3}{2} x_0^2 + (e^{x_0} - x_0)[(a^{x_0} - 1)e^{x_0} - 2x_0(a^{x_0} - \tfrac{1}{2})]\}$
$\qquad\qquad = x_0^2 T_0 e^{-x_0}(\tfrac{2}{3} - \tfrac{1}{8} x_0 \pm \cdots)$

(c') $\dfrac{e^{-x_0}}{r_i} \{\tfrac{5}{2} x_0^2 + (e^{x_0} - x_0)[(a^{x_0} - 1)(2(e^{x_0} - x_0) + a^{x_0})$
$\qquad\qquad - 3a^{x_0}x_0(a^{x_0} - x_0 e^{-x_0})]\} = x_0^3 T_0 e^{-x_0}(\tfrac{3}{8} \pm \cdots).$

On adding to T_1 the leading terms of (a'), (b'), (c') and expanding the exponential,

$$T_{\text{eff}} = T_1 + \tfrac{1}{2} x_0 T_0(1 + \tfrac{1}{3} x_0 - \tfrac{1}{3} x_0^2 \pm \cdots). \qquad (9.6.2)$$

Example. A 1000-channel spectrometer takes $T_1 = 100\ \mu$sec for processing a pulse that falls into channel 800 (we may assume that a peaked amplitude distribution is being measured), and is made insensitive during this time, as well as during all periods of length $T_0 = 5\ \mu$sec beginning whenever an input pulse arrives within combined blocking times. As a function of input rate, effective blocking time and fractional loss (using Eq. (9.6.1)) vary as follows:

$r_i =$	$T_{\text{eff}} =$	$(r_i - r_a)/r_i =$
10^3/sec	100 μsec	9.1%
3×10^3	100.0	23.1
10^4	100.1	50.0
3×10^4	100.4	75.1
10^5	101.3	91.0

Considering that for most pulse spectrometers T_1 depends largely on average pulse amplitude (and therefore on the particular spectrum being measured), the small contribution of T_0 may be neglected for practical purposes. A direct integration of "busy time" (or of its complement) by counting gated clock signals is in general more suitable for an accurate evaluation of true input rates than theoretical calculations.

9.6.3.3. Temporary Storage of One Pulse. In order to reduce deadtime losses, spectrometers can be equipped with a fast memory to store one or several pulses while a previous pulse is being processed. With a buffer capacity of one pulse, the first among an arbitrary number of pulses (say, $1 + \nu$) arriving, after any accepted pulse, within the spectrometer blocking time T_1, will be analyzed next as soon as the spectrometer is ready. The probability for receiving $1 + \nu$ pulses during T_1, namely[16]

$$P_{1+\nu}(x_1) = e^{-x_1} x_1^{(1+\nu)}/(1 + \nu)!, \qquad (9.6.3)$$

therefore applies to the loss of exactly ν pulses per accepted event. The total loss can be represented by an effective blocking time (to be used in the acquisition rate formula, Eq. (9.6.1)), neglecting T_0 as justified by the preceding numerical example:

$$T_{\text{eff}}^{(B1)} = T_b + T_1 e^{-x_1} \sum_{\nu=1}^{\infty} \frac{\nu x_1^{\nu}}{(\nu + 1)!} \qquad (9.6.4)$$

where T_b = minimum delay between an input pulse entering the buffer store and its transfer to the spectrometer. At low counting rates $(x_1 \equiv r_i T_1 \ll 1)$,

$$T_{\text{eff}}^{(B1)} = T_b + \tfrac{1}{2} x_1 T_1 (1 - \tfrac{1}{3} x_1 + \tfrac{1}{12} x_1^2 - \tfrac{1}{60} x_1^3 \pm \cdots) \qquad (9.6.5)$$

is mainly determined by T_b, the storage period, rather than by the analyzing time T_1. For $r_i > 2T_b/T_1^2$, this situation is reversed.

Example. Incorporating a one-pulse buffer circuit ($T_b = 5 \ \mu\text{sec}$) reduces the effective blocking times and fractional losses of the preceding example to the following values:

$r_i =$	$T_{\text{eff}}^{(B1)} =$	$(r_i - r_a)/r_i =$
$10^3/\text{sec}$,	$9.8 \ \mu\text{sec}$,	0.97%
3×10^3	18.6	5.29
10^4	41.8	29.47
3×10^4	73.3	68.75
10^5	95.0	90.48

9.6.3.4. Large-Capacity Temporary Storage. A buffer of arbitrary capacity (say, for k pulses) loses μ pulses whenever exactly $k + \mu$ pulses arrive during T_1. From the probability of such an occurrence in a random

---

[16] For a simple derivation of the Poisson formula, see B. Rossi, "High Energy Particles." Prentice-Hall, Englewood Cliffs, New Jersey, 1952.

sequence (Eq. (9.6.3)), again neglecting T_0, it follows that

$$T_{\text{eff}}^{(Bk)} = T_b + T_1 e^{-x_1} \sum_{\nu=1}^{\infty} \frac{\nu x_1^{\nu+k-1}}{(\nu+k)!}. \qquad (9.6.6)$$

The low counting rate expansion

$$T_{\text{eff}}^{(Bk)} = T_b + \frac{x_1^{k}}{(k+1)!} T_1 \left\{ 1 - x_1 \frac{k}{k+2} \left[1 - \frac{x_1}{2} \frac{k+1}{k+3} (\cdots) \right] \right\} \qquad (9.6.7)$$

shows that x_1 may now approach unity before T_1 becomes important, provided that $(k+1)! > T_1/T_b$. In the preceding example, $k = 3$ would yield

$r_i =$	$T_{\text{eff}}^{(B3)} =$	$(r_i - r_a)/r_i =$
10^3/sec,	5.0 μsec,	0.50%
3×10^3	5.1	1.50
10^4	7.3	6.83
3×10^4	27.4	45.11
10^5	75.0	88.24

From these results, it is evident that the performance of a slow measuring instrument in processing a *steady data flow* could not be improved much further by a larger buffer capacity.

The situation is very different when the pulses arrive in *bursts*. In this case, the buffer must have sufficient capacity, say $K - 1$, for storing all pulses of a burst, assuming that the analyzer system is too slow to process more than a few pulses during one burst. Counting losses are then essentially determined by the storage cycle of the buffer memory, using formula (9.6.1); however, *limited buffer capacity* may cause additional losses. These can be evaluated from the probability of receiving more than $K - 1$ events when the expectation value is N events per burst: putting $x_1 = N$ in formula (9.6.3),

$$P_{>K}(N) = e^{-N} \sum_{\mu=K}^{\infty} \frac{N^{\mu}}{\mu!} = e^{-N} \frac{N^{K}}{K!}$$

$$\times \left\{ 1 + \frac{N}{K+1} \left(1 + \frac{N}{K+2} (1 + \cdots) \right) \right\}$$

$$\approx \frac{e^{K-N}}{\sqrt{2\pi K}} \left(\frac{N}{K} \right)^{K} \left\{ 1 + \frac{N}{K+1} \left(1 + \frac{N}{K+2} (1 + \cdots) \right) \right\} \qquad (9.6.8)$$

or (somewhat less accurate)

$$\approx \frac{1}{\sqrt{2\pi N}} \left(\frac{N}{K}\right)^{(K+1-N)/2} \cdot \frac{K}{K-N}.$$

Example. If a burst consists, on the average, of $N = 30$ random pulses, the probability of exhausting a storage capacity of 49 pulses per burst is

$$P_{>50} = e^{-30}30^{50}\left[1 + \frac{30}{51} + \frac{30 \cdot 30}{51 \cdot 52} + \cdots\right]/50!$$

$$\approx e^{+20}(30/50)^{50}[1 + \cdots]/\sqrt{100\pi} \approx 5.2 \times 10^{-4}.$$

Most of the times that the buffer capacity is exceeded, only one or two pulses (out of 50) are actually lost so that the average fractional loss remains quite small, even at a higher fill-up level (e.g., $N = 40$ gives $P_{>50} \approx 7\%$, with 0.5% average loss).

Buffer systems having three[17] or more[18,19] words capacity (up to 16) with access times as short as 20 nsec[18] have recently been described in the literature; they are sometimes referred to as *derandomizers*.

9.6.3.5. Spectrum Distortion due to Peak Pile-Up. Statistical superposition of two or more pulses during the sampling period (T_s in Fig. 1) produces a secondary (or tertiary, etc.) spectrum whose shape depends on the detailed shape of individual pulses and on the response function of the sampler system. For a simplified illustration of what can happen, assume that the instrument responds to the peak amplitude occurring during a sampling period T_2. Consider the two idealized input pulse shapes given in the left part of Fig. 6, namely,

(1) rectangle of width T_2,
(2) linear rise of width T_2 (arbitrary decay).

At input rate r_i, the probabilities for any accepted pulse to be followed, within T_2, by no other, or by one, or two more pulses are, respectively:

$$P_0(T_2) = e^{-x_2}, \quad P_1(T_2) = x_2 e^{-x_2}, \quad P_2(T_2) = \tfrac{1}{2}x_2^2 e^{-x_2} \quad (9.6.9)$$

where $x_2 \equiv r_i T_2$ must be $\ll 1$ in order to avoid excessive distortion.

---

[17] J. B. S. Waugh, *Trans. IEEE* **NS-17**, No. 1, 345 (1970).

[18] J. Hahn, L. Cucancic, C. Gillmann, and A. Zidon, *Trans. IEEE* **NS-17**, No. 1, 405 (1970).

[19] K. B. Keller, *Trans. IEEE* **NS-17**, No. 1, 416 (1970).

In the case of a constant input pulse amplitude U_1 (Figs. 6-A1, A2), secondary, tertiary, etc., peaks appear whose area equals P_1, P_2, ..., respectively, when the undistorted primary peak is set $= P_0$. With pulse shape (1) these satellite peaks are sharp lines, while pulse shape (2) smears them into a rectangle and a triangle, respectively.

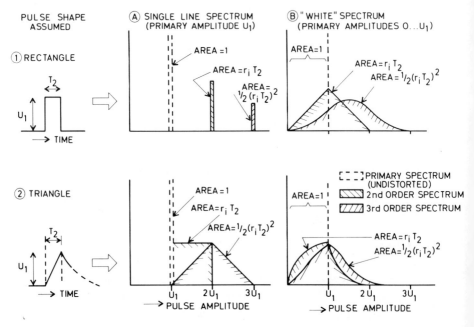

FIG. 6. Distortion of amplitude distributions caused by overlap of rectangular (top figures) or triangular (bottom figures) pulses: (A) for a single line spectrum; (B) for a flat continuum. Secondary and tertiary spectra have been normalized to unit area. Actual areas depend on the pulse rate r_1 and the overlap period T_2, as indicated.

The case of a continuous input spectrum extending from zero amplitude to U_1 is more complicated, as shown in Figs. 6-B1, B2. For instance, with pulse shape (2) the secondary spectrum follows the curves

$$\frac{dP_1}{dU} \sim \begin{cases} y(U/U_1) & \text{for } 0 \le U < U_1 \\ 1 - y(U/U_1 - 1) & \text{for } U_1 \le U < 2U_1 \end{cases} \qquad (9.6.10)$$

where $y(\alpha) \equiv \alpha(1 - \ln \alpha)$. Although all curves drawn in Fig. 6 were normalized so as to enclose equal areas, it is evident that ordinate scale factors can be specified only when $r_1 T_2$ is given, since secondary and tertiary contributions increase with different powers of the input rate r_1.

Example. $T_2 = 1$ μsec, input rate varies from $r_i = 10^4$ to 10^5 pulses/sec. At the lower rate, 1% of all events are falsified by 2-pulse superpositions (the tertiary spectrum is negligible). At the higher rate, only 90.5% of all pulses go into the true spectrum, whereas 9% and 0.5% are found in the secondary and tertiary peaks, respectively.

In practice, it must be recognized that different pulse shapes and different sampling methods can produce more complex distortion patterns, and that "sum peaks" may have physical significance. Experimental checks are done by superimposing uniform artificial pulses on an actual spectrum. The relative intensity of a parasitic satellite peak is determined from its dependence on the total input rate according to Eq. (9.6.9).

To avoid such distortions, a *time interval inspector* circuit[20] may be used to reject those events which are followed by the next one within T_2. The most elementary configuration capable of performing this function consists of a fast current pulse generator which is triggered by every input pulse and delivers standard voltage steps to a capacitor. The capacitor is discharged linearly towards its quiescent state, as fixed by a clamping device. The rejection criterion is derived directly from the clamp condition ("clamp nonconducting" forces rejection).

Alternatively, overlapping of consecutive pulses can be measured by a delayed autocoincidence technique. Depending on detector pulse shape, a delay of several microseconds may be needed. Instead of conventional passive delay lines (typically of the helical ferrite-loaded type, which allows several microseconds per meter), the use of one-shot generators now appears more practical.[14]

9.6.3.6. Tail Pile-Up. The clipping networks used in pulse amplifiers for limiting the width of individual pulses to a conveniently short duration T_0 will be discussed in Section 9.6.4.5. For simplicity, let us assume that the clipping circuit shapes the detector signal into a rectangular pulse with a perfectly accurate return to zero for $t > T_0$.

Conventional pulse amplifiers contain at least one "long AC coupling," with time constant $T_c \gg T_0$. The "longer" this coupling, the less visible is the baseline shift which it introduces in the transmission of *one* pulse. A large T_c will, however, increase the *statistical baseline fluctuations* occurring in the process of transmitting a sequence of pulses, although it

[20] G. G. Kelley, *Nucleonics* **10** (4), 34 (1952).

has no influence on the *average baseline shift* which is in all cases given by the duty cycle.

Example. $T_0 = 1\ \mu\text{sec}$, $T_c = 10^{-2}$ sec. For a typical pulse amplitude $A_m = +10.0$ V at the output terminal:

(a) a single pulse is followed by a slow tail of

$$A_u^{(1)} = - (T_0/T_c)A_m = -1\ \text{mV};$$

(b) the shift increases approximately linearly during a burst of duration $\ll T_c$; e.g., to -0.1 V when pulses arrive at a rate of 10^5/sec during a period of 10^{-3} sec;

(c) a stationary pulse rate r causes an average baseline shift of

$$A_u^{(r)} = - T_0 \cdot r \cdot A_m (= -1\ \text{V for } r = 10^5/\text{sec}).$$

Statistical *fluctuations of the shifted baseline* are, by analogy to those found in a counting rate meter, given by[21]:

$$\sqrt{\overline{\Delta A^2}} = \pm A_u^{(1)} \cdot \sqrt{\tfrac{1}{2}rT_c}$$

or ± 22 mV for our numerical example (at $r = 10^5$/sec). The corresponding dispersion of pulse amplitude measurements is sufficiently large to spoil the resolution of a modern γ-ray spectrometer, even if the enormous *average baseline shift* is accurately compensated.

It is easy to see why *bipolar pulses* are virtually unaffected by baseline shifts: let us assume a doubly rectangular pulse shape, $+A_m$ of duration T_0 followed by $-A_m$ of equal duration. Undershoots caused by transmission through the T_c circuit are calculated separately for each component by expanding the exponentials $(T_0 \ll T_c)$:

$$-(T_0/T_c)A_m \qquad\qquad \text{at } t = T_0 \quad \text{or}$$
$$-(T_0/T_c)A_m(1 - T_0/T_c) \quad \text{at } t = 2T_0 \text{ for the } +A_m \text{ pulse}$$

and

$$+(T_0/T_c)A_m \qquad\qquad \text{at } t = 2T_0 \text{ for the } -A_m \text{ pulse}$$

respectively. Superposition at $t = 2T_0$ gives: $A_u^{(\text{bipolar})} = A_m(T_0/T_c)^2$.

The baseline shift is small to a higher order and has changed sign, in comparison with the shift following a unipolar pulse. By reducing

[21] W. C. Elmore, *Nucleonics* 2 (4), 43 (1948).

the area of the negative-going part very slightly, baseline shift could theoretically be eliminated. In practice, amplifier nonlinearities tend to disturb the ideal balance depending on pulse amplitude; on the other hand, the "long" coupling time constant T_c could now be chosen much smaller $(T_c \gtrsim 10T_0)$ so that residual baseline shifts should not pile up over a large number of pulses.

Correction schemes for minimizing these effects in spectrometers using *unipolar pulses* are:

— dc restorers (Section 8.3.4)
— dc coupling throughout the amplifier (see Section 9.6.4.7)
— baseline inspection circuit.[14]

A well-designed instrument using these techniques maintains excellent resolution up to counting rates on the order of 10^5/sec: in tests using 1.33 MeV γ-rays,[15] the FWHM resolution could be kept as low as 2.64 keV at 10^5 counts/sec. From the low-rate FWHM of 2.15 keV, the rate-dependent part of the spectrometer contribution to the over-all resolution at 10^5 counts/sec would be extrapolated as

$$\sqrt{(2.64)^2 - (2.15)^2} = 1.53 \text{ keV} = 0.115\% \text{ of } E_\gamma.$$

The counting rate effect on the location of the peak is seen to be negligible.

9.6.3.7. Channel Width Errors ("Differential Nonlinearity"). In an ideal multichannel instrument, a "white spectrum" should produce a uniform channel count distribution, without humps or spikes. Deviations are sometimes expressed in terms of over-all spectrometer nonlinearity which may include imperfections of the amplifier chain. Let A_ν ($\nu = 0, 1, 2, \ldots, N$) denote the threshold amplitude for the $(1 + \nu)$th channel and $f(x)$ a smooth interpolating function such that $f(\nu) = A_\nu$. Channel width (except when formula (9.6.11b) applies) can be written as:

$$W_\nu = A_{\nu+1} - A_\nu \approx \left(\frac{df}{dx}\right)_{x=\nu+1/2}. \qquad (9.6.11a)$$

Any fine structure of $f(x)$ (such as irregularities associated with particular channels) is likely to cause large variations in W_ν and can therefore cause parasitic spikes in a uniform spectrum. After smoothing out the fine structure, the remaining expression

$$\bar{f}(x) = A_0 + W_0 x + c x^2 + d \cdot x^3 + \ldots \qquad (9.6.12)$$

may still contain nonlinear terms. These represent a systematic variation of channel widths, and can therefore cause parasitic humps in an otherwise flat spectrum:

$$W_\nu = W_0 + c(2\nu + 1) + 3d(\nu + \tfrac{1}{2})^2 + \cdots.$$

If the total number of channels is N, the average channel width becomes

$$\overline{W} = W_0 + cN + d \cdot N^2 + \cdots$$

so that the relative differential nonlinearity error may be written as

$$\varepsilon_\nu \equiv \frac{W_\nu}{\overline{W}} - 1 \approx \frac{c(2\nu - N) + d(3\nu^2 - N^2)}{W_0 + cN + d \cdot N^2}. \qquad (9.6.13)$$

This value can be measured as the relative variation of channel counts in a uniform test spectrum.

In practice, values for ε_ν of the order 1% must be considered as satisfactory, since the linearity of the most accurate "sliding pulsers" used for the test is not much better. In addition, there exist duty cycle effects: in general, the results depend not only on the ramp cycle of the pulser and the pulse rate, but also on the correlation between amplitudes of successive test pulses. Random triggering of the sliding pulser is a possible answer; however, the intricacies of the problem justify a general reference to the 1963 Monterey Conference[P-4] where nonlinearities in pulse spectrometers were discussed extensively.

Of the several known equalizing schemes devised for smoothing out spikes of the channel distribution in a pulse spectrometer, the "added-step" technique[22,23] has the remarkable feature that channel width is no longer related to the set of threshold amplitudes, A_ν. Consecutive channels are not contingent—there may either be a gap, or they may overlap.[†] On the other hand, Eq. (9.6.11a) is to be replaced by

$$W_\nu = \Delta A_{\text{control step}} = \text{const.} \qquad (9.6.11\text{b})$$

This method provides a channel width uniformity beyond the possibilities of experimental verification.

[22] E. Gatti and F. Piva, *Nuovo Cimento* **10**, 984 (1953).
[23] E. Gatti, *Nuovo Cimento* **11**, 153 (1954).

[†] There *must* be a gap when this technique is used for successive approximations (see Section 9.6.6.2).

9.6.3.8. Channel Position Errors ("Integral Nonlinearity"). With the aid of a high-precision test pulse generator, the channel position function (Eq. (9.6.12)) can be measured to an absolute accuracy of the order 10^{-3} of full range at any desired channel number. The resulting curve is fitted with a straight line, drawn so as to equalize the largest positive and negative deviations. At any channel number ν, the difference between $f(\nu)$ and the straight line is called *integral nonlinearity error*; if we neglect terms of order higher than cx^2, the errors turn out to be largest at $\nu = 0$, $N/2$, and N, where their absolute value is $= (c/8)N^2$. It is easy to see that such measurements are unsuitable for checking the desired channel width uniformity—on the contrary, differential nonlinearity is such a sensitive indicator of integral nonlinearity that it is rarely necessary to test the latter directly.

> *Example.* Assume that integral nonlinearity ascribed to the cx^2 term amounts to 0.5% of full scale (10 channels out of $N = 2000$):
>
> $$(c/8)N^2 = 0.005\,\overline{W}N, \qquad \text{or} \quad c = 0.04\overline{W}/N.$$
>
> By substituting this into Eq. (9.6.13), differential nonlinearity at both ends of the range is found to be relatively large, namely,
>
> $$\mid \varepsilon \mid_{\max} \approx cN/W = 4\%.$$

Inversely, if $\mid \varepsilon \mid$ is known to be $<1\%$ everywhere in the spectrum, we can be sure of an excellent integral linearity.

9.6.4. Analog Circuits in Pulse Spectrometers

9.6.4.1. The Basic Preamplifier Configuration. *Charge* pulses produced by nuclear detectors must be converted into *voltage* or *current* pulses in order to be processed by standard electronic equipment. As the charges involved are often quite small (e.g., some 10^3 to 10^6 electron charges) and have to be measured with high precision, it is essential that the conversion according to

$$\Delta U = \Delta Q/C_{\text{conv}} \tag{9.6.14}$$

result in a well-defined voltage step with a minimum of noise superimposed. This is the purpose of "charge-sensitive" preamplifiers: transfer of charge ΔQ from the detector (whose capacitance is shunted by wiring and amplifier input capacitances, totaling perhaps 20 to 50 pF) into a much smaller, nevertheless highly stable capacitor, C_f.

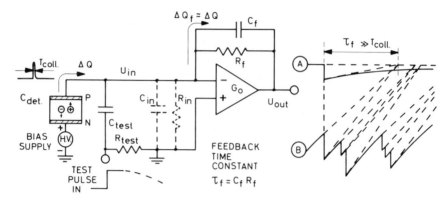

FIG. 7. Basic charge-sensitive preamplifier circuit. Output waveforms illustrate (A) the undisturbed exponential recovery following a single pulse; (B) the quasilinear saw-tooth pattern and average level shift resulting from a finite pulse rate.

The basic circuit (Fig. 7), which applies to tube-type as well as to transistorized preamplifiers, uses an inverting wideband voltage amplifier with a large gain factor, $G_0 > 100$, and a high input impedance (typically, $C_{in} < 10$ pF, $R_{in} > 100$ MΩ). In normal operation, only a small fraction of ΔQ, say ΔQ_1, is consumed in charging the combined input capacitance

$$C_1 = C_{det} + C_{test} + C_{in}. \qquad (9.6.15)$$

This part produces a step $\Delta U_{in} = \Delta Q_1/C_1$ at the amplifier input; the remainder, $\Delta Q_f = \Delta Q - \Delta Q_1$, goes into the feedback capacitor which is thus charged to

$$\Delta U_{in} - \Delta U_{out} = \frac{\Delta Q - \Delta Q_1}{C_f} = \frac{\Delta Q}{C_f} - \Delta U_{in}\frac{C_1}{C_f}.$$

On replacing ΔU_{in} by $-\Delta U_{out}/G_0$, we obtain

$$\Delta U_{out}\left(1 + \frac{C_1 + C_f}{G_0 C_f}\right) = \frac{-\Delta Q}{C_f}. \qquad (9.6.16)$$

A comparison with Eq. (9.6.14) shows that the effective *conversion capacitance* $C_{conv} \approx C_f$ if G_0 is chosen $\gg 1 + C_1/C_f$.

Similarly, R_{in} has very little effect on the *recovery time* if $G_0 \gg 1 + R_f/R_{in}$; the effective time constant becomes in this case $\tau_{rec} \approx \tau_f = R_f C_f$.

Example. $G_0 = 500 \mp 50$, $C_1 = (20 \pm 1)$ pF. By using a feed-back capacitor of $C_f = 0.958$ pF, we obtain $C_{conv} = (1.000 \pm 0.006)$ pF. A typical pulse of 3×10^5 elementary charges

($\sim$1.1 MeV γ-ray in a Ge detector) will produce a -50 mV step at the amplifier output, followed by a 1 msec exponential recovery if R_f is chosen to be 1000 MΩ. The same detector pulses can be simulated by applying, for example, a $+$ 150 mV step function across $C_{\text{test}} = 0.33$ pF.

Output *risetime* measured with sharp test pulses would critically depend on the G_0-frequency characteristic; a rough estimate based on a 3 dB cutoff at frequency f_0 indicates a closed loop amplifier risetime of

$$T_{\text{r0}} \approx \frac{0.35}{f_0 \mid G_0 \mid} \cdot \frac{C_1}{C_{\text{f}}}.$$

In actual application, however, output risetime will be given by the detector's charge collection time

$$T_{\text{r1}} \approx T_{\text{coll}}$$

which is generally longer than the response time of modern amplifiers. A typical two-stage amplifier provides $\mid G_0 \mid \cdot f_0 \sim 10^9$/sec so that T_{r0} could be made as short as 10^{-8} sec.

We may summarize the preamplifier function by its voltage response to a charge pulse ΔQ in terms of the complex frequency p as follows:

$$F(p) = \frac{\Delta Q}{p C_{\text{conv}}} \cdot \frac{a}{p+a} \cdot \frac{p}{p+1/\tau_{\text{rec}}} = \text{(I)} \times \text{(II)} \times \text{(III)}. \quad (9.6.17)$$

Here, (I) is the $\mathscr{L}$-transform of the ideal step function, while (II) and (III) specify high- and low-frequency cutoffs at

$$f_{\text{high}} = a/(2\pi) \approx 0.35/T_{\text{r0}}$$
$$f_{\text{low}} = 1/(2\pi\tau_{\text{rec}}) \approx 1/(2\pi C_{\text{f}} R_{\text{f}}).$$

Designing preamplifiers so that they handle an extremely wide frequency band (typically, 10^2 to 10^7 Hz) gives the advantage that any necessary filtering can be done in later amplifier stages (see Section 9.6.4.6) where this does not introduce additional noise sources.

9.6.4.2. Noise Sources in Preamplifiers. Except for applications where detector signals are quite large or need not be measured with great precision, noise considerations dictate the useful bandwidth, which corresponds to a certain signal shape. As shown below, the optimum choice is characterized by one single time constant, say τ_0. The following phenomena must be taken into account:

(1) leakage currents in detector and amplifier elements,

(2) shot effect of the operating current in the first (or the first two) amplifier stage(s),

(3) thermal noise of biasing resistors,

(4) flicker effect (in semiconductors as well as in vacuum tubes),

(5) nonohmic behavior of biasing resistors.

Apart from low leakage and low operating current levels, a small value for the time constant C_1/g_m is a decisive criterion for the choice of a suitable first-stage amplifying device (g_m = transconductance, C_1 = *total* input capacitance including detector + wiring).

For a very simplified discussion, we shall neglect the negative feedback via C_f, R_f (since it is known that feedback will not improve the signal-to-noise ratio), and evaluate noise sources (1) and (2) in terms of equivalent input charge fluctuations for an assumed measuring period of duration τ. Recalling that the number of elementary charges transported by a current I in a period τ is subject to an rms fluctuation $\Delta N_\tau^{(rms)} = \sqrt{N_\tau} = \sqrt{\tau I/e}$, it is seen that:

(1) the input leakage current I_1 causes charge fluctuations with a mean square of

$$Q_1{}^2 = (e\Delta N_\tau^{(rms)})^2 = e\tau I_1; \qquad (9.6.18)$$

(2) the operating current I_2 can be measured to an accuracy of $\overline{\Delta I_2{}^2} = (e\Delta N_\tau^{(rms)}/\tau)^2 = eI_2/\tau$.

On converting ΔI_2 into equivalent input voltage fluctuations across C_1, we obtain

$$Q_2{}^2 = \left(\frac{C_1}{g_m}\right)^2 \frac{eI_2}{\tau}. \qquad (9.6.19)$$

The combined noise $Q_{12}^2 = Q_1{}^2 + Q_2{}^2$ has a minimum for $Q_1{}^2 = Q_2{}^2$, that is, for a certain optimum measuring period given by:

$$\tau_{opt} = \frac{C_1}{g_m} \sqrt{I_2/I_1}; \qquad Q_{12,min}^2 = 2e\frac{C_1}{g_m}\sqrt{I_1I_2}. \qquad (9.6.20)$$

A complete, rigorous treatment[24,25] arrives at more complex formulas which essentially confirm the order-of-magnitude conclusions that can be

[24] A. B. Gillespie, "Signal, Noise and Resolution in Nuclear Counter Amplifiers." Pergamon, Oxford, 1953.

[25] E. Baldinger and W. Franzen, Amplitude and time measurement in nuclear physics, *Advan. Electron. Electron Phys.* **7**, 255 (1956).

drawn from (9.6.20). Let us denote the exact theoretical optimum values by τ_0 and Q_0.

Extensive experience with *tube amplifiers* has shown that certain tube types (such as 6AK5, 5591, E83F, E810F, 417A, etc., all triode-connected) can give noise levels as low as a few hundred electron charges, in accordance with Eq. (9.6.20). To achieve satisfactory amplifier recovery times, the actual pulse shaper time constant τ_1 is usually chosen somewhat smaller than τ_0. For example, at zero detector capacitance, measured $(\overline{Q^2})^{1/2}$ values range from about $500 \rightarrow 300 \rightarrow 200$ e as τ_1 is increased from $0.3 \rightarrow 1 \rightarrow 5$ μsec; at 100 pF added capacitance, noise levels are typically doubled. A review of practical applications is given in Fairstein,[26] along with a summary of tube caracteristics under special operating conditions. For additional data, see Kandiah.[27]

Around the year 1963, *field-effect transistors* (FETs) became competitive with tubes for low-noise applications. Using Van der Ziel's noise theory of the FET,[28] the over-all noise in a charge-sensitive amplifier was shown to be composed of the following terms[29]:

Channel thermal noise $\overline{Q_t^2} \approx 2.8kTC_1^2/g_m$

Channel $1/f$ noise $\overline{Q_f^2} = 4A_fC_1^2$

Leakage current shot noise $\overline{Q_l^2} = 2eI_1 \cdot \tau_0$

Input resistance noise $\overline{Q_R^2} = 4kT \cdot \tau_0/R_{in}$

From experimental data on the A_f parameter of then available FETs (1.3×10^{-11} V^2 for type FSP 401), a Q_0 of (200 e $+ C_{det} \times 60$ e/pF) with $\tau_0 \approx 7$ μsec was calculated, and confirmed by measurements. As the FET technology progressed further (types 2N3823; TIS-75; VX286; 2N4416; etc.) it finally yielded better signal-to-noise ratios than vacuum tube techniques. Relative freedom from microphonics and compatibility with a liquid nitrogen environment are additional advantages of FETs. Extraneous noise sources have been tracked down to lossy insulators[30] and adverse behavior of the high-valued bias resistors. The latter can

[26] E. Fairstein, Electrometers and amplifiers, *in* "Nuclear Instruments and their Uses" (A. H. Snell, ed.), Chapter 4. Wiley, New York, 1962.

[27] K. Kandiah, Ref. P–4, paper II-8, p. 65.

[28] A. Van der Ziel, *Proc. IRE* **50**, 1808 (1962).

[29] V. Radeka, Ref. P–4, paper II-9, p. 70.

[30] H. E. Kern and J. M. McKenzie, *IEEE Trans. Nucl. Sci.* **NS-17**, 260 (1970).

be eliminated by using an optoelectronic feedback system.[31] Selected FETs on ceramic mountings have a Q_0 of about 25 elementary charges, corresponding to a 200 eV FWHM resolution for x-ray spectra.[30]

State-of-the-art summaries have been given by Radeka[32] and by Elad.[33]

9.6.4.3. Overload Recovery Circuit. At the preamplifier output, individual pulses are normally $\ll 1$ V, i.e., much smaller than the maximum linear swing of a typical output stage (which we shall assume to be 10 V). As the recovery time constant $\tau_f = R_f \cdot C_f$ must be made very long in order to minimize thermal noise from R_f, many pulses can pile up in the preamplifier. In the numerical example discussed in Section 9.6.4.1, a steady rate of 200 average-size pulses per τ_f period would shift the average output level to -200×50 mV $= -10$ V. Thus, the preamplifier will be blocked most of the time whenever the pulse rate approaches $2 \cdot 10^5$/sec, unless a suitable correction circuit is used.

The obvious answer is a scheme which draws current from the amplifier input as soon as the output level shift exceeds a critical value, as determined by a conventional discriminator. The critical element is the current switch connected to the high-impedance input: it should not add a significant amount of noise to that of the charge amplifier. An FET was used in a practical solution described by Radeka,[34] resulting in an over-all recovery time of the order of 50 μsec after a large overload. Optoelectronic techniques[31] could obviously be applied for the same purpose.

9.6.4.4. Main Pulse Amplifier. The progress made in recent years in semiconductor technology, in particular in integrated amplifiers, has greatly simplified the design problems of equipment providing the necessary gain between the preamplifier (pulse amplitudes $< 10^{-1}$ V) and the recording and/or digitizing system (typical range 10 V). In fact, a configuration using three integrated circuits has proved adequate for an over-all maximum gain of $200 \times$ with extremely good linearity and stability[15]; a fourth IC was incorporated uniquely for convenience in changing signal polarity. A discrete-component buffer stage (typically a dual *NPN-PNP* emitter follower) may be required to drive a low-impedance load. A large amount of negative feedback is used in all linear amplifiers,

---

[31] F. S. Goulding, J. T. Walton, and R. H. Pehl, *IEEE Trans. Nucl. Sci.* **NS-17**, 218 (1970).

[32] V. Radeka, Ref. P–5, paper 46 (Vol. I).

[33] E. Elad, Ref. P–3, paper I-2, p. 21.

[34] V. Radeka, *IEEE Trans. Nucl. Sci.* **NS-17**, 269 (1970).

so that the gain depends only on the ratios of a few pairs of high-stability (metal film) resistors, as explained in Sections 6.4.2 and 12.1.1.3.

In certain applications, the over-all system gain (including the detector) may nevertheless be subject to drift. Various servo-systems for correcting such drifts have been described in the literature. The one most commonly used[35] with photomultipliers acts upon the high-voltage supply for controlling the over-all gain. Alternatively, variable-gain attenuators may be incorporated in the amplifier.[36-38]

9.6.4.5. Fast Clipping Circuits.

While pile-up of many pulses can and must be tolerated to a certain extent at the preamplifier level (see Section 9.6.4.3), the main amplifier should deliver well-separated pulses at its output. A pulse-shaping or -clipping network is therefore a vital part of any such amplifier.

The advantage of rectangular pulse shapes in assuring fast recovery after each event, and therefore good baseline stability at high event rates, has been demonstrated numerically in Section 9.6.3.6. The first delay-line clipping circuits[39] made use of the polarity change in the voltage signal reflected at a short circuit, as indicated in Fig. 8A. Exact ohmic termination of the line at the other end is critical if multiple reflections are to be avoided. Figures 8B and 8C illustrate more advanced configurations using differential amplifiers (high input, low output impedances). These offer great flexibility for tayloring particular pulse shapes: for example, single clipping can be obtained with a perfectly terminated line so that unwanted reflections are virtually eliminated. The pedestals (caused, e.g., by ohmic losses in the lines) can be compensated by potentiometer adjustments as shown in detail for the case 8B.

Instead of using two circuits of type 8A or 8B in cascade, double clipping can be achieved with a single unterminated line.[40] If T_d denotes the one-way electrical length of the lines B and C, then the pulse length is equal to T_d ($2T_d$ for double clipping), and recovery to 1% of peak amplitude is possible within $<1.5T_d$ ($<2.5T_d$ for double clipping). Parasitic capacitances or inductances must be kept at a minimum, or

[35] H. De Waard, *Nucleonics* **13** (7), 36 (1955).

[36] K. W. Marlow, *Nucl. Instrum. Methods* **15**, 188 (1962).

[37] A. Pakkanen and F. Stenman, *Nucl. Instrum. Methods* **44**, 321 (1966).

[38] H. Arque-Almaraz, *Nucl. Instrum. Methods* **32**, 283 (1965).

[39] W. C. Elmore, *Nucleonics* **2** (3), 16 (1948).

[40] D. Maeder, *Helv. Phys. Acta* **29**, 264 (1956); *IRE Trans. Nucl. Sci.* **NS-5**, 214 (1958).

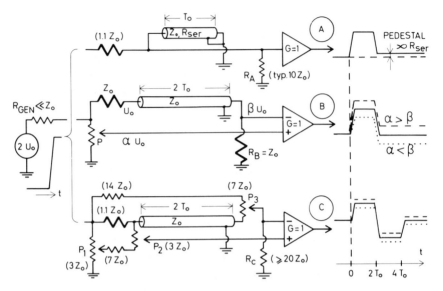

FIG. 8. Basic delay-line clipping and pedestal correction circuits. Terminating resistors are drawn in bold lines; R_A, R_B, R_C represent or include finite amplifier input
impedances. (A) single clipping with short-circuited line; (B) single clipping with
terminated line: P = pedestal adjustment; (C) double clipping with open-ended line:
P_1, P_2, P_3 allow arbitrary adjustments of symmetry and pedestal of output pulses
(for delay-lines having up to 3 dB attenuation, use resistor values given in parentheses).

should be compensated by complementary elements; otherwise they may
cause spikes of several percent at $t = 2T_d$, $3T_d$, etc. For a discussion
of practical design problems, see Fairstein.[26]

9.6.4.6. Optimum Pulse Shaping. Clipping circuits cut down the low-
frequency components of both the signal and the noise; whether a simple
RC or a delay-line "differentiator" is used, the time constant (or the
pulse width) should not be made shorter than τ_0 (see Section 9.6.4.2)
in order to maximize the signal-to-noise ratio ($\equiv S/N$).

   In addition, the high-frequency components should be cut off as well,
by a suitable low-pass filter: again, the time constant of a simple RC
"integrator" should be chosen $\approx \tau_0$. It can be shown that S/N is improved by using several (say, n) similar integrators in cascade. The pulse
shape resulting from a unit step input to an RC–RC circuit having equal
differentiator and integrator time constants τ is shown in Fig. 9 for $n = 0$,
1, 2, 3. For $n = 1$, the equivalent noise charge turns out to be 36%
higher than Q_0; with two integrators, it would merely drop to $1.22Q_0$.
Therefore, the simple RC–RC shapers have become quite popular in

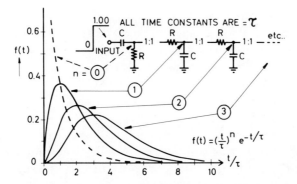

FIG. 9. Basic configuration and unit step response of RC–RC-type single-clipping/ multiple smoothing networks. The 1 : 1 decouplers may be regarded as unit voltage gain amplifiers having ∞ current gain. Note that the same even-n response functions can be obtained with half the number of integrator sections if series inductors are introduced ($C = 2\tau/R$, $L = \tau R/2$).

high-resolution spectrometry, although the exponential decay is un-satisfactory for high counting rates.

To obtain comparable S/N performance with any *delay-line clipper* (assuming $T_{\mathrm{d}} = \tau_0$), it must be combined with one or several, say n, integrators. For $n = 1$, τ would have to be chosen $> \tau_0$ according to[39]; a *unipolar rectangular pulse* would thus be distorted into one having a slow exponential tail (Fig. 10A) so that the advantage over the simple RC–RC shaper is lost. The alternative of using a large number of

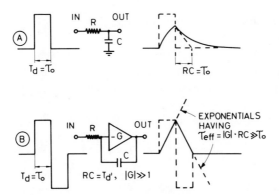

FIG. 10. Optimum high-frequency noise filtering for delay-line clipped pulses. (A) Unipolar pulse with single RC integrator of optimum time constant (note that *one RLC* section, equivalent to two RCs, would give slightly better S/N); (B) bipolar pulse with operational integrator; the triangular output comes close to the theoretical S/N limit.

integrators, each one having $\tau = \tau_0/n \ll \tau_0$, appears impractical, although the resulting Gaussian shape has a relatively short tail and gives satisfactory noise filtering ($Q_{\text{Gauss}} = 1.12Q_0$).

It is interesting to note, however, that a *bipolar rectangular* pulse is converted, by a single slow integration ($RC \gg \tau_0$), into a triangular waveform (Fig. 10B) which has no slow tail and yields excellent S/N, approaching the theoretical limit: $Q_{\text{triang}} = 1.08Q_0$. A complete review of the basic calculations was given in Baldinger and Franzen[25]; more recent studies by various authors[41,42] are indicative of the complexity of the problem. Time-variant filters, proposed in 1948,[43] are now approaching the stage of practical application.[44–48]

9.6.4.7. Pole-Zero Cancellation. It is easily seen that an "ideal differentiator" operating on the preamplifier output (Fig. 7) will produce clipped pulses superimposed on a shifting baseline: at a rate r of voltage step signals ΔU, stationary preamplifier operation implies the following slope (between steps) and level shift, respectively:

$$\text{average output slope} \quad \langle dU/dt \rangle_{\text{preamp}} = -r \cdot \Delta U \qquad (9.6.21)$$

$$\text{average output level} \quad \langle U \rangle_{\text{preamp}} = +r \cdot \tau_{\text{f}} \cdot \Delta U. \qquad (9.6.22)$$

A differentiator with time constant τ_1 converts the given slope into the following:

$$\text{average baseline shift} \quad \langle U_1 \rangle = -r \cdot \tau_1 \cdot \Delta U. \qquad (9.6.23)$$

To illustrate this effect more clearly, we have arbitrarily chosen $\tau_1 = 0.1\tau_{\text{f}}$ in constructing Fig. 11 (in practice, $\tau_1 < 10^{-3}\tau_{\text{f}}$).

The simple differentiator, Fig. 11A, does not transmit the input dc level, which is much larger than $\langle U_1 \rangle$, and is of opposite polarity. From Eqs. (9.6.22) and (9.6.23) it is evident that we can counteract the baseline

[41] C. H. Nowlin, J. L. Blankenship, and T. V. Blalock, *Rev. Sci. Instrum.* **36**, 1063 (1965).

[42] M. Bertolaccini, C. Bussolati, S. Cowa, I. De Lotto, and E. Gatti, *Nucl. Instrum. Methods* **61**, 84 (1968).

[43] D. Maeder, *Helv. Phys. Acta* **21**, 174 (1948).

[44] M. Konrad, *IEEE Trans. Nucl. Sci.* **NS-15**, No. 1, 268 (1968).

[45] K. Weise, *Nucl. Instrum. Methods* **61**, 241 (1968).

[46] H.-J. Schuster, *Nucl. Instrum. Methods* **63**, 342 (1968).

[47] V. Radeka and N. Karlovac, *IEEE Trans. Nucl. Sci.* **NS-15**, No. 3, 455 (1968).

[48] K. Kandiah, High Resolution Spectrometry with Nuclear Radiation Detectors. AERE-Rep. R6699 (1971).

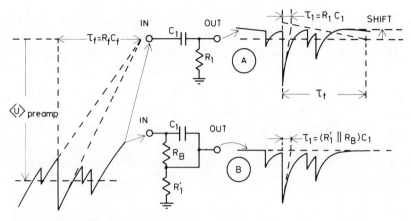

FIG. 11. Principle of pole-zero cancellation. (A) A conventional differentiator shifts the baseline in proportion to the dc pile-up of input pulses; (B) a modified differentiator compensates baseline shift, by transmitting a fraction of the input dc pile-up to the output.

shift by adding a fraction τ_1/τ_f of $\langle U \rangle_{\text{preamp}}$ to the differentiator output. Figure 11B shows how R_1 is replaced by a voltage divider (R_B, R_1') between input and ground; by choosing $R_B = R_1\tau_f/\tau_1 = \tau_f/C_1$, $\langle U_1 \rangle$ is cancelled.

For a formal demonstration, consider the input signal

$$F_{\text{in}} = \varDelta U/(p + 1/\tau_f)$$

which is transformed by the network B into

$$F_{\text{out}}(p) = F_{\text{in}}(p) \cdot \left(p + \frac{1}{C_1 R_B}\right)\left(p + \frac{R_1' + R_B}{R_1' R_B C_1}\right)^{-1}.$$

For $C_1 R_B = \tau_f$, the "zero" of this expression cancels the pole of F_{in}, and the resulting signal

$$F_{\text{out}}^{\text{(canceled)}} = \varDelta U/(p + 1/\tau_1)$$

represents an exact exponential with time constant

$$\tau_1 = C_1 \frac{R_1' R_B}{R_1' + R_B}.$$

Example. The preamplifier considered in Section 9.6.4.1 has $\tau_f = 1$ msec. A suitable $2\,\mu$sec RC clipper would consist of $R_1' = 1\,\text{k}\varOmega$, $R_B = 500\,\text{k}\varOmega$, $C_1 = 2000$ pF.

The same procedure must be used at every high-pass-type interstage coupling if baseline shift is to be avoided in over-all system performance.

In the case of delay-line clippers, equivalent shift cancellation is obtained by undercompensation of the pedestal (Fig. 8).

9.6.4.8. Timing and Gating Circuitry. In coincidence measurements, acceptance or rejection of pulses must be derived from accurate timing information. If pulses of varying amplitudes but constant delays are considered (referring to the 50% point of the leading edge in Fig. 1), it is obvious that a discriminator having a fixed upper threshold A_{disc} will trigger at different delay times depending on the peak amplitude A_{p}. For example,

$$t_{\mathrm{trig}}^{(\mathrm{up})} = \begin{cases} t_{0.1} & \text{for } A_{\mathrm{p}} = 10 \times A_{\mathrm{disc}} \\ t_{0.5} & \text{for } A_{\mathrm{p}} = 2 \times A_{\mathrm{disc}} \\ t_{0.9} & \text{for } A_{\mathrm{p}} = 1.11 \times A_{\mathrm{disc}}. \end{cases}$$

This time jitter, termed "walk" by Fairstein, can in principle be eliminated by the use of the *zero-crossover timing* technique[49,50]: if the discriminator hysteresis is made equal to A_{disc}—in other words, if the lower discriminator threshold is set at the baseline level—the backward transition occurs at time

$$t_{\mathrm{trig}}^{(\mathrm{down})} = t_{\mathrm{c}}$$

which is independent of peak amplitude in a linear system (it may still "walk" for overloading pulses, as indicated in Fig. 1).

Amplitude-independent *peak timing* can be obtained by applying the same technique after shaping the pulses (which must be relatively free of ringing) in a fast differentiator whose time constant $\tau_{\mathrm{diff}} \ll T_{\mathrm{r}} = $ rise-time of original pulses.

Although zero-crossover circuits actually allow a time resolution <10 nsec with slow inorganic scintillators, analysis of statistical fluctuations shows that *leading-edge timing* is basically more accurate.[51] To minimize walk in this case, the time information must be picked off the main signal path ahead of any noise filters, and processed in very fast circuitry (which does not need to be linear, as the output will be digital in nature). In the case of scintillation spectrometers, time and amplitude

[49] E. Fairstein, A Pulse Crossover Pickoff Gate for Use with a Medium Speed Coincidence Circuit. ORNL Instrum. and Contr. Div. Annu. Rep. (1 July 1957).

[50] W. Gruhle, *Nucl. Instrum. Methods* 4, 112 (1959).

[51] R. E. Bell, *Nucl. Instrum. Methods* 42, 211 (1966).

channels can be separated directly at the phototube (e.g., last dynode → slow output; anode → fast output). For semiconductor detectors, various splitting schemes have been described.[52-54]

Recent limiter-discriminators for the time channel using tunnel diode–transistor combinations achieve nanosecond resolution over a dynamic range of 10 : 1 or better (see Section 9.3.2). The discriminator output, typically 10 nsec wide with 1 nsec rise and fall times, is delay-line clipped before further processing in fast *coincidence* and other *gating circuitry* (which is sometimes referred to as "logic," although the vital importance of signal shape considerations at this level reveals its analog character). The development of modular transistorized gating circuits capable of nanosecond resolution began around 1960[55,56]; one of the most versatile configurations, based on current transfer between unsaturated long-tail pairs,[57] has since evolved into the fastest presently available integrated circuit family (ECL = "emitter-coupled logic"; see Section 2.3.2.3).

Linear gates are used to transmit analog-type input signals only when "open"; thus, a pulse source (e.g., the linear amplifier) can be disconnected from the measuring/recording system depending on rejection criteria as discussed in Section 9.6.3. Basic gate configurations are:

(1) a series switch making contact when "open,"
(2) a parallel switch making short circuit when "closed."

Combinations of (1) and (2) using high-gain transistors as switching elements give improved performance, with offset and feedthrough voltages as low as a few millivolts for a total swing of 10 V, and response times on the order of 100 nsec. Details are given in Section 8.3.2.

It is sufficient, in principle, to add a capacitive load across a series–parallel linear gate, in order to stretch an instantaneous input value over an arbitrary period during which both switches are nonconducting. *Pulse stretchers* are needed either as an integral part of measuring systems (for example, in a time-variant filter remembering the momentary baseline level preceding the pulse; in a conventional Wilkinson-type ADC;

[52] J. A. Scheer, *Nucl. Instrum. Methods* **22**, 45 (1963).
[53] C. W. Williams and J. A. Biggerstaff, *Nucl. Instrum. Methods* **25**, 370 (1964).
[54] I. S. Sherman and R. G. Roddick, *IEEE Trans. Nucl. Sci.* **NS-17**, 252 (1970).
[55] R. M. Sugarman and F. Merritt, Experimental Performance of High-Speed Limiters for Fast Coincidence Circuits. BNL Millimicro-Note No. 3 (1960).
[56] R. M. Sugarman, F. Merritt, and W. Higinbotham, Nanosecond Counter Circuit Manual, BNL 711 (1962).
[57] D. Maeder, Ref. P-4, paper VII-1, p. 325.

or in an analog level recorder), or as temporary analog stores for buffering a slow digitizing system. Apart from linearity, low offset, and low feedthrough (as in linear gates), the ratio of discharging ($\sim$ combined leak resistance) to charging ($\sim$ series switch resistance) time constant is crucial; with careful design, a $10^6 : 1$ ratio is possible so that a pulse having a 100 nsec rise (requiring $R_{series} \cdot C < 20$ nsec) can be stored about 200 μsec with $\sim 1\%$ droop; see also Section 8.3.5.

9.6.5. Analog Recording-Type Pulse Spectrometers

9.6.5.1. Individual-Pulse Recording. The *photographic method*, referred to in the introduction (see Section 9.6.1.4), is generally considered impractical due to the delay between film exposure and data analysis. Fast film reading machines are, in addition, quite expensive so that on the whole there is no economic advantage to be gained from the use of readily available film recording equipment. The photographic technique is, however, superior to other known procedures with respect to speed of data acquisition: since the necessary exposure time for a well-focused dot on a high-voltage CRT can be as short as 1 nsec, it is possible, in principle, to obtain individual markings, at ordinates y_1, y_2, $\ldots$ corresponding to peak amplitudes, from hundreds of pulses occurring in a burst of a few microseconds. In such cases, the mechanical film transport would be much too slow to separate successive events in the x direction; however, a fast, stepwise film transport between successive exposures can be simulated electronically, simply by applying the sawtooth-like output of an RC integrator (assumed to receive uniform pulses at its input) to the x deflection system of the CRT.

In continuous operation at 2 m/sec, a mechanical film transport can handle some 20,000 events/sec with negligible overlap if the RC step transport simulator is used simultaneously. Assuming an average spacing of 0.1 mm along the film axis, the storage capacity of a 120-m film is estimated at 10^6 events.

Entertainment-type *magnetic tape* units allow inexpensive analog recording, and immediate playback, of pulse distributions. Direct intensity recording is subject to considerable error ($\sim 6\%$ according to Cavanagh and Boyce[58]) due to the difficulty of reproducing a given magnetization level. Conversion of pulse amplitudes into duration of a standard signal yields records with excellent reproducibility (given by the "flutter"

---

[58] P. E. Cavanagh and D. A. Boyce, *Rev. Sci. Instrum.* **27**, 1028 (1956).

specification which is typically $<0.2\%$) but considerably slows down the acquisition rate. With the use of suitable analog buffer memories, this method appears feasible at pulse rates $\lesssim 10$/sec, for an amplitude resolution equivalent to about 200 channels. At the standard 19 cm/sec tape speed, a 540 m reel can thus store about 25,000 events.

Unless multichannel equipment is used for the playback of single-event records of any kind, total analyzing time is likely to exceed the acquisition time by an enormous factor.

9.6.5.2. Mechanical Curve-Plotting Spectrometer. The bias potentiometer of a single channel discriminator is coupled with a moving-paper chart recorder and the channel is connected to a counting rate meter whose output is plotted. The complete instrument is capable of a 1% accuracy, but subject, of course, to the speed limitations of the single channel procedure (see Section 9.6.2.2).

9.6.5.3. Analog Storage of Discrete Channel Counts. Any multidiscriminator type of spectrometer could have its counting circuits replaced by one capacitor for each channel to reduce some of its bulkiness.[59] The voltage on the capacitors may be made proportional to either the counts or the counting rates, and displayed sequentially on an oscilloscope. Cathode ray tubes with a series of collector electrodes[60,61] would appear ideally suited to replace the costly discriminator chain. In practice, it is very difficult to build such a device which satisfies the requirements of speed, channel width equality, and freedom from gaps or overlaps; and so far, none are in regular production. As an alternative, modern digitizing techniques have been used in the design of a recent 64-channel instrument of this type[62]: a standard pulse, generated after each event, is directed via a system of highly insulating gates to one out of 64 equal capacitors.

9.6.5.4. Gray-Wedge Technique. The light output from a cathode ray tube screen is sufficiently independent on spot position to be used for an analog integration of local count densities within a few percent. The simplest two-dimensional integrator system is a photographic plate or film, which is locally blackened in proportion to log(exposure), and thus to log(count density).

[59] R. E. Bell, Ref. P–1, paper II-6, p. 68.
[60] W. E. Glenn, *Nucleonics* **4** (6), 50 (1949); **9** (6), 24 (1951).
[61] D. A. Watkins, *Rev. Sci. Instrum.* **20**, 495 (1949).
[62] A. Westman, E. Petrusson, and P. A. Tove, Ref. P–5, paper 105 (Vol. II).

The inconvenience and inaccuracy of the exposure–density relation can be eliminated in the case of one-dimensional pulse spectra. The second dimension (e.g., the Y deflection; see Fig. 12) is then used for a variation of exposure per pulse, according to an intensity formula $I = f(y)$. The pulses $\delta N(x)$ delivered at amplitude x (within the spot width δx) now produce a fixed standard exposure B^* at a certain ordinate y^* which corresponds to $f(y^*) = B^*/\delta N(x)$. Therefore, any *isodensity curve* is a

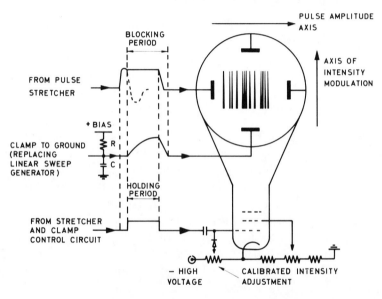

FIG. 12. Waveforms applied to the CRT in a gray-wedge spectrometer. The exponential sawtooth acts as an "electronic gray wedge," resulting in a linear intensity plot of the kind shown in Fig. 13.

graph of the spectrum with a (generally nonlinear) scale that depends on the function $f(y)$ and very little on the behavior of the photographic material.[63]

Plotting an isodensity line is done simply by taking a high-contrast print of the original photograph. If necessary, this procedure can be repeated or refined in the manner illustrated by Fig. 13 taken from Maeder.[64] Use of the Sabattier effect yields similar results.[65]

[63] D. Maeder, P. Huber, and A. Stebler, *Helv. Phys. Acta* **20**, 230 (1947).
[64] D. Maeder, *Nucl. Instrum.* **2**, 299 (1958).
[65] A. Rakow, *Atomkernenergie* **5** (3), 91 (1960).

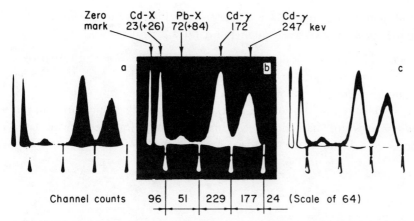

FIG. 13. Reproductions of a typical gray-wedge spectrogram (NaI scintillations from ^{111}In → ^{111}Cd): (a) original negative (contrast is exaggerated by the reproduction process); (b) first positive print; (c) double print, exposed through (a) and (b). [From D. Maeder, *Nucl. Instrum.* **2**, 299 (1958)]

The following two modulation formulas, between given end points y_0 and y_1, are easily obtained:

(a) *optical filter*, transmission $f(y) = T_0 e^{-a(y-y_0)}$, giving a logarithmic plot

$$\log \delta N(x) = \text{const} + ay$$

with a range factor $F = e^{a(y_1-y_0)}$,

(b) *electronic wedge*, constant intensity but variable sweep velocity according to $y = y_1 e^{-t/RC}$, giving a linear plot

$$\delta N(x) = \text{const} \cdot y$$

with a range factor $F = y_1/y_0 = e^{t_{max}/RC}$.

Since (a) absorbs most of the CRT light, (b) is more efficient by a factor $(F - 1)/T_0 \log F$, which amounts to ≈5 ... 30 for $F = 10$ ... 100. Method (b) has the further advantage that its parameters can be changed electronically.

Proper developing and calibration procedures[†] permit absolute intensity evaluations to 20%, while for intensity ratios the limiting accuracy is about 5%. The simplicity of the electronic system (essentially a con-

---

† Maeder,[64] Sect. 3.1.2.

ventional oscilloscope plus a pulse stretcher) makes the gray-wedge technique ideal for extremely high pulse rates.[66]

9.6.5.5. Two-Dimensional Analog Storage. Two coincident signals (e.g., pulse amplitudes) are applied to the X and Y plates of a CRT, whose beam is turned on shortly while both deflections are stationary. The (x, y) density distribution is integrated on a photographic film, and the x and y axes are marked by switching off the y (or the x respectively) deflection during a few hundred exposures.[†] It will be recognized that

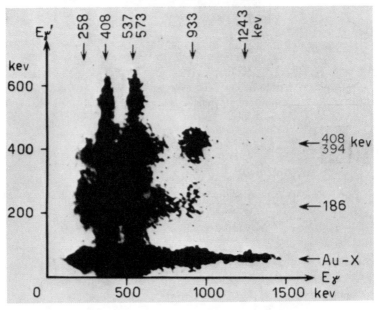

FIG. 14. Photographic record of two-dimensional distribution ($\gamma\gamma$-coincidences from ^{193}Hg in NaI scintillation crystals).

a quantitative, absolute intensity evaluation from such photographs would be quite time consuming; however, the coordinates of intensity peaks can be determined easily with a precision corresponding to at least 100×100 amplitude channels from the original half-tone pictures (note that gray-shade details were lost in the process of reproducing Fig. 14, an example taken from early $\gamma\gamma$-coincidence studies made at ETH,

[66] J. T. Flynn and F. A. Johnson, Rev. Sci. Instrum. 28, 867 (1957).

[†] Maeder,[64] Sects. 2.7 and 4.4.

Zurich). Isodensity contours obtained by high-contrast prints taken at various exposure levels allow a quick estimation of the relative intensities of different peaks. Typical applications were reviewed by Grodzins.[67]

The necessary equipment is extremely simple: the two "slow" linear signals (provided that they have time-coincident flat tops) are applied directly to the X and Y deflection systems of the cathode ray tube, and the fast coincidence output is used, with a suitable delay, to intensify the beam during the flat top. Pulse lengtheners are not normally required since a few nanoseconds of exposure time are ample for recording dots-on-film. Any number, from a very few (individual markings) to some 10^6 events (overlap produces different gray shades) can be stored on one frame.

The photographic 2D-integration technique thus yields a form of data presentation very similar to typical CRT displays of measurements made with the most elaborate digital-type pulse spectrometers (compare Fig. 18A) which are, however, much slower in operation and usually more limited in channel resolution.

With only minor additional circuitry, the basic simplicity of analog-type 2D-recording can be combined with the convenience of an instantly visible display: light integration (in the photographic emulsion) is replaced by charge integration (in the dielectric mosaic of a storage-type CRT). According to Stüber,[68] the type H-1033AP20 Tonotron tube has an (x, y)-resolution equivalent to 500×500 channels on a 150×150 mm² surface, and is capable of displaying accumulated charge densities by roughly proportional light intensity levels on the screen. Acquisition rate is limited by the minimum writing time (~ 10 μsec) and slow magnetic deflection circuitry. The display may be viewed for several minutes before the charge pattern is destroyed.

9.6.6. Digital Pulse Analyzer Techniques

9.6.6.1. Multidiscriminator Systems versus Analog–Digital Converters (ADC).
For systems requiring $K \gtrsim 100$ amplitude sorting channels, solutions based on an array of individual discriminators appear impractical even with modern IC technology, although this approach would be the most suitable one for achieving extremely fast operation (say, $>10^7$ pulses/sec). A classical "hybrid" solution, using two discriminator chains

---

[67] L. Grodzins, "Nuclear Electronics," Vol. II, p. 241. Intern. Atomic Energy Agency, Vienna, 1959.

[68] W. Stüber, Ref. P-5, paper II-149.

of only $1 + \sqrt{K}$ elements each ($+K$ gating elements),[69,70] has recently been proposed in a modified form.[71] Actually, the most widely used amplitude digitizing scheme is the Wilkinson-type ADC which is the basic principle of all high-precision digital voltmeters (see Section 9.5.1). As all amplitude measurements are made by one single discriminator, such a converter is very compact and offers excellent channel width uniformity; on the other hand, the channel numbers are determined by a counting process which takes an appreciable time (e.g., 100 μsec for assigning a pulse to channel # 1000 if a 10 MHz clock is used) during which the pulse spectrometer is blocked. Details on slow ADC circuits are given in Section 8.6.2.

9.6.6.2. Fast ADCs. The principle of successive approximations allows the number of digitizing steps to be considerably reduced, the minimum number being $A = \log_2 K$ using binary decisions. One design, using higher level decisions, arrives at a 3 μsec resolving time for a 2048-channel ADC.[72] Efficient channel width equalizing schemes are a necessity in any system based on successive approximations; otherwise certain channels would become useless, as can be seen by considering, for example, channel # 1023 in a binary step system.

> *Example.* The upper limit is defined by a single weighting resistor, the lower limit by a set of independent resistors. Even a $\pm 0.05\%$ accuracy for all resistors is insufficient to guarantee the mere existence of that channel—its width may turn out to be zero in the worst case.

Such errors can be eliminated, under certain restrictions,[†] by using a fixed added-step (compare Section 9.6.3.7) for the final channel assignment[73]; or in a more general way, by the sliding baseline scheme[74,75] whereby pulses belonging to, say, channel # 1023 are shifted on purpose

[69] D. H. Wilkinson, *Proc. Cambridge Phil. Soc.* **46**, 508 (1950).
[70] E. Gatti, *Nuovo Cimento* (9) **7**, 655 (1950).
[71] J. Leng and P. K. Patwardhan, Ref. P–4, paper II-29, p. 180.
[72] R. Kurz, Ref. P–3, paper 3-4, p. 179.
[73] K. Fränz and J. Schulz, Ref. P–4, paper II-27, p. 172.
[74] C. Cottini, E. Gatti, and V. Svelto, *Nucl. Instrum. Methods* **24**, 241 (1963).
[75] C. Cottini, E. Gatti, and V. Svelto, Ref. P–6, p. 309.

[†] The smallest channel width before equalization sets an upper limit to the equalized channel widths.

by known variable amounts so that some of them are first assigned to channels # 1024, 1025, . . ., (1023 + 64). The ADC output is corrected individually for each pulse by digital subtraction of the momentary baseline shift. As the effective channel width is averaged over 64 adjacent channels, irregularities are smoothed out, as demonstrated in Fig. 15.

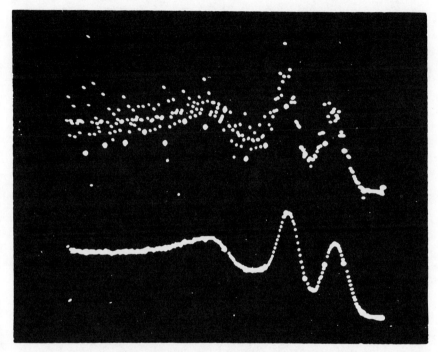

FIG. 15. Smoothing of channel width irregularities in a fast ADC by the sliding baseline method: Spectra obtained without (upper curve) and with (lower curve) smoothing [From C. Cottini, E. Gatti, and V. Svelto, Ref. P–6, p. 309].

9.6.6.3. Options in Digital Data Acquisition. The ADC output is a channel number, $k = 1, 2, \ldots, K$, presented usually in binary code on $A = \log_2 K$ address wires. A control signal on a separate wire indicates when the channel information is "ready." Further data processing can be organized in a number of ways:

(a) The "ready" signal is transmitted, via a system of logic gates, to one out of K individual counter circuits (active memory devices).

(b) A large capacity compact array of passive memory cells is used for storing the integrated counts of every channel. To register one count in the kth channel, its contents are read into a counter circuit, modified

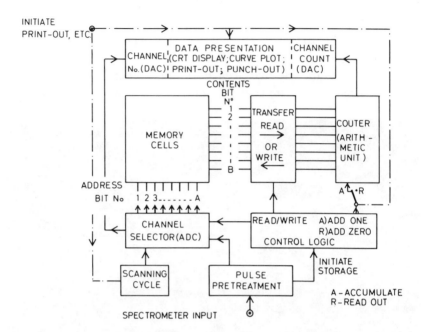

FIG. 16. Use of a passive memory as a pulse spectrometer.

by adding "one," and the result is written back into the memory (Fig. 16). Since only one such circuit is needed for the whole system, its complexity has little influence on the cost of the complete spectrometer; in particular, it can be arranged to subtract as well as add, or even to multiply incoming counts by a chosen power of 2, and therefore may be called an *arithmetic unit*. Additional circuitry is required for data display (Section 9.6.6.9).

(c) Direct communication with a computer follows the general concept of Fig. 16, with increased flexibility for arithmetics, multiplexing, and read-out.

(d) Individual events are stored in a large-capacity recording medium. Such "raw-data records" must be analyzed in a separate process, using equipment of categories (a), (b), or (c).

While method (a) is feasible only for applications for which a small number of channels is sufficient (say, $K < 100$), many type-(b) versions of multichannel instruments ($100 < K \lesssim 4000$) have been described in the literature (see Section 9.6.6.5). Solution (c) allows some further extension of K for a corresponding investment, whereas with (d) the total number of channels is limited only by a corresponding increase of analyzing time, as will be shown below.

9.6.6.4. Digital Raw-Data Recording. The necessity for a pulse distribution analyzing system with an extremely large number of channels, first realized in neutron time-of-flight work, initiated the development of digital tape recording systems.[76] Typical *instrumentation tape decks* use 25.4 mm wide tape on which a 24-bit word can be written about every 40 μsec at full tape speed.[77] Derandomizing buffer stores ($\gtrsim 10$ words) are indispensable for efficient use of tape space ($\sim 90\%$ utilized), both at high and low acquisition rates. At rates < 50 events/sec, a stepwise tape transport is recommended. Assuming that 4 tracks should be reserved for multiple parity checks, the remaining 20 tracks still allow 10^6 channels, which is ample for most 2D-and some 3D-experiments. At the rated writing density of 8 words/mm, a 2200-m length of tape can accommodate $16 \cdot 10^6$ events.

Modern *computer tape drives* using only 8 data tracks achieve essentially the same storage capacity for a given length of tape, due to a greatly increased packing density (up to 31 bits/mm). Even on a long record, occupying a full reel of tape, the average channel content will in general be very small (e.g., 16 events if 10^6 channel addresses are possible).

Fast and efficient playback is vital for any such measuring system to be practical: even with a totalizing device of 4000 channels capacity, complete analysis of all recorded data would take 250 scans, or about 2 days for a 2200 m reel at full tape speed. The problem of extracting the relevant information from tape records in a minimum of analyzing time was discussed by several authors at the 1964 EANDC conference.[78]

Paper tape punchers, used occasionally for the same purpose, have considerably lower capabilities: maximum acquisition rate is about 100 rows (of 6 bits)/sec—e.g., 50 events/sec classified into 4096 channels. One 250 m spool can store about 50,000 events.

9.6.6.5. Static Memory Arrays in Pulse Spectrometers. As active components are not required for maintaining the state (e.g., the direction of magnetization) of a memory cell, information can be stored indefinitely in a static memory, even when power supplies are shut off. On the other hand, there is no easy way of "looking at" the state of a static memory cell except by resetting it to 0—in other words, by destroying the information: if the cell happened to be in state 1, resetting produces a

---

[76] P. A. Egelstaff, Harwell neutron time-of-flight spectrometer, reported by F. H. Wells, *Nucl. Instrum.* **2**, 165 (1958).

[77] A. B. Idzerda, Ref. P–7, paper IIb-4, p. 104.

[78] P. A. Egelstaff and E. R. Rae, Ref. P–7, paper IIb-1, give a general survey.

signal which can be "sensed" and displayed. Nondestructive read-out of static memories therefore requires special circuitry to rewrite the original information into each cell immediately after it was sensed. These techniques are covered by standard textbooks on computers,[79-82] and may be characterized by a typical memory access time, $T_{RW}(=$ complete read–write cycle). Suitability of different types of static memories for application in pulse spectrometers (where the total bit capacity must be about 10 to 20 times the number of channels) is illustrated by the following list:

(a) *Magnetic cores* wired in a 3-dimensional matrix are immediately accessible as soon as the address (X, Y) is known (*random access*). All cores of a given channel are interrogated simultaneously, by applying coincident current pulses to one X-address wire and one Y-address wire. In the first spectrometers using magnetic core storage,[83,84] a complete read–write cycle took 16 μsec; actually, commercial instruments now offer capacities between 100 and 4096 channels, with $T_{RW} \sim 1$ μsec.

(b) *Magnetic film* allows faster access, typically within $T_{RW} \sim 200$ nsec.[85] Such a memory was used in a very fast 256-channel spectrometer.[86]

(c) *Electrostatic storage*[87,88] is not truly "static," in that the charge pattern is subject to leakage and must therefore be regenerated periodically, usually a few times per second. Since this regenerating cycle can be interrupted temporarily, a desired address is nevertheless immediately accessible for adding (or subtracting) a count. In selected Williams tubes (5ADP1), a 64×64 dot array can be stored, and the memory cycle takes a few microseconds per bit. Barrier grid tubes (RCA 6499) have better spot definition than a Williams tube and permit writing within 0.1 μsec.[89]

[79] B. Kazan and M. Knoll, "Electronic Image Storage." Academic Press, New York, 1968.

[80] A. I. Kitov and N. A. Krinitskii, "Electronic Computers." Pergamon, Oxford, 1962.

[81] A. P. Speiser, "Digitale Rechenanlagen." Springer, Berlin, 1965.

[82] A. Profit, "Structure et technologie des ordinateurs." Armand Colin, Paris, 1970.

[83] P. W. Byington and C. W. Johnstone, *Inst. Radio Eng. Nat. Conv. Record* **3**, Part 10, 204 (1955).

[84] R. W. Schumann and J. P. McMahon, *Rev. Sci. Instrum.* **27**, 675 (1956).

[85] B. Alexandre, G. Antier, and G. Grunberg, Ref. P–6, p. 657.

[86] T. L. Emmer, *IEEE Trans. Nucl. Sci.* **NS-12**, No. 1, 329 (1965).

[87] F. C. Williams and T. Kilburn, *Proc. Inst. Elec. Eng. (London)* **96**, Part II, 183 (1949).

[88] J. P. Eckert, H. Lukoff, and G. Smoliar, *Proc. IRE* **38**, 498 (1950).

[89] F. H. Wells, *Symp. Fast Pulse Tech.* Paper B1. Univ. of California Radiat. Lab., Rep. UCRL-8706 (1959).

The main advantage of electrostatic stores is the facility with which the memory contents can be displayed: it is either directly visible on the storage screen, or on a monitor CRT operating in parallel with the storage tube. Figure 17 reproduces the display of a decimal-coded 64-channel spectrometer[90] using a Williams-type store.

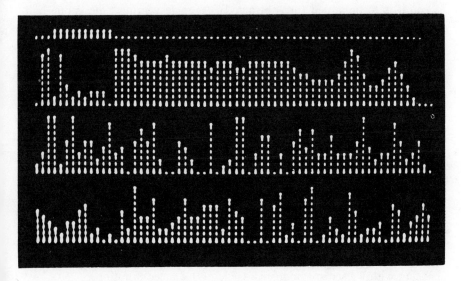

FIG. 17. Cathode ray tube display of ^{60}Co scintillation spectrum stored in a decimal electrostatic memory (courtesy of Dr. R. Müller, ETH Zürich). Readings are, for example, 938 and 786 counts at the two photopeaks.

(d) *MOS devices* are not truly "static," although the operating currents needed to maintain the state of a memory cell have been considerably reduced in recent years. Owing to rapid progress in LSI technology, MOS flip-flop arrays may become, in the future, competitive for pulse spectrometer applications with capacities of a few hundred channels; the possibility of continuous nondestructive read-out would offer great advantages for data display.

9.6.6.6. Circulating-Type Dynamic Memories. Very economical solutions result when individual storage cells are replaced by a continuous medium in which information circulates in a time sequence. Since attenuation and dispersion are unavoidable, the pattern must be regenerated periodically, typically at such a rate ($>$1 kHz) that the memory cycle can be utilized

[90] R. Müller, Z. *Angew. Math. Phys.* **13**, 13 (1962).

directly for address selection without introducing prohibitively long waiting periods:

(1) *Pulse amplitude* is compared, by means of a voltage discriminator, with a linear sawtooth generated in synchronism with the memory cycle[91]; the timing of the discriminator output determines the channel in which a count is to be added (or subtracted). Average waiting time is one-half of the memory period.

(2) *Time-of-flight* is assigned to memory channels by using a synchronized particle source.[92] Detector pulses (possibly several per cycle) are counted shortly after their occurrence, with a waiting time averaging one-half of the channel period.

In pulse spectrometers using such a store, the transfer and arithmetic circuits (Fig. 16) becomes basic memory constituents, while the bulk of storage cells reduces to a single component called a *delay-line*. System design is thus greatly simplified, although the technical problem of highly reliable regeneration (e.g., $\sim 10^{11}$ bits during a day's run) must be solved. Characteristics of actual instruments depend on the type of delay-line chosen, for example:

(a) *A mercury tube* was originally used[91] to provide 1200 bits (60, 80, or 120 channels of 20, 15, or 10 binary digits) as ultrasonic wave packets with 1 μsec spacing, in a cycle of 1.2 msec duration.

(b) *Ferromagnetic wires* (nickel; steel; Fe-Co alloys such as Vacoflux), excited by a short coil, propagate longitudinal shock waves at about 5000 m/sec or 5 mm/μsec. Dispersion causes initially short pulses (e.g., 0.3 μsec) to spread gradually over many wire diameters so that storage capacity is practically limited to ~ 2000 bits (in a 3-msec cycle). Limiting factors and parasitic effects are summarized by Gutmann *et al.*[93]

(c) Low-attenuation bulk material (e.g., *glass or quartz plates*) can be shaped to guide an ultrasonic wave through many reflections from a transmitting to a receiving face, so that total delay can amount to 1 msec in a physically small unit. Carrier frequencies of the order of 40 Mc/sec permit a bit spacing as small as 0.1 μsec. The first published time-of-flight spectrometer of this kind provided 500 channels of 10 bits[92] (2-μsec channel spacing, 1-msec total cycle). A modern design with ICs, con-

[91] G. W. Hutchinson and G. G. Scarrott, *Phil. Mag.* (7) **42**, 792 (1951).

[92] H. L. Schultz, G. F. Pieper, and L. Rosler, *Rev. Sci. Instrum.* **27**, 437 (1956).

[93] J. Gutmann, K. Jauch, A. Kunz, D. Maeder, M. Peter, and R. Von Felten, *Proc. 1966 Int. Conf. Instrum. High Energy Phys.* paper A-8, p. 60. Stanford, California (Sept. 1966).

ceived for measuring amplitude spectra, has a 0.13 msec average access time for its 128 channels of 16 bits,[94] and uses derandomizing buffers (see Section 9.6.3.4) so that effective deadtime for random events at rates below 8 kHz is reduced to about 20 μsec.

The general field of delay-line techniques was reviewed by Eveleth.[95] *Electromagnetic delay-lines* have not been used so far as digital multi-channel stores, although extremely short circulation periods are possible in a coaxial system.[96] However, unless the lines are made of super-conducting material, total capacity is practically limited to about 100 bits. With only one pulse circulating in a line, coaxial memories are actually used (in a semianalog manner) in the "vernier chronotron."[97]

Magnetic drums and disks, although not affected by the problem of continuous regeneration of the stored information, resemble the recirculation-type systems as far as memory access is concerned: even the fastest drums are in fact quite slow (1 cycle $\approx$10 ms) and therefore require elaborate buffering when used for random pulse sorting.[98]

9.6.6.7. Associative Memories. Due to the general scarcity of channel counts in multidimensional experiments, there is a possibility of performing such measurements with high resolution using equipment of rather limited capacity. The following hypothetical example illustrates the basic idea.

Example. A 1000×1000-channel coincident pulse distribution exhibits a strong correlation by which 90% of all pulse pairs are confined to a small selection (say 100×100) out of the 10^6 possible channel combinations. At the end of a run of 10^6 events, it is found that 10^5 events ($= 10\%$ of 10^6) are spread over 990,000 possible channels, giving insignificant average intensities (0.1 counts/channel). The remaining 10,000 channels contain the relevant information, with an average of 90 counts/channel. The selection of "most probable channels" is not known a priori; it must be determined automatically by the measuring equipment.

[94] J. B. S. Waugh, *IEEE Trans. Nucl. Sci.* **NS-17**, No. 1, 345 (1970).
[95] J. H. Eveleth, *Proc. IEEE* **53**, No. 10, 1406 (1965).
[96] D. Maeder, Ref. P–7, paper IVb, p. 379.
[97] H. W. Lefevre and J. T. Russel, *Rev. Sci. Instrum.* **30**, 159 (1959).
[98] R. L. Chase, *IRE Nat. Conv. Record*, Part 9, 196 (March 1959); or "Nuclear Electronics," Vol. II, p. 223. Int. At. Energy Agency, Vienna, 1959.

In an "associative analyzer,"[99] event parameters (a linear combination of channel numbers, $k_1 + k_2 K_1 + \cdots$) are called *descriptors* and are stored individually in the available memory cells, together with a *count* number which indicates how often a given descriptor occurred during an experiment. For the above example with $10^3 \times 10^3$ channels, a descriptor would occupy the first 20 bits of a memory location. Only those descriptors which actually occur as pulses in the experiment will be assigned memory space, following a "first come–first served" procedure. When the memory is filled with descriptors, those which have occurred only once are eliminated by clearing the respective memory locations, so that new (and possibly more probable) descriptors can be accepted during the remainder of the experimental run.

This technique, when implemented by software on a conventional computer, takes considerable searching time to determine the memory location where an incoming pulse should be stored. Various schemes[100] including content-addressable hardware[101] have been devised in order to speed up the operation of associative analyzers.

9.6.6.8. On-Line Computer Use for Pulse Spectrometry.
Any type of digital pulse spectrometer may be considered as a special-purpose computer, "hard-wired" for data sorting and totalizing applications, with a minimum of arithmetic capabilities. In contrast, a general-purpose computer, equipped for real-time data processing, can be operated as a "stored-program pulse spectrometer" with high flexibility allowing involved spectrum corrections and transformations.[†]

Even a small computer, although its random-access memory may not be larger than that of a comparable hard-wired pulse spectrometer, has an enormous total channel number capability when operated as a buffer between fast ADCs and a slow-access system of large capacity, e.g., a magnetic drum or disk. Various operation modes, such as:

— transfer of a spectrum from a hard-wired instrument into the computer (for analysis, display, print-out, etc.)
— direct entry of individual events (from an ADC through an "add-one-to-storage" unit) to the core memory

[99] I. N. Hooton, Ref. P–7, paper IVb-2, p. 338.
[100] B. Soucek, *Nucl. Instrum. Methods* **36**, 181 (1965); *Rev. Sci. Instrum.* **36**, 750 (1965).
[101] I. N. Hooton, *IEEE Trans. Nucl. Sci.* **NS-1**, No. 3, 553 (1966).

[†] This is one of the major topics treated in Ref. P–7.

— direct entry of individual events and transfer to a disk memory
for totalizing the spectra,

have been described in the literature.[102]

Recent trends are towards "multi-user systems" in which several
ADCs and other data sources share a large disk memory via a multi-
plexer interface.[103–107] In a typical system of this kind,[103] a PDP-8 computer was programmed to allocate buffer storage to incoming events from
a maximum of 14 independent ADCs (2048 channels each), and to
schedule disk access to achieve high sorting rates (e.g., up to 4000 counts/
sec simultaneously from each of 2 ADCs). Other features of on-line
computers are their versatility as multidimensional analyzers (Section
9.6.6.7) and their interactive data display (Section 9.6.6.9). The poten-
tialities of software-controlled data acquisition systems are fully exploited
in the complexity of high-energy physics experiments where real-time
feedback based on analyzed results, or at least a sample of such, is vital
for adjusting and checking the experimental parameters.[108,109] Relative
merits of large and small computers and of their combined use in such
applications are reviewed by Lidofsky.[110]

9.6.6.9. Data Display. Unless the data (event descriptors and associated
counts) are stored in an analog-type memory, they generally need to be
converted into analog voltages (DACs are discussed in Section 8.6.3)
in order to be applied to the deflection plates (X, Y) and/or the beam
intensifier (Z coordinate) of a CRT display. An example of a digital
storage system requiring no DAC for producing a quasi-analog display,
conserving full digital accuracy, was quoted in Section 9.6.6.5 (Fig. 17).

In contrast, Fig. 15 reproduces linear, truly analog-type displays;
obviously, relative channel count accuracy depends on the display scale:

[102] J. Zen, A. Muser, J. D. Michaud, and F. Scheibling, Ref. P–3, paper 5-8, p. 307.

[103] D. C. Uber, *IEEE Trans. Nucl. Sci.* **NS-17**, No. 1, 390 (1970).

[104] J. W. Reynolds and N. A. Betz, *IEEE Trans. Nucl. Sci.* **NS-17**, No. 1, 383 (1970).

[105] H. V. Jones, *IEEE Trans. Nucl. Sci.* **NS-17**, No. 1, 398 (1970).

[106] J. P. Adam, J. C. Brun, L. Faucher, and C. Victor, Ref. P–5, paper 110 (Vol. II).

[107] J. Moisset and M. Barthelémy, Ref. P–3, paper 5-5, p. 293.

[108] S. J. Lindenbaum, On-line computer techniques in nuclear research. *Ann. Rev. Nucl. Sci.* **16**, 619 (1966).

[109] S. J. Lindenbaum and S. Ozaki, *Proc. Int. Conf. Data Handling Syst. High Energy Phys.* Cavendish Lab., Cambridge, England, (March 23-25, 1970); Vol. 2, p. 921 (D. H. Lord and B. W. Powell, eds.). CERN Publ. 70-21.

[110] L. J. Lidofsky, Ref. P–3, Supplement, Panel Discuss. on Future Trends in Phys. Measurements and Nucl. Instrumentation, p. 11.

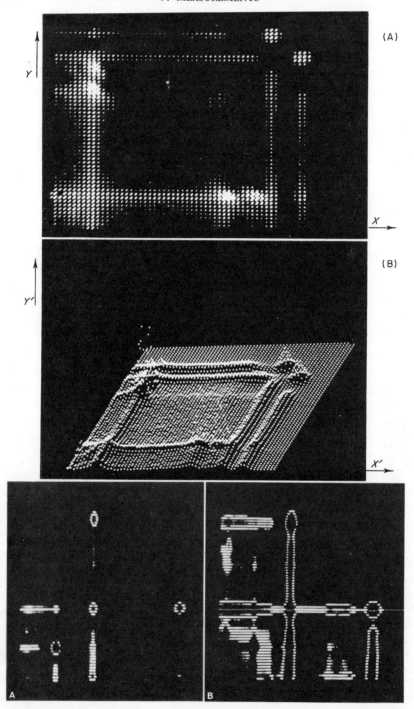

(A)

(B)

it must be chosen differently for viewing various portions of a spectrum covering a large intensity range. The more advanced data acquisition systems allow a variety of display modes: for example, logarithmic scales, automatic range adjustments, intensification of every nth channel, or of selected regions in a spectrum, etc. If intensities are displayed on a linear scale with an automatic scaling adjustment such that the highest channel count in the spectrum just reaches the top of the range, the gross features of a spectrum become apparent as soon as a few hundred events have been analyzed (note that in a fixed-scale display, the spectrum would hardly become visible before at least $\frac{1}{10}$ of the experimental run had been accomplished). In practice, the same effect is obtained by software implementation of a "mean rate analyzer."[111]

Multidimensional distributions pose special display problems; so far, a suggested 3D display device[112] is not commercially available. 2D spectra are most simply represented as a brightness distribution on a flat CRT screen, with X and Y deflections directly derived from the two detector pulse amplitudes. The resulting pictures resemble Fig. 14 (or its photographic complement) except that in digital instruments the total number of channels tends to be rather limited so that the discrete structure of the distribution becomes visible, as in Fig. 18A taken from Dimmler and Krüger[113] (64×64 channels). To obtain a more quantitative reading of intensity levels, various techniques have come into use:

— by gating the CRT brightness control, black "isodensity lines" corresponding to certain Z-windows are cut into an otherwise continuous distribution[114,115]

— the complementary procedure (bright isodensity lines) is illustrated in Figs. 19A,B (from O'Kelley[116])

[111] I. N. Hooton and G. C. Best, *Nucl. Instrum. Methods* **56**, 284 (1967).

[112] D. A. Mack, Ref. P–4, p. 320.

[113] G. Dimmler and G. Krüger, Ref. P–7, p. 393.

[114] G. Bianchi, C. R. Corge, and J. P. Meinadier, Ref. P–7, p. 174.

[115] J. F. Whalen, Ref. P–7, p. 399.

[116] G. D. O'Kelley, Ref. P–7, p. 385.

FIG. 18. Displays of two-dimensional distributions ($\gamma\gamma$-coincidences from a ^{60}Co source). (A) Map with continuous brightness modulation; (B) isometric mode (x', y' defined by Eq. (9.6.24)). [From G. Dimmler and G. Krüger, Ref. P–7, Fig. 3]

FIG. 19. Contour map displays of $\gamma\gamma$-coincidences recorded with two 3 in. $\times$ 3 in. NaI(Tl) detectors at 180° from an ^{88}Y source. The intensifier window is set: (A) wide, at a high level of channel counts, (B) narrow, at a low level of channel counts. [From G. D. O'Kelley, Ref. P–7, Figs. 1a and 3.]

— in an "isometric" display, the two CRT deflections are derived from all three coordinates, for example, as

$$x' = x + \alpha y$$
$$y' = y + \beta z.$$

$$(9.6.24)$$

The illusion of a perspectivic view of the $z = z(x, y)$ topography, reinforced by a simultaneous brightness modulation proportional to z, is illustrated by Fig. 18B.

In addition, plane sections through 2D-distributions can be displayed in the conventional (single-spectrum) manner, or can be intensified in an isometric display. The use of a "light pen" greatly facilitates display manipulations such as the marking of a particular channel (peak or valley) or group of channels, or the fitting of a theoretical curve to experimental data.[117] Direct-view storage tube displays[118] can operate with smaller computers but do not provide the light pen and intensity modulation facilities of a "refresher"-type display system.

For a general survey of the role of various display systems in the progress of complex experiments, see Brun.[119]

[117] R. L. Heath, Ref. P–4, p. 313.
[118] H. Klessmann and J. Zahn, Ref. P–3, paper 5-11, p. 321.
[119] J. C. Brun, Ref. P–5, paper 142 (Vol. II).

9.7. Magnetic Resonance*

9.7.1. Introduction[1-3]

In general terms, magnetic resonance involves inducing and detecting quantum transitions between energy levels of nuclear, atomic, and molecular systems by the application of alternating (radio frequency) magnetic fields. It is essential that the system under study possess a magnetic dipole moment to interact with the applied field. This dipole moment may be associated with a nuclear spin (nuclear magnetic resonance or NMR) or with an electronic spin or angular momentum (electron spin resonance, ESR, or electron paramagnetic resonance, EPR).

The energy quantum, $\hbar\omega$, of the applied radio frequency (rf) magnetic field must satisfy the usual condition that

$$\hbar\omega = E_2 - E_1 \qquad (9.7.1)$$

where ω is the angular frequency of the rf field and $E_2 - E_1$ is the energy difference between a pair of levels of the system. When the system energies result solely from the interaction of the magnetic dipole moment μ with an applied magnetic field H_0, there will be $2I + 1$ equally spaced energy levels with spacing $\Delta E = \gamma\hbar H_0$ between adjacent levels. Here I is the angular momentum quantum number and γ is the magnetogyric ratio of the nucleus (electron). In this simple case, transitions occur only between adjacent levels so that there is just a single resonant frequency

$$\omega_0 = \gamma H_0. \qquad (9.7.2)$$

Order-of-magnitude indications of ω_0 may be inferred from the two basic representative magnetogyric ratios:

$$\gamma_{\text{proton}} = 2\pi(4.26 \text{ kHz/Oe})$$

$$\gamma_{\text{electron}} = 2\pi(2.80 \text{ MHz/Oe}).$$

[1] A. Abragam, "The Principles of Nuclear Magnetism." Oxford Univ. Press, London and New York, 1961.
[2] C. P. Slichter, "Principles of Magnetic Resonance." Harper, New York, 1963.
[3] C. P. Poole, Jr., "Electron Spin Resonance." Wiley, New York, 1967.

* Chapter 9.7 is by Richard Barnes.

9.7.1.1. **Magnetic Resonance in Atomic Beams.**[4,5] Experiments utilizing neutral atomic or molecular beams represent the most fundamental type of nuclear magnetic resonance, and were historically the first such experiments. The typical experimental arrangement involves three regions of magnetic field, A, B, C, through which a highly collimated beam of atoms or molecules travels without change of speed after having been evaporated from a suitable source or "oven." The A and C magnets have pole faces shaped so as to produce large transverse gradients which deflect the nuclear moments and separate the beam into its $2I + 1$ energy states. As the beam passes from source through the A and C magnets to an appropriate detector, its path is a very elongated S-shape due to the opposing gradient directions of the A and C magnets.

Within the central (homogeneous) B magnet the field has the constant value H_0 and the rf exciting field is applied at frequency ω. Changes in the orientation of the total angular momentum occur in the B magnet as either H_0 or ω is scanned through the resonance condition. When this occurs the atom is "flopped" out of its normal path, and the detected beam intensity decreases. Apparatus requirements for beam experiments are elaborate, but the experimental results can be interpreted without the complications which arise from interatomic (and intermolecular) interactions in bulk matter. A second point is that very small samples may be used, and this has led to considerable application of the beam technique to the measurement of the spins and moments of relatively short-lived radioactive nuclear species.

9.7.1.2. **Magnetic Resonance in Bulk Matter.** In bulk matter, the magnetic moments interact with the applied field H_0, with one another, and also with the random thermal motions of the lattice. Since, except at very low temperatures, the thermal energy $kT \gg \gamma\hbar H_0$ for both electronic and nuclear moments even in moderately strong magnetic fields, the lattice functions effectively as an infinite thermal reservoir at temperature T. In thermal equilibrium, and assuming that interactions between the moments are also weak, the equilibrium magnetization is proportional to H_0 and is given by the Curie formula for the magnetic susceptibility:

$$M_0 = \chi_0 H_0 = \frac{N\gamma^2\hbar^2 I(I+1)}{3kT} H_0 \qquad (9.7.3)$$

where N is the density of magnetic moments per unit volume.

[4] N. F. Ramsey, "Molecular Beams." Oxford Univ. Press, London and New York, 1955.
[5] I. Rabi, S. Millman, P. Kusch, and J. R. Zacharias, *Phys. Rev.* **55**, 526 (1939).

In the resonance experiment the applied field is the sum of the fixed field H_0 and a weak field H_1 perpendicular to H_0 and oscillating at frequency ω. In arriving at an equation of motion to describe the behavior of magnetic moments in such a field, Bloch[6] assumed that the torque equation

$$dM/dt = \gamma M \times H \qquad (9.7.4)$$

which describes the precessional and nutational motions of the magnetization could be added to equations describing the transient damping out of nonequilibrium components of M. Thus, with H_0 in the z direction,

$$dM_z/dt = -(M_z - M_0)/T_1 \qquad (9.7.5)$$

defines the longitudinal, or spin-lattice, relaxation time T_1 with which M_z approaches its equilibrium value $M_0 = \chi_0 H_0$. Similarly,

$$dM_{x,y}/dt = -M_{x,y}/T_2 \qquad (9.7.6)$$

defines the transverse relaxation time T_2 with which M_x and M_y are reduced to their equilibrium value of zero. T_2 is effectively the time for different groups of spins to become dephased in their precessional motion.

Under these assumptions the motion of the magnetization may be described by the equation

$$\frac{dM}{dt} = \gamma M \times H - \frac{M_x i + M_y j}{T_2} - \frac{M_z - M_0}{T_1} k \qquad (9.7.7)$$

where H is the net field including both H_1 and H_0, and i, j, k are unit vectors in the x, y, z coordinate directions in the laboratory frame.

The general solution of this equation of motion, for fixed values of the parameters, is a combination of decreasing exponential terms and of a steady-state solution. The steady-state solution for the magnetization has a component which is in-phase with the applied oscillatory field H_1 and a component which is out-of-phase. This phase relationship is customarily treated by writing the part of the magnetization parallel to H_1 as a complex quantity M_1. In fact, this part of the magnetization is usually written in terms of a complex susceptibility, $M_1 = (\chi' - i\chi'')H_1$.

---

[6] F. Bloch, *Phys. Rev.* **70**, 460 (1946).

The components of the susceptibility may be identified from the steady-state solution as

$$\chi' = \tfrac{1}{2}\chi_0\omega_0 \frac{\Delta\omega T_2^2}{1 + (\Delta\omega)^2 T_2^2 + \gamma^2 H_1^2 T_1 T_2} \tag{9.7.8}$$

$$\chi'' = \tfrac{1}{2}\chi_0\omega_0 \frac{T_2}{1 + (\Delta\omega)^2 T_2^2 + \gamma^2 H_1^2 T_1 T_2} \tag{9.7.9}$$

where $\Delta\omega = (\omega_0 - \omega)$. Clearly, the susceptibility exhibits a resonance character similar to that of the (complex) impedance of a tuned circuit. However, when the third term in the denominators becomes significant, both χ' and χ'' decrease in magnitude. This so-called "saturation" term $\gamma^2 H_1^2 T_1 T_2$ is usually made smaller than the other terms in steady-state experiments, and if one defines $a = \tfrac{1}{2}\chi_0\omega_0 T_2$ and $x = (\Delta\omega)T_2$, the equations above in that case take the form

$$\chi' = ax/(1 + x^2) \tag{9.7.10}$$

$$\chi'' = a/(1 + x^2) \tag{9.7.11}$$

which are of the Lorentzian line shape and may be shown to verify the Kramers–Kronig relations.[2] Neither of these statements is true if the saturation term is not negligible.

In the usual experimental arrangements, the spins are located in a resonant circuit element, either a lumped circuit inductor or a cavity. The changes in the quality factor Q and resonant frequency ω of the resonant circuit in passing from off-resonance to on-resonance (e.g., by varying the applied field, H_0) may be expressed in terms of the complex susceptibility by[7]

$$-2\Delta\omega/\omega + i\Delta(Q^{-1}) = 4\pi(\chi' - i\chi'')f. \tag{9.7.12}$$

In this expression, f represents the ratio of magnetic energy within the volume occupied by the spin sample to the total magnetic energy stored in the resonant element and is known as the filling factor. Magnetic resonance can thus be detected by the change in Q or by the change (shift) in the resonant frequency of the circuit.[1] These signals are conventionally referred to as the absorption and dispersion modes, respectively. Unless otherwise specifically indicated, it is the absorption-mode signal which is the one ordinarily observed.

[7] J. C. Slater, *Rev. Mod. Phys.* **18**, 441 (1946).

Finally, the average rate of energy absorption per unit volume by the spin system at resonance due to the H_1 field is given by

$$P = \frac{\omega}{2\pi} \int_0^{2\pi} [-\mathbf{M} \cdot (d\mathbf{H}/dt)]\, dt = 2\omega\chi''H_1^2. \qquad (9.7.13)$$

In addition, as first shown by Bloembergen et al.,[8] the signal-to-noise voltage ratio to be expected under given experimental conditions is expressed by the relation (cgs units):

$$\frac{V_s}{V_n} = \left[\frac{N\hbar^2\gamma I(I+1)}{3kT} \left(\frac{T_2}{T_1} \right)^{1/2} \right] \frac{Af}{16} \left[\frac{QV_c\omega_0^3}{kTBF} \right]^{1/2}. \qquad (9.7.14)$$

In applying this relation, the rf power level is assumed to have its optimum value from the standpoint of signal detection, namely that which reduces the absorption component χ'' to half maximum value by saturation. The factor A is dimensionless and of order unity, V_c is the volume of the resonant circuit element (coil or cavity), B is the noise bandwidth of the detector, and F is the noise figure of the entire system.

The first bracket in Eq. (9.7.14) contains those factors which pertain directly to the nature of the sample itself and are essentially unalterable. The experimenter has some control over the remaining factors. In accord with the assumption that sufficient rf power is available to saturate the resonance, the importance of the other factors goes as: ω_0; A and f; Q, V_c, B, and F. Clearly, it is desirable to choose ω_0 (and hence H_0) as great as possible. At constant ω_0, V_s/V_n varies directly as γ, the magnetogyric ratio. However, at constant H_0, it varies as $\gamma\omega_0^{3/2} \sim \gamma^{5/2}$; hence, the sensitivity is inherently better by orders of magnitude in the ESR case compared to the NMR situation. It should be noted that the filling factor f is more important than the Q of the coil, although at low rf levels (avoiding saturation) these become of equal importance. Finally, it should be noted that T_1, T_2, Q, and F may be functions of ω_0, and that on this account these factors in Eq. (9.7.14) may be somewhat interdependent.

9.7.2. Nuclear Magnetic Resonance

9.7.2.1. Steady-State Methods. In typical steady-state (cw) measurements, the resonance is scanned slowly in a time long compared with T_1, keeping the amplitude of the H_1 field constant. The spin system then remains in thermal equilibrium with the lattice.

[8] N. Bloembergen, E. M. Purcell, and R. V. Pound, Phys. Rev. 73, 679 (1948).

The coupling of the radio-frequency energy between the external circuitry and the ensemble of spins in the sample may be accomplished in two general ways. (a) In the induction experiment, or crossed-coil method, the spin system is irradiated with rf energy from one coil, producing spin transitions, and the resultant induced signal is obtained at the terminals of the second coil.[9] Figure 1 shows the basic block schematic of this arrangement. (b) In the resonance method a single coil is used, sometimes in a balanced bridge configuration which permits the spin-induced signal to be separated from that of the rf source.[10] Figure 2 shows a general schematic of this arrangement.

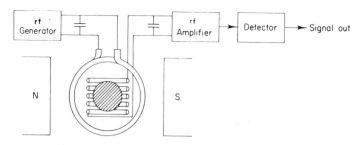

FIG. 1. Basic crossed-coil arrangement. Both coils have axes perpendicular to the applied magnetic field and to each other.

In these methods the signal is characterized by two parameters, its amplitude and phase. Alternatively, these may be chosen to be the amplitude components in-phase and out-of-phase (quadrature) with a reference signal derived from the rf source. What amounts effectively to phase detection of the rf signal is normally accomplished with ordinary amplitude detection and a small degree of unbalance of the bridge or induction (crossed-coil) circuit. In this way, either the χ'', absorption-mode, or χ', dispersion-mode signal may be selected for detection.

A common version of the single-coil method employs a vacuum tube or transistor coupled directly to the resonant circuit, thereby forming an oscillator. The simplest of these are referred to as *marginal* oscillators since the rf level must be maintained at a very low level to avoid saturation effects on the resonance, particularly when samples with large $T_1 T_2$ products are investigated. Alternatively, an amplitude-stabilizing feedback loop can be added to maintain the amplitude constant at a low level.

[9] F. Bloch, W. W. Hanson, and M. Packard, *Phys. Rev.* **70**, 474 (1946).
[10] E. M. Purcell, H. C. Torrey, and R. V. Pound, *Phys. Rev.* **69**, 47 (1946).

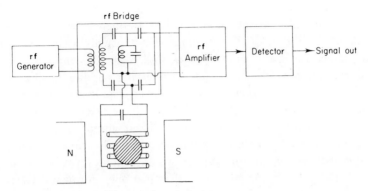

FIG. 2. Basic single-coil arrangement. A variety of other bridge circuits may be employed in addition to the one shown.

The relative merits of single-coil and crossed-coil systems depend greatly upon auxiliary considerations dictated by the conditions of the particular experiment. In certain cases, for example, experiments at high pressure or high temperature, the difficulties of introducing numerous rf leads into the sample chamber argue in favor of a single-coil arrangement with just one rf lead. For experiments at low temperatures, single-coil circuits are typically immersed in the cryogenic bath, thereby preserving the filling factor. A widely-used commercial crossed-coil probe, by contrast, requires that the cryostat tip be inserted in the receiver coil, thereby greatly reducing the filling factor. However, a variety of miniaturized crossed-coil heads have been described in the literature, suitable for immersion in the cryobath. These are also suitable for use in superconducting solenoids.[11]

More generally, although marginal oscillator circuits are simple and relatively insensitive to microphonics, they operate best at only one level of H_1, making it difficult to vary H_1 or to choose it on the basis of the sample T_1. Marginal oscillators are also not well-adapted to crystal-frequency stabilization (but such circuits have been described). They have the advantage of selecting the absorption-mode signal χ'' automatically and unambiguously. Bridge circuits (single-coil) are difficult to use over wide frequency ranges, but operate well at all levels of H_1. However, they do become more difficult to control at very high H_1 levels. The operating frequency may be readily crystal-stabilized. Induction probe arrangements (crossed-coil) work well at all H_1 levels and are readily crystal-stabilized. They can be operated over wide frequency

[11] J. B. Mock, *Rev. Sci. Instrum.* **41**, 129 (1970).

ranges, but usually cannot be scanned in frequency without upsetting the balance condition. This last point is not usually a drawback since the resonance can be observed just as well by field scanning. The disadvantages inherent in the use of flux-steering paddles in crossed-coil balancing may be avoided at least in part by electronic balancing schemes.

Many transistorized marginal oscillators have been described in the literature since the first was reported.[12] Their circuits are very similar and belong generally to the Colpitts type. Figure 3 shows an improved and updated version of this type,[13] which employs two controls (for feedback and emitter current) rather than just one. In this way the circuit is suitable for samples of many different relaxation time characteristics. A somewhat similar FET circuit has also been described which may be operated either marginally or as a superregenerative oscillator for pure quadrupole resonance detection.[14]

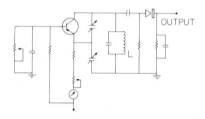

FIG. 3. Transistorized marginal oscillator circuit [from Shih-yu Feng, *Rev. Sci. Instrum.* **40**, 963 (1970)]. The sample coil is L, and the transistor is either 2SA103, 2N1178, 2N371, or 0C170. Values of the other components are given in Shih-yu Feng.

Superregenerative oscillators, which constitute a particular subset of single-coil system, have been used as magnetic resonance detectors since almost the very beginnings of the field. Such oscillators are characterized by a quenching (or squegging) action at a frequency typically on the order of one one-thousandth of the oscillator frequency. This quenching leads to a frequency modulation of the carrier at the quench frequency, and the consequent appearance of a sideband spectrum. A single resonance is then detected at each of the sideband positions as well as at the carrier frequency.

Application of this type of detector to conventional NMR problems has been slight because of the complications introduced by the sidebands

[12] J. R. Singer and S. D. Johnson, *Rev. Sci. Instrum.* **30**, 92 (1959).
[13] Shih-yu Feng, *Rev. Sci. Instrum.* **40**, 963 (1969).
[14] F. Bruin and H. Kunaysir, *Rev. Sci. Instrum.* **42**, 1480 (1970).

and because the nonconstant nature of the oscillation amplitude may lead to distorted line shapes. On the other hand, most nuclear quadrupole resonance work has been carried on with this type of detector because the relatively high rf level attained, combined with the typically short spin-lattice relaxation time of pure quadrupole resonances, leads to excellent sensitivity.

The same remarks apply also to the case of nuclear resonance in magnetically ordered materials (ferro or antiferromagnets). In both the pure quadrupole and magnetically ordered cases, resonances are observed in zero applied magnetic field, so that the detector must frequently be scanned over a relatively wide frequency range during the search operation. The total range of frequency encountered in these types of resonance study is also very great, the lowest frequency being about 1 MHz and the highest about 1000 MHz.

The original circuits employed vacuum tubes, of course, but transistorized versions of practically equivalent sensitivity have been described.[14] Oscillators for the higher frequency ranges still typically employ vacuum tubes.[15] When used with appropriate recording equipment for automatic frequency scanning and resonance searching, it becomes necessary to readjust continually the quench condition of the oscillator proper. Several arrangements have been described for this purpose, which consist basically of a means of monitoring the noise level of the oscillator and correcting the quench condition so as to maintain the noise level constant.[16]

Actual display or visualization of the resonance condition is most simply accomplished by sweeping or modulating the applied field H_0 with a sinusoidal component, $H_m \cos \omega_m t$, which is also applied to the horizontal plates of an oscilloscope while the output signal from the resonance detector is applied to the vertical plates. The modulation amplitude H_m must be several times greater than the width of the resonance being observed. In most cases ω_m can be chosen in the low audio-frequency range. Evidently, if the resonance shape is to be displayed with reasonable fidelity, the passband of the receiver (detector) must include the modulation frequency as well as many of its harmonics. Thus, the noise bandwidth of the system is large, and the sensitivity is usually rather low. This method of detection and display is appropriate to strong resonance signals.

For increased sensitivity, as well as faithful reproduction of resonance

[15] E. D. Jones and K. B. Jefferts, *Rev. Sci. Instrum.* **36**, 983 (1965).
[16] G. E. Peterson and P. M. Bridenbough, *Rev. Sci. Instrum.* **35**, 698 (1964).

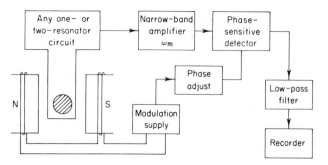

FIG. 4. Schematic arrangement of the differential sampling scheme. The modulation supply furnishes a weak sinusoidal magnetic field component H_m at the frequency ω_m. The combination of narrow-band amplifier, phase-sensitive detector, phase adjust, and low-pass filter is briefly referred to as a *lock-in detector*.

shapes, the most common approach involves the differential sampling scheme, shown schematically in Fig. 4. Now the amplitude H_m is reduced to a small fraction of the resonance linewidth, so that the major component of the output signal from the detector is at the modulation frequency ω_m. This signal is amplified by a narrow-band amplifier at the frequency ω_m and phase-detected (lock-in detected), the reference signal for the phase detection also being derived from the modulation supply. The output at this stage is essentially dc, since its frequency spectrum extends from zero up to a low value determined by the rate at which the resonance condition is being scanned. A suitable low-pass filter can then be used to further reduce the noise bandwidth and increase the effective signal-to-noise ratio. Clearly, as this low-pass filter is made increasingly restrictive, the resonance must be scanned (by varying either H_0 or ω) ever more slowly. Scanning times of the order of hours may be used; however, the long-term stability of the entire system then becomes an important factor in determining the actual signal-to-noise ratio which can be achieved.

A technique now in widespread use in all types of magnetic resonance is that of signal enhancement by continuous time averaging.[17] The basic scheme involves the use of a multichannel memory device to sum repetitively the signals from successive scans (in frequency or field) across the spectrum. This may be done by converting the output signal from the spectrometer proper, i.e., the output signal from the lock-in detector stage, into a pulse train, where the instantaneous pulse frequency is

[17] M. P. Klein and G. W. Barton, *Rev. Sci. Instrum.* **34**, 754 (1963).

proportional to the dc voltage level of the lock-in output. The memory channels are opened consecutively to these pulses at fixed time invervals, usually based on a crystal-controlled oscillator, and the field or frequency variable of the spectrometer system is scanned in synchronism with the same timing pulses. As the data accumulate, the signal amplitude increases directly with the number of complete spectra stored, but the noise amplitude grows only as the square root of the number of scans (passes). The signal-to-noise ratio then improves as the square root of the number of scans.

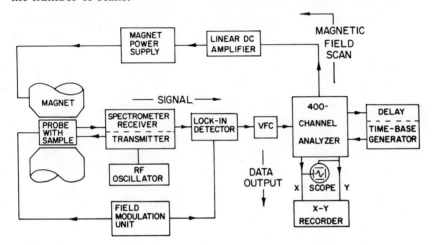

FIG. 5. Schematic arrangement for continuous time averaging, utilizing a multi-channel memory device. The VFC unit is a voltage-to-frequency converter whose instantaneous output frequency is proportional to the dc voltage level of the lock-in output.

Figure 5 shows the block diagram of a typical arrangement of this type. In this example the magnetic field is directly controlled by the channel advance pulse. However, the same instrumentation may be used to scan frequency, as, for example, in nuclear quadrupole resonance or in zero-field NMR. The function of the time base generator delay unit in the diagram is to permit the magnetic field to return to its initial value again following the return of the active channel from its final value to channel zero.

And lastly, an additional further step which may be taken to increase spectrometer sensitivity is to cool the electronics, at least the first stage or stages thereof. FET and MOSFET circuits are particularly appropriate for this, but only a few systems have been described in the literature.[18,19] Both of these employed MOSFET rf preamplifiers operating at liquid helium temperature.

9.7.2.2. Transient Methods.[20] The transient solutions to Eq. (9.7.7) were first investigated by Torrey[21] and subsequently by Hahn[22] who initiated the techniques of "pulsed NMR" based on the use of intense pulses of the H_1 field of very short duration. Such pulses are generally separated by much longer intervals during which $H_1 = 0$. In this way, the magnetization vector responds solely to the external (steady) field H_0 and to the various relaxation mechanisms during these intervals. A wide variety of pulse sequences are possible, incorporating as many as eight pulses of various widths and separations. Many of these sequences are intended to measure the basic relaxation times T_1 and T_2 of the Bloch equations, but some have other, additional applications.

Experimental arrangements for pulsed methods are similar to those utilized in steady-state work. An important difference is that the rf exciting source must typically provide hundreds of watts of peak power in contrast to the few milliwatts required for steady-state detection. These intense pulses usually overload the receiver, in spite of careful balancing adjustments and other schemes, so that the latter must be designed for rapid recovery from such saturation.

Phase-sensitive detection of the rf signal offers better sensitivity (and linear detection) than straight amplitude detection. Signal averaging techniques can also be utilized in pulse experiments. A suitable multi-channel memory device sums repetitively the signals from successive pulse sequences. In this application, since the width (in time) of the transient response is typically measured in milliseconds, a very fast digitizing rate is called for. Memory channel dwell times in the range 0.1–1.0 μsec are frequently employed. Appropriate timing and gating circuits are required to control the length and spacing of the rf pulses. In the most exacting cases, these intervals are derived from the (crystal-controlled) master oscillator for the resonance frequency by digital logic circuitry. In this manner, all of the rf pulses can have complete phase coherence. The block diagram of a typical basic pulse spectrometer system is shown in Fig. 6.

[18] D. W. Alderman, *Rev. Sci. Instrum.* **41**, 192 (1970).

[19] D. S. Miyoshi and R. M. Cotts, *Rev. Sci. Instrum.* **39**, 1881 (1969).

[20] T. C. Farrar and E. D. Becker, "Pulse and Fourier Transform NMR." Academic Press, New York, 1971.

[21] H. C. Torrey, *Phys. Rev.* **76**, 1059 (1949).

[22] E. L. Hahn, *Phys. Rev.* **80**, 580 (1950).

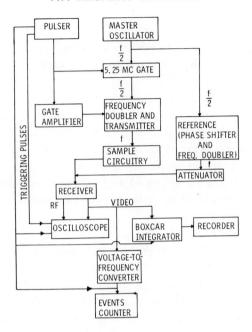

FIG. 6. Block schematic of a basic, general-purpose pulse spectrometer. The use of frequency doubling and pulse gates at both frequencies minimizes rf leakage to the receiver between pulses. Phase-sensitive detection of the rf signal is provided for, and the box-car integrator is employed as a single channel memory device in detecting weak signals.

A simple pulse sequence for measuring T_2 by means of the "spin-echo" experiment is indicated in Fig. 7. The magnetization **M** has its equilibrium value in the direction of $\mathbf{H}_0$ initially, and at $t = 0$ an rf pulse of such amplitude H_1 and duration t is applied to tip **M** through an angle of 90°. This follows according to the relation $\theta = \gamma H_1 t$, which describes the precession of **M** about $\mathbf{H}_1$ through the angle θ in time t in a coordinate frame rotating at the resonant frequency ω_0. Following such a 90° pulse, the magnetization precesses about $\mathbf{H}_0$ and decays in

FIG. 7. Pulsed NMR signals. The free induction decay (FID) follows the 90° pulse, and the spin-echo signal appears at 2τ.

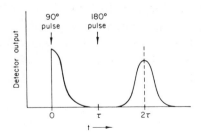

amplitude as the phase difference between different groups of spins (spin packets) grows with time. The resulting decaying signal picked up by the receiver is referred to as the *free induction decay* (FID).

When a second pulse is applied at time $t = \tau$, such that the magnetization of each spin packet is reversed (a $180°$ pulse), the accumulated phase differences are reversed in sign (since the direction of the precession about $\mathbf{H}_0$ always remains the same). Hence, after an equal time interval τ, the phase differences all return to zero, and an "echo" signal is detected by the receiver. The amplitude of the echo is a function of the pulse spacing τ, and for a spin system described by the Bloch equations, the echo amplitude decays exponentially, having the dependence $\exp(-2\tau/T_2)$ on the transverse relaxation time T_2.

In contrast to the basic $90°$–$180°$ pulse sequence, much more sophisticated techniques involving as many as eight pulses have been developed by Waugh and his co-workers.[23] These sequences effectively average to zero the spin-spin (or dipolar) interactions between nuclei, so that the broad resonance lines characteristic of NMR in solids are "narrowed." This narrowing makes possible precise measurements of the small shifts in resonance frequency which reflect, for example, chemical bonding in solids and which are otherwise difficult to observe.

Finally, the pulse technique which most nearly corresponds to conventional steady-state NMR is Fourier transform spectroscopy. This approach is based on the fact that the spectral representation in the frequency domain $f(\omega)$ is the Fourier transform of the free induction decay (FID) in the time domain $G(t)$:

$$f(\omega) = k \int_0^\infty G(t) \cos \omega t \, dt \qquad (9.7.15)$$

where k is a constant of proportionality.[24]

In order for $G(t)$ to contain the desired information, the width of the exciting pulse must be small compared to the reciprocal of the highest frequency separation in the spectrum, and the FID signal must be detected for a sufficient length of time, consistent with the smallest frequency separation in the spectrum. Of course, the exciting pulses must not be repeated too frequently compared to the spin-lattice time T_1 or partial saturation of the spectrum will occur. In effect, the entire spectral region of interest is excited and detected simultaneously with the Fourier

[23] J. S. Waugh, L. M. Huber, and U. Haeberlen, *Phys. Rev. Lett.* **20**, 180 (1968).
[24] I. J. Lowe and R. E. Norberg, *Phys. Rev.* **107**, 46 (1957).

components of the pulse. By contrast, in steady-state spectroscopy the resonance frequency is scanned through the region of interest so that the response from only one frequency at a time is observed. The Fourier technique therefore affords a very significant advantage in measurement time required for given sensitivity, which can approach factors of 100–1000. In typical applications,[25] the FID signal is stored in a multichannel memory device, and the Fourier transform operation is accomplished by computer using the so-called "fast Fourier transform" (FFT) algorithm.[26]

9.7.3. Electron Spin Resonance.

9.7.3.1. Steady-State Methods. There is no fundamental difference between electron spin resonance (ESR) and nuclear magnetic resonance insofar as either basic theory or methods of detection are concerned. However, due to the fact that electronic gyromagnetic ratios are typically roughly 1000 times larger than those of nuclei, the resonance frequency domain for ESR falls in the microwave region (3–30 GHz) for magnetic fields produced by conventional laboratory magnets. The lumped circuit elements of NMR are replaced by waveguide and resonant cavity elements. Of course, in sufficiently weak applied fields, the resonance frequency for ESR will also occur in the NMR region of the rf spectrum. This is not a very sensitive arrangement, but it can be readily utilized for demonstration or instructional purposes with a simple NMR spectrometer and a very small quantity ($\sim 10^{-6}$ mole) of an organic free radical such as diphenyl-picryl-hydrazyl (DPPH).

A second consequence of the stronger electronic moments is that the relaxation mechanisms contributing to both T_1 and T_2 are also stronger. Hence, electron spin systems typically have shorter T_1 and T_2 values, and in consequence of this the resonance linewidths are generally much greater than in the nuclear case. Indeed, in most cases ESR specimens are prepared in magnetically dilute form in order to reduce the interaction between spins, thereby yielding significantly narrower resonances. For the same reason, ESR furnishes a particularly powerful tool for the investigation of paramagnetic defects and impurities in materials.

Just as in the case of NMR, in designing an ESR spectrometer for high sensitivity the operating frequency ω_0 should be chosen as large as possible and practicable, consistent with Eq. (9.7.14) and with magnetic

[25] R. R. Ernst and W. A. Anderson, *Rev. Sci. Instrum.* **37**, 93 (1966).
[26] J. W. Cooley and J. W. Tukey, *Math. Comput.* **19**, 297 (1965).

field strengths available. Klystrons furnish the usual source of rf power for such spectrometers, and these are available to cover frequency ranges from about 5 to 50 GHz, including the most used ranges at 10 GHz (X band) and 24 GHz (K band). Certain solid-state diodes may also be used as microwave sources (see below).

Shown in Fig. 8 is the block schematic of a relatively simple, yet practical, ESR spectrometer. The power supply for the klystron must be very well-regulated, and the klystron itself may be operated in an oil bath for thermal stability. Usually the klystron frequency is stabilized (locked) to that of the sample cavity or alternatively to one of the high harmonics of a crystal oscillator. The gyromagnetic properties of ferrite are utilized in the isolator which is a standard commercial waveguide

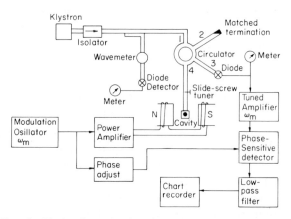

FIG. 8. Block schematic of a microwave ESR spectrometer.

component that permits power transfer from klystron to waveguide but absorbs virtually all power reflected back toward the klystron from the waveguide system (see Section 10.2.5). This largely eliminates so-called "frequency pulling" of the klystron by any of the reactive circuit elements, including the cavity. The klystron frequency and power are monitored by means of the auxiliary circuit employing the directional coupler, wave meter, and diode detector and its meter. Where greater accuracy is required, the signal from the directional coupler may also be extracted and the frequency measured with a frequency counter.

A solid-state alternative to the klystron as the microwave frequency power source is available in the Gunn diode,[27] which has been utilized

[27] J. B. Gunn, *IBM J. Res. Develop.* **8**, 141 (1964).

in several spectrometer applications in the 7–10 GHz range. In a particularly simple design, the Gunn diode is placed directly in the resonant cavity with the paramagnetic sample, and the entire spectrometer consists simply of the cavity together with a short section of waveguide terminated by a detector diode and tunable short.[28] Although this spectrometer has significance primarily for instructional purposes, the application of the Gunn diode as a microwave source is especially attractive in view of the extremely simple power supply and frequency stabilization requirements relative to klystrons[28] (see also Section 10.3.2).

The key element of the spectrometer is the circulator which yields the sensitivity of a bridge circuit by enabling the detector to respond to the signal from the sample cavity without interference from the incident rf power. Qualitatively, the circulator acts in the following manner: The spacings between the ports, measured in units of λ_g, the wavelength in the guide, are $\lambda_g/4$ between ports 1 and 2, 2 and 3, and 3 and 4, and $3\lambda_g/4$ between ports 1 and 4. The power incident at 1 circulates around the ring in both directions, leaving the different ports if the phase relations allow this. At ports 2 and 4 the two waves have the same phase, whereas at 3 they have opposite phase. Hence, the power incident at 1 is equally divided between ports 2 and 4, half being lost in the matched load at 2 and half going to the cavity. None of the incident power appears at 3, provided that the cavity, which at its resonant frequency behaves as a purely resistive load, is properly matched to the characteristic impedance of the guide by means of an impedance-matching element— usually a slide-screw tuner or three-screw section.

When paramagnetic resonance occurs, sideband frequencies $\omega_0 \pm \omega_m$ are produced by the interaction of the rf energy with the spins in the sample (ω_m is the modulation frequency). The sideband power enters the circulator at 4, divides as before, and the two waves reach ports 3 and 1 in-phase. That which reaches 3 furnishes the output signal, that which reaches 1 is absorbed in the isolator. In practice, a 4-port circulator differs little from a hybrid ("magic") tee; a 3-port circulator on the other hand can replace both the 4-port circulator and matched load, thereby doubling the signal voltage at the detector (assuming no saturation of the resonance).

The crystal diode detector requires a certain rf biasing power level (at ω_0) to detect the sidebands with good signal-to-noise ratio, and this bias power is usually obtained by mismatching the cavity slightly with

[28] L. W. Rupp, Jr. and W. M. Walsh, *Amer. J. Phys.* **38**, 238 (1970).

the slide-screw tuner. This element also serves to select either the absorption or dispersion-mode signal by controlling the relative phase between the klystron frequency ω_0 and the sidebands $\omega_0 \pm \omega_m$ at the detector.

The differential sampling scheme, already described with respect to its application to NMR, furnishes a very significant improvement in both signal-to-noise and fidelity of line-shape reproduction. Since the electronic gyromagnetic ratio is so large, the widths of even relatively narrow ESR lines ($\sim$0.2 Oe) correspond to rather high frequencies ($\sim$600 kHz). Consequently, much higher modulation frequencies may be used, 100 kHz being a fairly standard choice. This greatly reduces the (frequency)$^{-1}$ noise characteristic of crystal detectors relative to that which would result from using a low audio frequency. The requisite amplifiers, lock-in detectors, and associated circuitry (transistorized) have been described in detail in the literature.[29] Applications of continuous averaging methods to ESR signal enhancement proceeds exactly as in the NMR case.

9.7.3.2. Transient Methods. Analogous to the transient NMR methods described in Section 9.7.2.2, pulse methods have been utilized in the study of ESR relaxation times (particularly T_1) since as early as 1954.[30] A wide variety of experimental arrangements have been described in the literature. In place of klystrons, magnetrons which are capable of delivering high-power, short-duration pulses provide the usual source of microwave power in these spectrometers.

One relatively simple spectrometer[31] employs a 4-port circulator and low-power klystron (not pulsed) in an arrangement like that of Fig. 8. A pulsed magnetron is also coupled into the microwave system in parallel with the klystron power via a high-power attenuator. The reflected klystron signal from the sample cavity is detected and monitored in the conventional steady-state manner. With the spectrometer tuned to resonance, and before applying a pulse, the detected power is proportional to the steady-state rf susceptibility χ'' in the absence of saturation. Immediately following the application of a pulse from the magnetron, χ'' rises virtually instantaneously to an effective initial value, $\chi''(0)$. The decay of $\chi''(t)$ from this $\chi''(0)$ back to χ''(steady-state) follows an exponential dependence on t/T_1 for spin systems behaving in accord with the Bloch

[29] T. H. Wilmshurst, "Electron Spin Resonance Spectrometers." Plenum Press, New York, 1968.

[30] N. Bloembergen and S. Wang, *Phys. Rev.* **93**, 92 (1954).

[31] C. F. Davis, Jr., M. W. P. Strandberg, and R. L. Kyhl, *Phys. Rev.* **111**, 1268 (1958).

equations, (9.7.7). The fast transient recovery of the rf susceptibility may be recorded directly from an oscilloscope photographically or alternatively by means of an appropriate fast multichannel memory device as described for NMR applications in Section 9.7.2.2.

9.7.3.3. Electron Nuclear Double Resonance (ENDOR). The methods of ESR and NMR have been combined in a variety of experimental arrangements. One such arrangement, often referred to by the acronym ENDOR (Electron Nuclear Double Resonance),[32] capitalizes on the high sensitivity of ESR to detect and measure interactions between nuclei and electronic moments.

A typical experiment employs a coil within the ESR resonant cavity, by means of which the nuclear spin system may be irradiated with rf energy. Sufficient microwave power is applied to saturate the electronic resonance (keeping the static field H_0 fixed at the resonance value), and in this condition the rf power source is scanned in frequency. When the frequency matches the resonance condition corresponding to the electron-nuclear interaction, the nuclear spin transition which occurs provides additional relaxation mechanisms for the electron spin, and the electronic resonance recovers from saturation. In this manner a spectrum of resonances is traced out on the chart recorder, corresponding to the different electron-nuclear interaction strengths present. This technique has proven especially valuable in the study of impurity ions in semiconductors,[32] and of color centers in alkali halide crystals.[33]

[32] G. Feher, *Phys. Rev.* **114**, 1219 (1959).
[33] W. C. Holton, H. Blum, and C. P. Slichter, *Phys. Rev. Lett.* **5**, 197 (1960).

9.8. Computers*

9.8.1. Analog Computers[1-3]

9.8.1.1. Introduction. Analog computation was used by very early scientists and engineers in surveying and map making and the Greeks made extensive use of analogs in their early work in astronomy. The common slide rule and planimeter were more modern analog devices but the development of the ball and disk integrator by James Thomson and the interconnection of these integrators to solve differential equations as suggested by Lord Kelvin in 1876 are considered to be the true beginning of the differential analyzers or analog computers of today. Vannevar Bush made the mechanical differential analyzers a practical computational tool and these analog computers were given considerable use through 1950. The electronic analog computer was developed during and after World War II and the techniques of that analog signal system are now used extensively. New configurations are continually being designed for special-purpose or limited-purpose processors and a number of general-purpose machines of great flexibility are available in computational centers.

The general-purpose analog computer finds greatest use in the simulation of complex physical systems described by differential equations. With an accurate simulation, the system can be better understood, the effects of changes in system design can be determined, and component tolerances can be realistically specified. These analog simulations operate faster, generally, than the digital computers and such a system simulation may often be operated simultaneously with a physical system, an accomplishment usually denied to digital simulations. In simulations where speed is not such a great factor, digital techniques are very competitive.

[1] G. A. Korn and T. M. Korn, "Electronic Analog and Hybrid Computers," 2nd ed. McGraw-Hill, New York, 1972.

[2] A. Hausner, "Analog and Analog-Hybrid Computer Programming." Prentice-Hall, Englewood Cliffs, New Jersey, 1971.

[3] J. G. Graeme, G. E. Tobey, and L. D. Huelsman, "Operational Amplifiers, Design and Application." McGraw-Hill, New York, 1971.

* Chapter 9.8 is by **Paul E. Russell.**

Probably the most desirable configuration for design through simulation is a hybrid arrangement with the basic dynamics simulated in analog fashion and with the analog under the direct control of digital logic. Such hybrid computers are finding considerable use.

Special-purpose electronic analog signal systems are widely applied in instrumentation, interpretation, and control functions. Analog signal processors are manufactured in module form and the newer technologies in electronic circuit fabrication have made these modules relatively cheap, small, and reliable. The processors may perform very complex manipulations of signals, but comparison devices, simple addition or subtraction circuits, operating in a place or manner unsuited to human beings, make up a very large part of the application. Special-purpose analog arrays are available also to solve certain classes of problems such as algebraic equations or partial differential equations of a limited form.

An analog computer is a parallel machine with all computing elements operating simultaneously and with almost all elements providing outputs that are voltages varying with time. Thus, time is the independent variable and the output voltages of the computing elements bear a known relationship to the dependent variables of the problem. The solutions appear as continuously available voltage signals for recording, display, or perhaps for further processing in follow-on hardware. In repetitive analog computers, the solutions appear in repeating patterns and may be easily displayed on an oscilloscope.

9.8.1.2. Components of Analog Computing Systems.

Problem variables are most often represented by corresponding voltages of magnitude between ± 100 or ± 10 V. These voltages, machine variables, are constrained by interconnection of the computing elements to obey the equations that describe the system to be simulated. The usual elementary operations implemented in the computing elements are (1) multiplication of a voltage by a constant usually related to a coefficient in the describing equations, (2) addition of two or more voltages, perhaps with amplification, (3) integration of a voltage with respect to the variable time, (4) generation of a voltage that is a function of an input voltage, and (5) multiplication of two time-dependent voltages. The notation for these operations is illustrated in Fig. 1. Fundamental to the elements and their proper array are operational amplifiers, linear and nonlinear circuit elements, power supplies, read-out devices, control relays and other switches, and means for proper interconnection of the elements. Interconnection on general-purpose machines is arranged through patch-

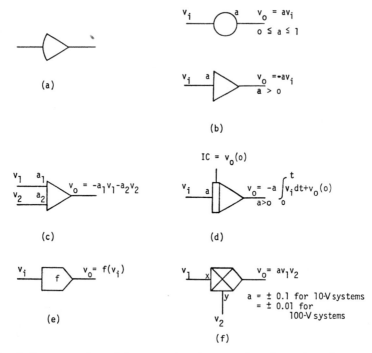

FIG. 1. Representations of the computing elements of an electronic analog computer. (a) High-gain operational amplifier (inverting); (b) multiplication by a constant; (c) addition of voltages; (d) integration with respect to time, t; (e) function generation; (f) multiplication of voltages.

boards which may appear to be extensive and elaborate to provide for the flexibility required there. Likewise, read-out and control functions are much more complex on general-purpose computers. However, the elementary operations on all analog computers are generally implemented as described below.

An attenuator or potentiometer provides multiplication of an input voltage by a constant less than unity. The potentiometer must be properly calibrated and loading effects of impedances that are connected to the arm must be taken into account. Though potentiometer inductance is negligible for all except extremely high computation frequencies, the stray capacitance of the potentiometer and interconnected lines and components may require special attention in the selection of best hardware and layout, even at moderate or low frequencies, if high accuracy is required.

Though voltages can be manipulated with only passive elements, an operational amplifier is most commonly used in conjunction with fixed circuit elements so as to obtain amplification as well as the elementary manipulation. Operational amplifiers are discussed in Chapter 6.5. For ± 10-V computers, the operational amplifiers will have dc gains higher than 10^4, have full output capability to at least 50 kHz, input impedances higher than 0.25 MΩ, and noise voltages less than 10 μV. Somewhat higher noise is expected in ± 100-V amplifiers. As commonly applied, the operational amplifiers are inverting amplifiers though differential amplifiers with noninverting as well as inverting inputs may be used. With careful matching of gains and high common-mode rejection, differential amplifiers offer a great many possibilities.

A general arrangement of the operational amplifier is shown in Fig. 2. With the amplifier gain and input impedance very high,

$$v_g = \frac{v_o}{\text{amplifier gain}} \simeq 0$$

and

$$\frac{v_1}{Z_1} + \frac{v_2}{Z_2} + \cdots + \frac{v_i}{Z_i} + \frac{v_o}{Z_f} = 0$$

so that

$$v_o = -\left[\frac{Z_f}{Z_1} v_1 + \frac{Z_f}{Z_2} v_2 + \cdots + \frac{Z_f}{Z_i} v_i \right]$$

If all impedances are made wholly resistive, then

$$v_o = -\left[\frac{R_f}{R_1} v_1 + \frac{R_f}{R_2} v_2 + \cdots + \frac{R_f}{R_i} v_i \right]$$

and a summing function with amplification on each input signal has been implemented. Commonly, the resistors are fixed and are of the order of

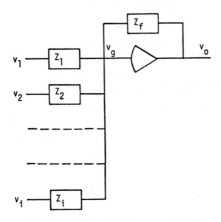

FIG. 2. An operational amplifier with input and feedback impedances.

megohms, proportioned so as to provide gains in the range 0.1 to 10.0. With potentiometers in tandem with these amplifying capabilities, any reasonable gain can be set to 3-digit accuracy.

If all input impedances of Fig. 2 remain resistive and the feedback impedance is made to be purely capacitive, then the input voltages are not only amplified and summed, they are integrated with respect to time, t, computer time. If, through a simple switching arrangement, the capacitor is charged to some initial value prior to the beginning of the integration, the initial condition on v_0 can be inserted. The output in that event is

$$v_0 = - \left[\frac{1}{R_1 C_f} \int_0^t v_1 \, dt + \frac{1}{R_2 C_f} \int_0^t v_2 \, dt + \cdots + \frac{1}{R_i C_f} \int_0^t v_i \, dt \right] + v_0(0).$$

For problems using integration, the independent variable is then inherently computer time. The resistors are commonly megohms, capacitors microfarads, so that channel gains can be set in the range from 0.1 to 10.0 [i.e., $dv_0/dt = (0.1 \text{ to } 10.0)v_i$].

Differentiation of input voltages with respect to computer time may be accomplished with a purely resistive feedback impedance and with capacitive input impedances, though this is seldom done since differentiation emphasizes noise and may also produce an overload on the amplifier output when input waveforms are steep. Approximate differentiation is sometimes inplemented so as to avoid these difficulties by replacing the capacitive input impedance with a series combination of a resistor and capacitor, but the approximation is very inaccurate for input voltages with components of frequency larger than the reciprocal of the time constant of the series combination. Usually, the requirement for differentiation can be avoided by proper manipulation of the system equations.

Complex networks can be arranged in the place of the input and feedback impedances of Fig. 2 to effect complex transfer functions. More will be said on these active filters in the next section.

Function generation, the generation of an output voltage as a nonlinear function of an input voltage, is required for many problems. Logarithmic amplifiers, limiters, sine–cosine resolvers, and comparators are special cases of function generation. Cathode ray tubes and servo driven potentiometers have been used for many applications, and certain nonlinear resistors, as varistors, have served to simulate specific functions. Combinations of multipliers, summers, and pots readily serve to generate polynomial functions, but higher order polynomials require large numbers of multipliers.

Biased diodes in resistance networks are the most widely used elements, providing suitable approximations to many classes of nonlinear functions. In these networks the diodes serve as voltage-sensitive switches, selecting gain values determined by the resistive elements, so that the desired function is approximated by straight-line segments. Breakpoints of the line segments are set by the bias voltages and the slopes are determined by the gain values. A number of useful arrangements are shown in Figs. 3

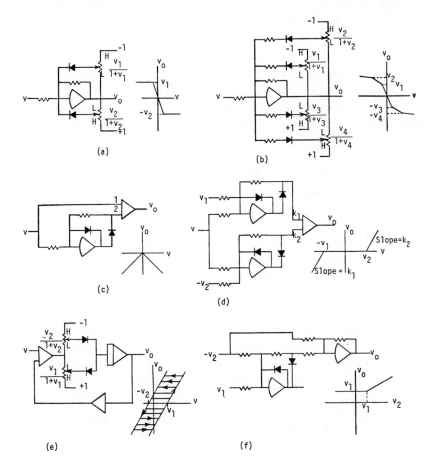

FIG. 3. Diode function generators. In these circuits, the potentiometer resistances are small compared to the fixed resistors shown. (a) Feedback limiter, hard limiting; (b) feedback limiter, soft limiting; (c) absolute value; (d) dead space; (e) backlash; (f) maximum selector.

and 4. Diode networks may appear in the input or feedback systems. It should be noted that input summing resistors can provide for the addition of signals before function generation, if the diode networks are placed in the feedback circuit. Figure 4a, however, illustrates function generation before addition with networks in the inputs. In a general-purpose computer, the diode function generator circuits allow for many line segments with wide variation of break point and gain. Best practice uses the function generator to supply only corrections to analytic approximations to the required function. Functions that are not monotonic are derived by adding a nondecreasing function to a nonincreasing function. Diode characteristics are temperature-sensitive and considerable care must be exercised in their use to prevent break-point drift, a consideration accounted for in the circuits of Fig. 3. Break points for these circuits tend to be sharp rather than smooth, a characteristic that is not always desirable.

Diode function generators can be used to generate functions of several variables as may be required in instrumentation and control applications or in dynamic simulations. Combinations of functions of two variables can serve since $f_1(v_1, v_2) = f_2[f_3(v_1), v_2]$. That approach is quite tedious and is usually avoided. A sometimes substitute approach generates $f_1(v_1)$ for a number of fixed v_2 and then interpolates for other values of v_2.

Analog signal multiplication can be implemented with servo driven pots with the servo driving a number of ganged pots, some linear and providing multiplication, some perhaps nonlinear and providing a nonlinear function. The relatively slow response times for these servos coupled with maintenance difficulties have caused them to be replaced almost completely by all-electronic multipliers. Though most electronic multipliers will use quarter-square, variable transconductance, or time division techniques, logarithmic function generators, triangle averaging arrangements, or special modulation systems may be encountered.

In the quarter-square multiplier, diode function generators are used in squaring circuits to develop the nonlinear parts of $v_1 v_2 = \frac{1}{4}[(v_1 + v_2)^2 - (v_1 - v_2)^2]$. Ten segments in a squaring circuit can develop the squared relation to within 0.1% of full scale and these in tandem with absolute value generators in an arrangement as in Fig. 4a can serve to provide four quadrant multiplication with less than 0.5% (of full scale) error. True error voltage may be large, however, at small signal levels. The circuits are fast and with only small dynamic error.

The triangle-averaging multiplier of Fig. 4b develops the product utilizing a triangle wave voltage source as a kind of carrier which is removed later in a low-pass filter. The sum and difference of the inputs

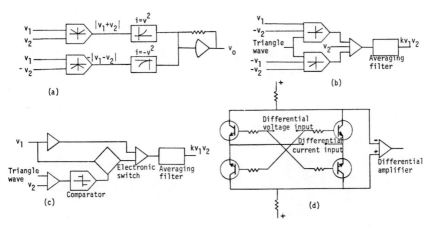

FIG. 4. Analog multipliers. (a) Quarter-square multiplier; (b) triangle-averaging multiplier; (c) time division multiplier; (d) variable transconductance multiplier.

separately bias the triangle wave which is then clipped and averaged. The triangle wave must be linear, with sharply defined peaks. The amplitude must be constant and the waveform must have a zero average value. Inaccuracies in the waveform of the carrier limit the over-all accuracies obtainable, but 1.0% multipliers are available at low cost. Bandwidths are limited by carrier frequency and by the low-pass filter.

In the time division multiplier, a triangle wave source is added to one input voltage and this sum is applied to a comparator, thereby developing a square wave with pulse-width modulation. As noted in Fig. 4c, this square wave can control an electronic switch that gates the second factor, which after proper smoothing is proportional to the product of the two input voltages. Such multipliers are low cost and display reasonable accuracy and frequency response.

Transconductance multipliers exploit the fact that emitter current can control the gain of a transistor amplifier. One factor of the product generates the emitter current, and the product is formed in a very straightforward amplification of the second factor of the product. In Fig. 4d, the transistors must be perfectly matched, but the method serves extremely well in integrated circuit construction and is widely used. Accuracy can be made as good as 1.0% at frequencies up to 10 kHz. Some designs provide better than 5.0% accuracy even to 100 MHz (see also Section 8.5.1).

Multipliers may be used to compute the square or square root of a voltage waveform or to divide, though division is avoided if at all possible since the output of a divider can saturate with low levels of denominator

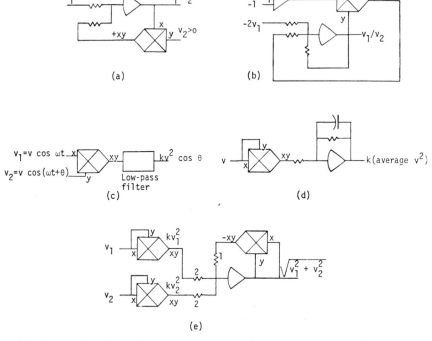

FIG. 5. Circuits with multipliers. (a) Standard division circuit; (b) improved division circuit; (c) phase detector; (d) mean square computation; (e) computation of vector magnitude.

voltage. Since rms calculations, power calculations, and other similar operations are so commonly required, modules are available incorporating multipliers to perform many of those functions. See Fig. 5.

9.8.1.3. Analog Techniques in Simulation and Signal Processing. Applications of analog computing devices were mentioned in a broad way in Sections 9.8.1.1 and 9.8.1.2 as the general systems and components were described. The subject is now treated in greater detail.

The fundamental notions are the same whether the application be signal processing or simulation, only the attitude of the user changes. In Fig. 6, a computing array is indicated for solution of a second order linear differential equation as the system might be forced by an input c, or as it might respond to initial conditions. The array is a simulation of the second order system but it is also useful in processing a signal c through a second order filter with an undamped natural frequency $\sqrt{b}$, and a damping ratio $0.5a/\sqrt{b}$.

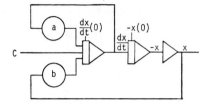

FIG. 6. Solution of $d^2x/dt^2 + a(dx/dt) + bx + c = 0$. A second order filter.

The dynamics represented by linear differential equations common to physics and engineering are usually simulated using only the summing amplifiers and integrators, and the scale-changing potentiometers. The basic techniques are those illustrated in the second order simulation already mentioned. If the dynamics are represented by a single, possibly high order differential equation, a simple mechanization follows from solving the equation for the negative of the highest order derivative. Integrators arranged in tandem provide successively lower derivatives (alternating in algebraic sign) if the first integrator is driven by a signal equivalent to the negative of the highest order derivative. The inputs to the first integrator are derived from the outputs of the integrators in the chain, properly sized and signed according to the constraining equation. Illustrating this technique, a second order array is shown in Fig. 6; however, the extension to higher order equations is straightforward.

If the system is represented by a set of linear equations, each equation in the set can be simulated as noted, with coupling terms in the equation being developed from the simulations of the other equations in the set as required. If the system is represented with state variables and a set of first order equations, the mechanization is extremely direct, with each equation in the set requiring a single integrator. A second order system in state variable form is shown in Fig. 7. Any linear system, any linear filter, can be formed, provided the noise, saturation, and frequency limits of the analog components and their interconnection are not abused.

If a transfer function notation (see Part 12) is used in describing the linearized dynamics, the notation can be carried directly into the analog

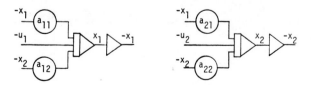

FIG. 7. Solution of $[\dot{x}] = [A][x] + [u]$ for a second order system. Connections between points with the same label are not shown.

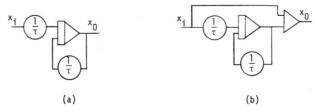

FIG. 8. Simulation of a first order lag (a) and lead (b). (a) $x_0 = -x_i(1/(\tau s + 1))$; (b) $x_0 = -x_i(\tau s/(\tau s + 1))$.

array. A first order lag, interpretable through a first order linear differential equation if desired, is shown in Fig. 8. Higher order transfer functions may be constructed similarly. Figure 9 shows another second order array, this time constructed from the transfer function notation. The transfer function can also be simulated through use of more complex input and feedback impedances in modifications of the circuit of Fig. 2 of Section 9.8.1.1. The circuit mentioned in Section 9.8.1.1 for approximate differentiation is an example. Other first order transfer functions are simulated as noted in Fig. 10. Note that Figs. 10a and 8a are alternate mechanizations. Some more complex networks are shown as examples of higher order simulations in Fig. 11. The operational amplifiers provide gain and isolate the parts of the networks in these circuits, but transfer function simulation can be accomplished with passive elements alone if these advantages are not important.

Nonlinear operations require function generators or multipliers or both. Figure 5 noted some elementary uses of multipliers. Figure 12 illustrates signal processing through nonlinear function generators. The solution of nonlinear differential equations or linear differential equations with time-varying coefficients follows the format noted for linear equations generally. The classic Van der Pol equation is illustrated in Fig. 13. Note that the first multiplier could have been a function generator. Nonlinear system

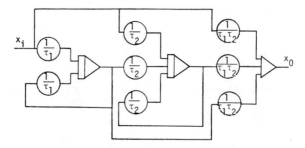

FIG. 9. Simulation of $-s^2/(\tau_1 s + 1)(\tau_2 s + 1)$.

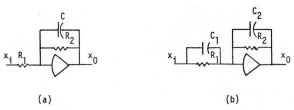

FIG. 10. Simulation of first order transfer functions using complex networks around the amplifiers.

(a) $\quad x_0 = -x_i \left(\dfrac{R_2}{R_1} \right) \left(\dfrac{1}{R_2 C s + 1} \right);$ (b) $\quad x_0 = -x_i \left(\dfrac{R_2}{R_1} \right) \left(\dfrac{R_1 C_1 s + 1}{R_2 C_2 s + 1} \right).$

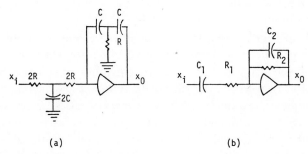

FIG. 11. Simulation of higher order transfer functions using complex networks around the amplifiers.

(a) $\quad x_0 = -x_i \left(\dfrac{1}{4R^2 C^2 s^2} \right);$ (b) $\quad x_0 = -x_i \dfrac{R_2 C_1 s}{(R_2 C_2 s + 1)(R_1 C_1 s + 1)}.$

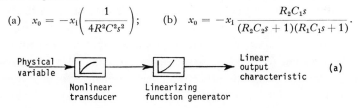

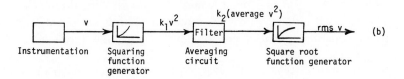

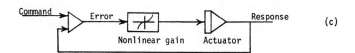

FIG. 12. Applications of function generators in analog signal systems. (a) Linearization of a transducer output; (b) rms calculation; (c) simulation of a nonlinear actuating servo.

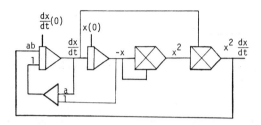

FIG. 13. Solution of $d^2x/dt^2 = -[a(bx^2 - 1)\,dx/dt + x]$.

simulation and the solution of nonlinear differential equations are accomplished in a very impressive way on the analog computer, especially when the computer simulation must be fast as in realtime simulation.

Attempts at solution of partial differential equations or simulation of distributed parameter systems usually begin by conversion to an approximating set of ordinary differential equations or lumped parameter systems. After conversion, the simulation follows the forms noted earlier. The technique is extremely valuable in special cases, as in simulating certain wave equation phenomena, but generally the excessive equipment requirement makes the solution more practical on the digital computer.

Analog studies of random processes mechanize the calculation of mean square error, maximum or minimum error, probabilities, etc., using standard computing modules. The input signals may be from the original source or may be simulated. They may be made available from tape records and these tapes may then be used as input a large number of times. If recorded playback is utilized, time scaling may be applied to permit processing at speeds other than the original recording speed.

Analog techniques lend themselves very effectively to the solution of many forms of algebraic equations. Summation and multiplication of variables are the principal manipulations required, with the mechanization being used either to provide automatic setting in accord with the algebraic equation constraints or to allow for a manual iteration to effect those constraints.

In a different type of application, fairly complicated signal manipulations can be performed rapidly with analog computation techniques. For example, the particle identifier of Goulding et al.[4] can determine the mass number A and the charge z of light particles emitted in nuclear

[4] F. S. Goulding, D. A. Landis, J. Cerny, and R. H. Pehl, A new particle identifier technique for $z = 1$ and $z = 2$ particles in the energy range >10 MeV. *IEEE Trans. Nucl. Sci.* **NS-11**, No. 3, 388 (1964).

reactions in a few microseconds. The particles are made to pass through a thin detector in which they deposit an energy ΔE and come to rest in a thick counter, dissipating the rest of their energy, E. Using logarithmic function generators, the computer calculates the quantity $(E + \Delta E)^{1.73} - E^{1.73}$, which is independent of the energy of the particle and proportional to $z^2 A^{0.73}$, uniquely identifying the particle. With increasing use of on-line digital computers, though, this class of analog calculations will become less important except where speed is essential.

9.8.2. Digital Computers[5-8]

9.8.2.1. Introduction. In a manner of speaking, a computer is any calculating device. The term can as well be applied to a counter, an abacus, or other adding machine, as to a modern-day large-scale processing machine. Our discussions in this treatment center on those computers in which the signal systems are primarily electronic. Thus, there are two fundamentally different types of computers to be considered, analog and digital. The analog computers represent physical system dynamics as voltages or currents in electronic arrays, and some detail on those configurations has been provided in the preceding section. Digital computers of concern in this section employ electronic operations fundamentally much less complicated, often only modifications of counting processes. The electronic circuits required are treated in Part 8, so that this section can focus on the digital systems, their elements, and their configurations.

The development of the modern digital computer is often said to have surfaced in the later half of the 1940s. Many of the concepts date much earlier, however. In the early 1800s, Charles Babbage conceived an "engine" to perform a programable sequence of operations on digital quantities. In 1812 he built a fixed program machine to assist in the determination of logarithms and trigonometric functions. The symbolic logic, so important in the logical systems design, was developed by George Boole in the 1850s. The punched card that plays such an important part

[5] F. J. Hill and G. R. Peterson, "Digital Systems: Hardware Organizations and Design." Wiley, New York, 1973.

[6] H. V. Malmstadt and C. G. Enke, "Digital Electronics for Scientists." Benjamin, New York, 1969.

[7] C. Belove, H. Schachter, and D. L. Schilling, "Digital and Analog Systems, Circuits, and Devices: An Introduction." McGraw-Hill, New York, 1973.

[8] G. B. Davis, "Introduction to Electronic Computers," 2nd ed. McGraw-Hill, New York, 1971.

in the input and output functions of many digital configurations was developed by Herman Hollerith in the 1880s. Howard Aiken designed what was basically a large mechanical calculator, the Harvard Mark 1, in 1937.

The machine was completed in 1944 and is considered to be an important lead into the automatic electronic digital computers of today. Mechanical relays, switches, and other electromechanical devices provided storage and control functions. Though slow and unreliable by today's standards, many of the features of programable electronic computers were incorporated. The ENIAC, completed in 1945, at the University of Pennsylvania used electronic components and was programed by plugs and switches. It was far faster than the Mark I and is thought of as the first electronic digital computer. The EDSAC followed, using an internally stored program. The EDVAC appeared soon also, being the first electronic computer designed with internal instructions in digital form and with binary arithmetic units. The EDVAC was a serial machine— information was transferred serially, one digit at a time—but the concept of a parallel machine was noted at about the same time.

The first commercial installation, a UNIVAC I in 1951, ushered in the "first generation" of digitals. The IBM 701 in 1953 and the IBM 650 in 1954 were typical of the period, using vacuum tubes and therefore being big and only moderately fast. Transistors brought about the development of a second generation, available commercially from about 1959 and typified by the IBM 1620 and IBM 7094. Further trending toward microelectronics produced faster, smaller, more reliable circuits. Third generation computing systems using integrated circuits were available in 1965. Advances continue and large-scale integration of circuitry and MOS techniques offer a great deal for the immediate future. Cryogenic devices have considerable potential, but a fourth generation has no clear definition at this time.

9.8.2.2. Computer Systems. The digital processing system, generally, separates into hardware and software, though the design of one determines and is determined by the other. Hardware involves the equipment for preparation or development of the data, the input of the data to the system, the processing of the data, the storage of data, and the output from the system. All of these elements may be continuously interconnected into the system or may perhaps be operated "off-line," for flexibility is a major feature in many systems. Many separate users may "time share" a central system if each is provided an input channel and suitable control is exercised.

Software includes computer programs and routines to direct the built-in flexibility, to cause the computer operation to be that required for the current processing of data. The function provided by software can be built solidly into the hardware, thereby committing the machine to a limited range of use. However, flexibility requirements usually dominate and software is commonly the very versatile communications channel to the many elements of the hardware system. Software programs may provide for very routine processing as sorting or updating of records, or for very complex manipulations in the numerical solution of high order equations. The larger and perhaps specialized programs may be combinations of many rather standard subroutines which have application in processing of a large number of different problems.

In the hardware system, the central processing unit (CPU) provides the arithmetic unit, the main storage elements, and the means for control. All arithmetic manipulations are performed in the arithmetic unit on data drawn from the storage unit, the data transfer and the numerical manipulations being sequenced and otherwise directed by the control unit. Supplemental storage units to the primary storage unit are used to hold additional programs and data. This supplemental storage is often of very large capacity relative to that in the CPU, and most often takes the form of magnetic tape or disk. Magnetic drums, punched cards, or paper tape may also be used. As operated, a word or more often a group of words are read from the supplemental memory into the main memory, manipulated, and perhaps modified and then returned to the secondary unit. Main memory is most likely in magnetic cores.

The input data are produced from documents or instrumentation and made available to the system in the form of punched cards, magnetic or paper tape, magnetic characters on a paper base, or similar machine readable medium. Many machines accept direct typewriter input. The computer input device "reads" this input data from the medium, stores it in the internal or the secondary memory, or if so directed, may move it to the output system. Input devices then are card readers, tape readers, magnetic or optical scanners, or perhaps a digital encoder from a communications or instrumentation channel.

The output may appear as punched cards, tapes, graphical plots, print on tab sheets, visual displays as on a cathode ray tube, or perhaps as a digitally-coded electronic signal input to control a process or function. Line printers have been the most common output mechanism, but visual displays have become very popular in systems where parts of storage are to be read out often and no permanent record is required.

Computers are often classified according to size. Word length, memory capacity, cycle time, and cost are factors in size. A word is a group of binary digits (bits) which are manipulated as an entity. An 8-bit word can represent one of up to 256 binary numbers, a 16-bit word up to 65,536 numbers, etc. The bits of the word must provide sign bits for algebraic sign and parity bits for error detection in most systems. The word length determines data resolution, address codes, operation codes, etc., and is often selected in deference to the data format of the input. Word lengths to 24 bits are very common.

Capacity indicates the number of words that can be stored in internal memory, usually core memory, as noted earlier. Capacity requirements are set by the amount of data required in internal memory at one time, which is set by the complexity of the problem. Memory elements are binary so that memory capacity is an integral power of 2, with common values over 4000 and up to a million or more.

Cycle time is the time required to obtain a word from memory, restore the word in memory, and prepare the memory for the next operation. Most computers manipulate data much faster than the input and output devices can generate or accept signals so that speed limitations for most users is determined more by input–output equipment than by cycle time.

Larger computers find use as large data banks, gathering data and processing large sequences of data, and in supervising smaller computers. Their large memories and longer word length permit rapid data handling. Multiple words can be processed in single operations. The large memory allows contact with a number of processes, operators, or other computers simultaneously.

Small computers may be general purpose or for dedicated applications. "Minicomputers" with up to 8000 word core memory and up to 18-bit word length, augmented with secondary storage, are having a decided impact in applications formerly time-shared on larger computers. Some can control several different data acquisition devices simultaneously, and with the wide variety of peripheral equipment for input, storage, and output that is now available, the number of new installations is very significant. Software for these smaller machines may be more expensive than the hardware. Medium size computers with 64,000 word memories and 24-bit words can manipulate data faster than small machines and supervise several. Such a computer could control a process and simultaneously run scientific or engineering calculations. Executive programs can be run, too, to coordinate or schedule multiple tasks on a priority basis. Programming has been developed progressively to satisfy most needs.

9.8.2.3. Applications.[9] In experimental physics, digital computers are used either "off-line" or "on-line." In the first case, after the desired run of experimental data has been collected and encoded in a suitable input medium, it is submitted together with the appropriate program to the computer for treatment and analysis. Most often, this is done by batch processing in a central computation facility. This procedure has the considerable advantage that the experimenter profits from the experience of the computing staff and does not have to worry about any computer hardware. The disadvantage is, of course, the time lag between data taking and evaluation which prevents any immediate feedback from the computer output to the ongoing experiment.

An on-line computer, on the other hand, can receive input from monitors in the experimental system itself and also from the data obtained in the experiment. The data may arrive at controlled intervals or at random. In the latter case, the computer will stop its program for a few machine cycles ("data interrupt"), accept the input and store it, and then continue the program. The computer can thus control the functioning of the apparatus and at the same time perform an evaluation of the results, providing a basis for decisions on the continuation of the experiment. These may be made by the experimenter or, again, by the computer.

On-line computations can either be done by time sharing in a large central facility or by using a separate laboratory computer of the appropriate (more modest) size. The choice will depend on the amount of data to be taken and on the complexity and needed speed of response. The control of a high-energy accelerator, for example, will always be done by a laboratory computer. Compound systems should also be considered, where part of the control and data taking may be done by a small laboratory computer, while large-program evaluation proceeds simultaneously by time sharing in a large computer. The majority of on-line data acquisition and control systems are found in nuclear and particle physics laboratories,[10,11] but a rapid increase of their use in other fields is to be expected.

---

[9] See, for example, Digital computers in physics research. *Phys. Today* **23**, No. 7 (1970).

[10] On-Line Data-Acquisition Systems in Nuclear Physics, 1969. Nat. Acad. Sci., Committee on Nucl. Sci., Washington, D.C., 1970.

[11] Yearly reports on IEEE nuclear science symposium, in the first number of each volume; e.g. *IEEE Trans. Nucl. Sci.* **NS-20**, No. 1, 334–374 (1973).

9.9. Equipment Testing*

9.9.1. Classification of Tests

9.9.1.1. Introduction. It is a fact that electronic equipment can be tested in different ways depending on the aim of the person dealing with the equipment. The instrument designer, for example, needs a complete insight into the operation of the apparatus including the dependence of its performance on individual components; the manufacturer wants to test the finished equipment under severe combinations of adverse circumstances, unlikely to occur in practice; the average user is mainly interested in being protected against false measurements, and therefore is likely to check the performance under actual running conditions only. In very complex experiments, automatic tests are necessary, in order to detect developing faults before actual breakdown of the equipment. Finally, maintenance and repair work involves still another approach. The basic ideas underlying known test procedures will be outlined in this chapter.

9.9.1.2. Routine Checks and Adjustments. Many electronic instruments have built-in autotesting facilities which should be used about as often as the instrument is turned on. Typical examples are the end-of-scale adjustment of ohmmeters (to allow for battery voltage variations), the autocalibration push button of a digital voltmeter, or the sensitivity and overshoot adjustments of oscilloscope probes.

Thus, whenever an oscilloscope probe is taken from one oscilloscope to another, it is essential to check its response with the built-in calibrator before attempting to use it for any measurements. The crucial points of such self-testing features are:

(a) reliability of the built-in standard;
(b) simplicity of the procedure; in particular, adjustment buttons or screws must be easily accessible;
(c) indications on front panel should be self-explanatory, without reference to an instruction manual.

---

* Chapter 9.9 is by **Daniel Maeder.**

9.9.1.3. Internal Adjustments and Calibrations. In complicated circuits where several parameters may need readjustment, it may be necessary to proceed in a prescribed order, or to use special calibration equipment. For these kinds of tests, detailed instructions found in the manufacturers' manuals should be followed carefully. As these operations are not expected to occur too frequently, the corresponding buttons and screws are usually accessible only after opening the instrument. Such recalibrations should be done according to a regular maintenance schedule, or else whenever a deviation from equipment specifications is noticed.

9.9.1.4. Marginal Testing. Rather than wait until the user notices excessive inaccuracies which cannot be eliminated by normal adjustments, one may attempt to spot developing weaknesses before they cause real trouble. This can be done by running the equipment temporarily under slightly abnormal operating conditions. The following elementary examples illustrate this idea.

> *Examples.* (a) A regulated power supply designed to furnish 12.00 ± 0.06 V dc at its output when powered from a 220 V $\pm 10\%$ ac line could be run temporarily at ac voltages outside this range, say 190 or 250 V. Knowledge of the ac voltages at which the output begins to change abruptly allows a direct evaluation of the safety margin which exists in a given condition of operation.
> (b) An amplifier designed for operation from a 12.00 ± 0.06 V regulated dc supply is run temporarily at 11 or 13 V while checking the gain. From the percentage of gain variation observed, one can evaluate the gain stability to be expected in practical use.
> (c) An amplifier designed to provide a linear gain for input signals of a few millivolts is temporarily operated with much larger signals, and the resulting nonlinear response observed. From the measured distortion, one can estimate the margin of linearity in normal operation.

In it evident that in all these cases great precaution must be taken in order to avoid damaging the equipment by overloading.

9.9.1.5. Fault Tracking. Once a malfunction has been discovered, repair work is directed towards restoring the original state of the instrument in every detail: by this we mean that one should avoid any kind of readjustment in which the over-all effect of a faulty element is compensated by a voluntary deviation of another element from its nominal value.

Examples. (a) An attenuator giving a wrong output level could be modified by changing any one of its three resistors to produce the correct response into a nominal load. However, changing a good resistor for compensating an error in one of the others would result in a wrong input or output impedance.

(b) If several components have deteriorated, all causing gain reduction in an amplifier, it would not be acceptable to repair the trouble by changing just one of them, leaving the others with wrong values. Bandwidth or linearity might be seriously impaired if this were to be done.

The general method of locating faulty parts consists of following the flow of signals through the equipment until a deviation from normal behavior first becomes noticeable. This procedure merits a detailed discussion (see Section 9.9.6) with emphasis on the possibilities of component diagnostics *in situ*, i.e., with a minimum of disassembling–reassembling operations.

The usual assumption of only one component in a system being defective may be a reasonable starting hypothesis, but this should be checked as soon as a faulty component has been found. This is done by reconstructing the whole chain of interactions that led to the failure, and by checking all elements which could possibly have been involved.

9.9.1.6. Automatic Testing during Experiments.

The complexity of advanced experiments, for example, in high energy physics, suggests the use of automatic test routines, in order to reduce the probability of false measurements. In experiments involving a small number of critical parameters (e.g., the over-all gain in a high-resolution pulse spectrometer), automatic testing may be developed into a servo system providing for automatic corrections.

In more complex experiments, relying on many different information channels, automatic correction may prove too difficult; however, failures or drifts in any channel should be detected promptly and brought to the experimenter's attention.

Whenever experimental data are received in periodic bursts at a low duty cycle, it is possible to use the remainder of the cycling time for testing. For example, it has become common practice in high energy physics experiments to generate "artificial events" (simulating the signals that would be produced by certain types of particles) in the periods between accelerator bursts. By suitable gating of the recording channels, separate records of actual and simulated events can be obtained. Thus, even in the case that some measuring channels may have been drifting away from desired settings during a run, the necessary corrections may be determined so that the data can be saved.

In the practical application of this idea, it is essential that the test cover the complete chain of equipment. Thus, the classical method of superimposing test pulses on the input of a pulse spectrometer would not check the phototube or semiconductor detectors preceding the electronics system. Test particles impinging on the detectors with known energies and timing would be ideal for checking any particle physics experiment, but such particles are not always available, or might cause confusion with the "unknown particles" to be measured. Test systems using pulsed light sources have therefore been developed,[1-5] which allow a completely automatic testing program of scintillation and Cerenkov counter arrays to be carried out in a time-sharing mode during an experiment. Modern data acquisition systems such as CAMAC[6-8] allow an immediate reaction under computer control, whenever the test results suggest that experimental conditions should be modified. Automatic testing in space research experiments is described in a series of publications[9] edited under the auspices of SETE (Secretariat for Electronic Test Equipment, N. Y. University, Research Division, Building No. 2).

[1] Q. A. Kerns and G. C. Cox, A nanosecond triggered light source using field emission. *Nucl. Instrum. Methods* **12**, 32 (1961); also Lawrence Radiat. Lab. Rep. UCRL-9269 (1960).

[2] F. A. Kirsten, The application of automatic testing to complex nuclear physics experiments. *IRE Trans. Nucl. Sci.* **NS-7**, No. 3, 333 (1962).

[3] G. Amsel and C. Zajde, Use of fast infrared emitting diodes with semiconductor detectors. *Rev. Sci. Instrum.* **35**, 1538 (1964).

[4] J. Johnson and D. Porat, Nanosecond light source, XP-20. *Rev. Sci. Instrum.* **38**, 1796 (1967).

[5] D. Maeder and M. Sabev-Galé, *Colloq. Int. Électron. Nucl., Versailles* paper No. 57 (Sept. 1968).

[6] CAMAC: A Modular Instrumentation System for Data Handling—EUR 4100 (March 1969). Can be obtained, in English, French, German, and Italian editions, from EURATOM, I-21020-ISPRA (Va), Italy. See also CAMAC Tutorial Issue, *IEEE Trans. Nucl. Sci.* **NS-18**, No. 2, 1 (1971).

[7] *Proc. ISPRA Nucl. Electron. Symp.*, session on CAMAC Instruments, p. 379–401 (May 1969).

[8] "Proceedings of the First International Symposium on CAMAC in Real-Time Computer Applications, Luxembourg, December 4–6, 1973." Commission of the European Communities, in collaboration with the ESONE Committee (suppl. to CAMAC bulletin April 1974).

[9] Automation in Electronic Test Equipment, Vol. III (D. M. Goodman, Director of project SETE), New York (1969).

9.9.1.7. Fault Isolation by Semiautomatic Techniques (FIST). Project FIST[10] was undertaken with the aim of reducing malfunction diagnosis time in very complex electronic equipment, particularly for such equipment which is to be used by personnel having limited technical knowledge. Though actual repair work will always be carried out by a well-trained technical staff, the unskilled user should be able to ascertain by himself whether the system is operating correctly or not, and to replace faulty parts immediately.

Modular construction is essential for allowing an easy exchange of defective sections in complex apparatus; in fact, this construction principle has been widely accepted, for example, in the computer industry, as well as in nuclear instrumentation.[11] Recently, international collaboration has progressed towards standardization, both electrically and mechanically, of a general-purpose modular circuit system,[6] well adapted to computer-controlled applications. More or less automatic fault diagnostic procedures are indispensable for systems of this kind.

The FIST conception is based on the modest assumption that the user knows how to read a test instrument in terms of "good"/"no good," but cannot even be asked to manipulate function and range selector switches or to connect measuring probes to certain test points in a circuit. Therefore, the test instrument is simply plugged into a multipole connector (typically 24 pins) installed on the module which is to be checked. In order to test various characteristics of different modules (ac and dc voltages, gain, pulse risetime, etc.) on the same instrument, special interface circuits, called "transformation networks," must be permanently installed on every module.[12]

It is clear that existing electronic instruments would have to be completely redesigned if a minimum-skill testing facility of this kind were to become mandatory for future instrumentation.

9.9.1.8. Environmental Testing. Manufacturers of electronic instruments usually specify a set of environmental tests to which their products are subjected. U.S. military requirements[13] probably go beyond the most

[10] G. Shapiro, O. B. Laug, G. J. Rogers, and P. M. Fulcomer, NBS-Monograph 83 (Sept. 1964).

[11] L. Costrell, Standard Nuclear Instrument Modules (NIM). USAEC Document TID-20893 (Rev. 3, Dec. 1969).

[12] G. J. Rogers, M. Sigman, and D. P. Stokesberry, FIST transformation network design manual. NBS Rep. 9613 (Sept. 1967).

[13] U.S. Dept. of Defense Military Std., for example: MIL-STD 446B, 454C. Washington, D.C. 20360.

adverse operating conditions found in any physics laboratory. More conventional sets of factory tests include:

— Temperature: typically $-28°C \dots +65°C$
— Humidity: up to 95% at 40°C
— Air pressure: down to 400 mmHg (corresponding to 5000 m of altitude)
— Vibration: up to 50 Hz with 0.25 mm excursion, during several minutes
— Power line: typically $\pm 10\%$ voltage variation, 50 to 400 Hz

These tests are usually more than sufficient to guarantee satisfactory operation under laboratory conditions. Some of the larger research laboratories submit homemade as well as commercial equipment to certain tests according to self-established standards, e.g.[14]:

— Temperature: $0°C \dots +65°C$
— Vibration[15]: cyclic, from 5 Hz at 1.3 g peak acceleration through 500 Hz at 5 g and back to 5 Hz
— Supply voltages: dc $\pm 2\%$, ac $\pm 10\%$

A comprehensive checking of dynamic input–output characteristics and internal calibrations is usually performed separately for each type of variation, other conditions being kept normal.

With respect to magnetic stray fields, two complementary aspects can be important, depending on the application:

(a) Equipment must not by itself produce noticeable stray fields; typical limits are:

$<10^{-6}$ G at $f > 0.1$ Hz
$<10^{-7}$ G at $f > 1$ Hz
$<10^{-8}$ G at $f > 10$ Hz

etc., as suggested by ESRO.[16]

(b) Equipment must operate correctly in a strong magnetic field (e.g., 10 kG, at any orientation).

[14] B. Righini et al., Internal rep. of the CERN-NP Division, Test and Instrumentation group. Rep. No. 12 through 35, issued in the years 1968-74, are concerned with pulse shapers, power supplies, discriminators, coincidence and strobing circuits, limiters, ADCs, attenuators, and fan-out circuits.
[15] Following the Specifications for Ground Equipment, MIL-STD-810 A, USAF (25 June 1964).
[16] ESRO, Noordwijk, Holland. Circular ESTEC/CO/RM/CM/24.542 (Feb. 1970).

Requirement (b) is typical for nuclear and high energy physics instrumentation.

9.9.2. Basic Test Instrumentation

Although the list of instruments needed in the electronics shop depends on the range of applications—digital/analog/continuous waves/pulse techniques/noise problems—it is likely to include a few items from each of the following categories, even in the case of a relatively small laboratory.

9.9.2.1. Component Testers.

Ohmmeter. This widely used feature, incorporated in the traditional laboratory meter, allows a quick check of resistors to a precision on the order of 5%; near midscale, a comparison of resistors may be possible even to $\pm 1\%$ with a good instrument.

Care must be exercised in checking nonlinear devices, such as semiconductor diodes or transistors: the "resistance" value read is by no means identical with the dynamical impedance of the device; the source voltage on the high-resistance range may in some cases be high enough to cause avalanche breakdown of a normally nonconducting device.

To check the source voltage (including its polarity), connect a second meter across the ohmmeter terminals, shunted by a suitable resistor so that the ohmmeter reading is midscale. The external meter then indicates $\frac{1}{2}$ the source voltage of the ohmmeter, and the combined external resistance equals the internal impedance. Typical multitesters have three or more resistance ranges with midscale values (R_0) of, say, 20 Ω, 2 kΩ, and 200 kΩ, and source voltages (U_0) ranging from 1 to 10 V.

In testing an "unknown" device, a reading R_x should be interpreted correctly as follows: at a voltage $U_x = U_0/(1 + R_0/R_x)$, the device draws a current $I_x = U_0/(R_0 + R_x)$. Thus, readings obtained on different ohmmeters will, in general, establish different points of an "unknown" (U, I)-characteristic curve.

Bridges. Moderately priced classical-type instruments allow the measurement of complex impedances in a range extending from a fraction of an ohm to several megohms, at a precision on the order of 2%. If the operating frequency is fixed at the usual value of 1000 Hz, the impedance limits correspond to capacitance ranges of roughly 100 pF to 1000 μF, and to inductance ranges of 10 μH to 100 H.

Since many C's and L's of smaller size are used in present-day circuitry, a direct-reading LC-meter for low values is a necessary complement of the conventional *RLC* bridge, unless the latter is of such a design that it may be operated as well at much higher frequencies without loss of accuracy (e.g., at 1 MHz). Care must be used in interpreting direct *LC*-meter readings for components whose values are slightly beyond the nominal range, or which are very lossy. In case of doubt, a check with a conventional bridge is advisable.

The main drawback of conventional *RLC* bridges is the necessity of a simultaneous two-parameter adjustment, especially tedious when testing high-loss components. A mechanical coupling, incorporated in certain instruments, facilitates this task; however, the use of a phase-sensitive detector (such as an oscilloscope) in place of the traditional magnitude indicator is more efficient. The most advanced instruments have a built-in phase-sensitive feedback system which automatically compensates the resistive component of the device under test within a fraction of a second ("electronic autobalance"). Digital versions have recently become available which allow an accuracy of 1% for automatic measurements of *R*, *L*, and *C*.

Characteristic curve tracers. Although transistor (or tube) differential parameters can be checked with less expensive small testers (usually comprising two moving-coil meters), a curve tracer seems preferable even for a small electronics shop. The display reveals in a single operation all the nonlinearities of the component being tested; these can be compared in detail with those of a similar device, chosen as a reference, or with published curves.

The "transistor curve tracer" is equally useful for testing other non-linear elements, such as

— VDRs (voltage dependent resistors)
— CLDs (current limiting diodes)
— Conventional diodes
— Zener diodes
— Tunnel diodes, etc.

Integrated circuit (IC) testers. Input–output as well as transfer characteristics of any 2-port device at low frequency can, in principle, be checked on a curve tracer. However, a test of digital ICs usually involves several inputs; these have to be activated simultaneously in various combinations (some must be "on", others must be "off"), and the effect of propagation times must be checked. An IC tester therefore consists essentially of a programable system of standard pulse generators and variable loads. Universal testers on-line with a computer are a necessity for IC manufacturers. Although smaller testers have recently become commercially available, it appears that the average electronics shop would have no difficulty in assembling, from its own stock of ICs, a simple test setup for automatic checking of a given type of IC, should the necessity arise.

9.9.2.2. Power Supplies.

Variable supplies. Variable ac (Variac) and dc are needed, for example, for marginal testing (see Section 9.9.1.4). Three-phase ac may become more common as the trend towards low-voltage/high-current circuitry continues, since a 3-phase system allows more compact rectifiers and filters for dc supplies.

Fixed supplies. Modular circuit systems require fixed, well-regulated dc voltages, in order to avoid interference between modules. Multiples of 6 V have been chosen for an internationally standardized low-voltage supply system, namely +24 V (+12 V), +6 V, −6 V, (−12 V), −24 V.[6,11] Some modular circuits would suffer damage if all the required supply voltages were not furnished simultaneously; therefore, a protection feature cutting off the negative supplies when one of the positive voltages is missing (and viceversa) is very desirable for any multivoltage power supply in a test laboratory, in addition to the conventional short-circuit protections.

Both NIM and CAMAC specifications call for an absolute tolerance of ±0.5% of the dc supply voltages (long-term stability); regulation should be better than 0.1% (including ±10% line voltage variations), noise and ripple smaller than 3 mV$_{pp}$ (where pp stands for peak-to-peak). More details are given in Refs. 6 and 11.

Integrated circuits can generally be operated directly from the standard 6 V supply, except in cases of extreme package density, where the use of a silicon series diode is recommended so that voltage across the ICs may drop to the nominal 5.3 V.

General-purpose instrumentation. Modern dc supplies can be used alternatively as voltage sources with extremely low internal impedance (typically 10^{-2} Ω, current-limited for protection), or as current sources (>10 kΩ, voltage-limited); remote sensing is available for applications where voltage drops along leads must be automatically compensated. Special master-slave connections are required when several supplies are to be operated in parallel (or in series), in order to equalize their current (or voltage, respectively) loading for a given total current (or voltage) output.

Protections. The fast-acting electronic current-limiting feature is in general preferable to the traditional slow-blow cutout protection, except for such applications where large current surges are unavoidable (filaments, motors). Externally caused voltage surges are not necessarily absorbed in the regulation circuitry, unless a special "crowbar" feature is provided (typically set at 1 V above the nominal regulated output level).

9.9.2.3. Multimeters.

Analog meters (moving coil system). The most widely used category of inexpensive test instruments covers dc and ac voltages from about one volt (full-scale) to several kilovolts, and dc currents from about 50 μA (full-scale) to several amperes, with an accuracy (and a resolution) of the order $\pm 2\%$ (resp. $\pm 0.2\%$) of full-scale deflection. Ac accuracy is somewhat impaired due to nonlinearities of the rectifier elements, especially on the lowest voltage range, and is in general limited to an rms indication of sine waves in the audio range of frequencies. Most of these instruments can be used as ohmmeters (see Section 9.9.2.1); some of them also have ac current ranges. While internal resistance of the dc voltage ranges is satisfactory for most applications (typically 20 kΩ/V; see Section 9.9.3.3), it is usually somewhat low for ac voltages ($<$3 kΩ/V). On the other hand, the internal resistance for the current ranges is seldom specified, but is usually too high to allow straightforward measurements, especially on the ac and the higher dc ranges (typical voltage drop is of the order of 1 V; see Section 9.9.3.4).

More expensive moving coil meters, equipped with amplifiers, choppers, elaborate protection systems, etc., are no longer competitive with recent digital instruments.

Digital meters. The simplest models of this class provide a 3-digit read-out (i.e., a 0.1% resolution) with a $\pm 1\%$ accuracy covering, for example, dc and ac (audio-frequency) voltages from 1 V to 1 kV, dc and ac currents from 1 mA to 1 A, and resistances from 100 Ω to 1 MΩ. Input impedances are improved over those of analog meters; an electronic overload circuit protects the instrument and serves as a guide for switching to the correct measurement range.

Apart from improved accuracy, the main advantages of a digital meter are readability from a distance, automatic polarity selection and indication, and elimination of errors caused by scale interpolations and conversions.

More advanced models feature automatic range setting; 4- or even 5-digit read-out with correspondingly improved accuracy; a wide extension of measuring ranges down to microvolts and picoamperes, up to gigaohms; increased bandwidth for ac ranges; differential input; etc.

Accessories. Instead of acquiring a number of specialized instruments to provide a complete range of test facilities, it is more economical to standardize on a general-purpose multimeter whose range of applications can be extended by means of:

— high-voltage probes (typical attenuation 100 to 1, input impedance $10^9 \ \Omega$)
— rf and vhf probes (with diode or thermocouple)
— high-current shunts (dc measurements)
— high-current transformers (ac measurements at low voltage drop)
— low-current clip-on probes (for the convenience of measuring dc and ac[†] currents without disassembling the circuit under test; see Section 9.9.3.4).

Some of these accessories can also be used in connection with oscilloscopes.

9.9.2.4. Signal Generators.

General-purpose oscillators. Sine wave oscillators of the RC type combine low harmonic distortion with reasonable frequency stability over a wide range, typically from a few hertz to about 10 MHz. Low-frequency versions, making use of operational amplifiers, extend the range down to 10^{-3} Hz and provide several output functions, such as square, triangular, and sine oscillations. Calibrated output voltages are independent of the frequency within a few percent, and are typically on the order of 10 V_{rms} into 600 Ω.

LC-type oscillators are useful for producing high-frequency sine waves, typically a few volts peak-to-peak into a 50 Ω coaxial load. They can be tuned with excellent frequency resolution over wide ranges, up to about 10^9 Hz. Techniques used at higher frequencies are described in Part 10.

Any inverting wideband amplifier with sharply defined saturation characteristics may be used as a square wave oscillator, by connecting

---

[†] While the ac probe is a straightforward clamp-on transformer, its use as a dc probe requires auxiliary circuits which generate a dc compensating field when the probe is clipped around a current-carrying conductor. Automatic adjustment of the compensating current provides a practically instantaneous meter reading proportional to the unknown current.

its output to its input via an adjustable delay line. Using fast ECL integrated circuits, for example, a 0.8 V output with rise and fall times of about 1 nsec is obtained. Oscillators composed in this manner offer great flexibility for testing integrated circuits in triggered or gated operation. Frequency stability is satisfactory if T_1, the delay-line length, is chosen $\gg T_0$ (see Fig. 1).

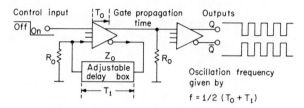

FIG. 1. A simple square wave oscillator for test purposes, assembled from fast logic gates. Using a MECL III circuit (OR gate) with 75 Ω coaxial lines, rise and fall times of about 1 nsec are obtained ($R_0 = 82\ \Omega$ biased to -1.8 V).

Fast pulse generators. For testing transistor switching characteristics, wideband amplifier response, and for time domain reflectometry, a sharply rising wavefront between well-defined initial and final levels is the most appropriate signal shape. A repetition rate of a few kilohertz is quite sufficient for convenient viewing on a sampling oscilloscope.

The shortest possible risetime is at present about 20 psec, with very limited amplitude (0.2 V into 50 Ω). Ten volts into 50 Ω is available at 1 nsec risetime from commercial transistor circuits operating at high repetition rate. Still higher amplitudes at 1 nsec risetime are available from avalanche transistors, at considerably lower repetition rates ($<$10 kHz). Clean step functions with adjustable, accurately calibrated amplitude up to $\sim$100 V can be obtained using mercury relays, though at a very limited repetition rate ($\lesssim$200 Hz).

With vacuum tubes, a 10 nsec rise of several kilovolts has been produced across a 50 Ω load, at repetition rates up to 1 kHz.[17]

If the wavefront generator provides sufficient output power, the signal can be easily transformed according to the particular needs of the user. It can be:

— split into several parallel, quasi-independent signals
— shaped into a reactangular pulse
— shaped into a pulse pair.

[17] D. Maeder, *ZAMP* **19**, 524 (1968).

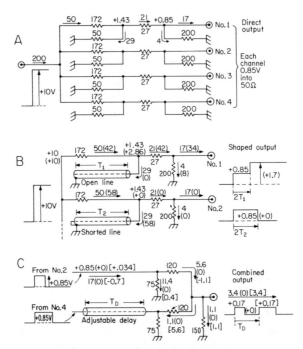

FIG. 2. Network for adapting a high-power single-output step function generator for various applications (currents in mA): (A) A given source is split into four "independent" identical channels of 50 Ω output impedance. Attenuation has been chosen sufficiently high so that short- or open-circuiting of any channel changes the three other output voltages by less than 1%. (B) Any of the four outputs can be delay-line-shaped without changing the flat-top amplitude, upon replacement of one of the 50 Ω resistors in (A) by an open- or short-circuited line. With attenuating resistors as indicated, crosstalk to the other channels is at most 2%; if short and open lines of equal length should be used on two channels respectively, their effects on the two remaining channels cancel each other. (C) Any of the four outputs can be superimposed, with a freely chosen relative delay and individual shaping, by an attenuating T network. Values in parentheses: after end of No. 2 pulse, but before arrival of delayed No. 4 pulse; values in square brackets: during delayed No. 4 pulse. Numerical values were chosen to reduce reflections from a shaping line (circuit B) to less than 2% of output amplitude. Indications in square brackets show a 34 mV signal sent into the other channel (where a certain fraction may be reflected).

Figure 2 illustrates the simplicity of the necessary circuits. Most of the fast step function generators must be operated at a low duty cycle; within certain limits the duration of the "ON"-level can be continuously adjusted. This adjustment may also serve to shape or duplicate the signal by proper use of the trailing edge (the transition to the "OFF"-level).

The more expensive fast pulse generators provide several outputs which can be adjusted individually in amplitude, duration, and relative delay.

General-purpose test pulser. This kind of instrument serves to simulate slow or medium-fast pulses of various shapes occurring, for example, at different points in an amplifier chain. Independent adjustments of rise-time, flat-top duration, and fall time are essential features. Minimum rise and fall times need, in general, not be shorter than 10 nsec, repetition rates should be continuously variable from, say, 1 Hz to a few megahertz. An output of a few volts into 50 Ω is ample for this kind of instrument; on the other hand, it should have a high duty cycle capability (50%) and provide outputs of both polarities. Many of these pulsers have gating facilities so that a pulse train can be produced.

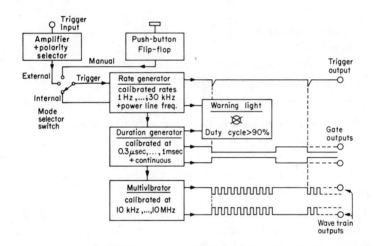

Fig. 3. Functional diagram of a versatile wave train generator for test purposes.

Square wave train generator. The full flexibility of a general-purpose test pulser is not always required for tests using pulse trains. The block diagram of an inexpensive wave train generator, built from standard integrated circuits, is shown in Fig. 3. Trigger and repetition rate facilities were designed to cover most applications encountered in the author's laboratory.

Elaborate wave train generators are available for special purposes, such as computer simulation ("word generator"), TV equipment testing ("pattern generator"), and the like.

9.9.2.5. Advanced Measuring Instruments.

Oscilloscopes. The obvious extension of the analog-type multimeter into the time domain allows the following measurements:

(a) Voltage as a function of time to be read on a doubly calibrated scale. Typical accuracy is on the order of $\pm 3\%$ for both voltage and time over a wide selection of measuring ranges. For an ideal instrument of this kind, there is no distinction between dc and ac measurements.

(b) Current as a function of time, either by observing the voltage differences across a dropping resistor (ac + dc), or by using a clip-on current probe (ac signals only).

(c) Direct reading of complex impedances, through time domain reflectometry displayed on a calibrated screen (see Section 9.9.4.5).

Proper choice of an oscilloscope for a given test is determined by requirements on

— input impedance
— sensitivity
— waveform or bandwidth
— trigger and sweep facilities
— observation and recording procedures.

As these instruments tend to be costly, a test laboratory needs a well-balanced selection of oscilloscopes which covers every foreseeable application by means of a suitable set of accessories. A survey of the most versatile combinations and some hints for their practical manipulation are given in Section 9.9.4.

Counters, timers, and frequency meters. An extension of digital-type multimeters into the frequency domain adds relatively little to their complexity: a highly stable oscillator (the "clock"), some trigger and gating circuitry, and a scaler with read-out facilities are the main constituents of every digital measuring instrument, as illustrated by the various applications indicated in Fig. 4. The basic configuration can be used as an events counter (switches in position A; counting periods defined by external "start" and "stop" signals), or as a timer for controlling other equipment. Combining this with a ramp generator and a level comparator turns the instrument into a digital volt-or amperemeter (position B). Alternatively, incorporating a second scaler (with a preselection facility, but without display) converts the ensemble into either a frequency meter (position C), or a frequency ratio meter (clock replaced by one of the unknown frequency sources).

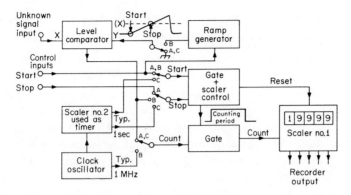

FIG. 4. Basic configuration of a universal digital test instrument. Of the various possible combinations, only those corresponding to (A) an events counter, (B) a digital voltmeter, (C) a frequency meter, are shown explicitly.

To obtain good accuracy from a finite count, special interpolation circuits have been incorporated in the most advanced frequency meters. An alternative solution, preferable for low-frequency work, consists in counting the clock pulses falling into a preselected number of periods of the unknown frequency. Start–stop pulses are derived from a zero-crossing discriminator in the period-measuring configuration. By an obvious extension of this system, the phase angle between two signals of equal frequency can be digitized.

Frequency limits of typical "fast" counters are presently around 100 MHz; this limit can be extended by means of binary "prescalers" to about 1 GHz, or by heterodyne converters to over 10 GHz.

Universal digital multimeter. While a 4-digit resolution is in general more than sufficient for digital multimeters, time and frequency measurements can be made with much higher resolution: a typical 10 MHz crystal clock in a thermostat is stable within 1 part in 10^{10} over periods of several hours.

The disparity in accuracy levels appears to be the main reason why counters/timers were not originally conceived as additional operation modes of a universal multimeter. Only very recently have a few digital instruments incorporating such a conception appeared on the market.[†] These are, in the first place, 7- or 8-digit counters, with a DVM option offering 5-digit resolution and roughly 4-digit accuracy. Presently,

---

[†] E.g., HP 5326 B; alternatively, any counter combined with a voltage-to-frequency converter.

current, resistance, and ac measurements with this type of instrument seem to require nonstandard accessories.

Calibration problems. It is good practice to check similar test instruments against each other frequently—not only when test results begin to look abnormal. Even if no inconsistencies are found, the calibration of each measuring instrument should be checked periodically (say, once a year) against a high-reliability standard. In principle, one single high-precision universal digital multimeter could serve as such a standard, covering practically all needs of a small electronics laboratory. In addition, a set of precision capacitors and inductors for recalibrating impedance bridges and meters should be available. The calibration of pulse generators can be checked against the wavefronts obtained from an accurately known dc supply voltage through a mercury relay.

Accuracy of the high-precision digital laboratory standard should be maintained through regular servicing by the manufacturer. Stability of the most accurate DVMs is on the order of $\pm 0.01\%$ (dc) or $\pm 0.05\%$ (ac) over a six month period; a well-behaved quartz crystal time base may drift by less than 10^{-8} per month. Reliable impedance standards (R, L, C) keep their values within about $\pm 0.01\%$ per year.

9.9.3. Testing Steady-State Conditions

9.9.3.1. Significance of Steady-State Data. Practical use of electronic equipment implies that operating conditions are related in a specific way to some input variable: In general, proper performance of a given circuit requires an output change within specified limits when the input is varied.

In this section, we define as *steady state* a set of operating conditions with negligibly small ac components ("zero signal" level). In the absence of ac components, reactive elements have no effect, so that equivalent circuits are simplified as follows:

— inductors become short circuits
— capacitors become open circuits.

Checking of steady-state potentials and currents therefore does not guarantee proper operation of an electronic device; however, a deviation of these conditions from the design values does indicate circuit trouble.

Example. An amplifier designed to give an open loop gain $A = 3000 \pm 1000$ is stabilized by negative feedback ($\beta = 0.01$) to an over-all gain factor between 95.2 and 97.5. If, by some shift developing in the biasing network, one of the transistors is made to operate either near cutoff or near saturation, A may drop to 1000. The slight drop in over-all gain (from 96.7 to 91) might escape attention in a signal flow test, although gain *stability* is seriously affected. However, dc testing will reveal such a condition immediately.

As demonstrated in this example, routine checks of steady-state conditions on individual components of a circuit often show the development of circuit trouble earlier, or more conspicuously, than tests of over-all dynamic performance. Morever, dc checking can be done with simple, inexpensive equipment, and the results are correspondingly easy to interpret. In particular, short circuits, open circuits, and faulty semiconductor devices or tubes are located immediately; on the other hand, it must be kept in mind that dc measurements neglect the functions of nondissipative circuit elements.

9.9.3.2. Circuit Requirements for Significant Dc Tests. Equipment having dc coupled input and/or output should be connected to the designed source and/or load with the correct bias voltages to permit normal operation and to prevent damage. For example, an emitter follower with its input left floating is bound to deviate from its normal operating point; removal of the plate resistor from an output pentode might result in excessive screen grid dissipation; and so forth.

Even for dc tests, ac coupled input/output terminations should in general be returned to ground along the shortest possible path. Again, deviations from the nominal source and load impedances may in some cases result in an abnormal regime: for example, a supposedly stable circuit may break into high-frequency oscillations when its ac-coupled output is short-circuited (or open-circuited), although this would seem compatible with normal dc conditions.

Similar accidents can happen when additional loads (for example, a multimeter probe lead) are tied to internal circuit nodes. Therefore, a *current meter* should be bypassed by a sufficiently large capacitor, and should always be inserted at the "cold end" of a circuit element; while a *voltmeter* probe should be decoupled by a sufficiently large series resistor of small physical size (e.g., 10 kΩ, $\frac{1}{8}$ W) if it has to be connected to a "hot point." A normally oscillating circuit can be dc tested without complications on its "cold points" only: dc supply lines, and bias networks as far as these are bypassed to ground. Connection of any measuring instrument to an oscillating "hot point" is likely to detune the circuit and to modify the oscillating regime unless the probe tip decoupling resistor is chosen so large that it may affect the meter calibration.

Circuits having more than one stable state may have a tendency to flip when a dc meter is connected. However, by means of an external dc biasing lead to the proper transistor (or tube), one can enforce any stable (or even a metastable) state temporarily. In order to measure a stable state correctly, the external biasing lead should be removed before actual readings; a metastable state, on the other hand, cannot be dc tested without leaving the enforcement on, and therefore will be disturbed to some extent.

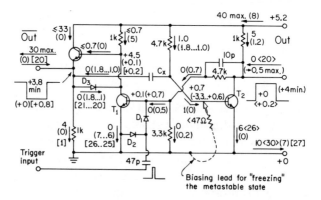

FIG. 5. Multivibrator. Metastable states in switching circuitry can be rendered stable temporarily for the purpose of dc testing. Without the need for an oscilloscope, stationary operating potentials and currents can be checked for each circuit state (for the metastable one as long as the biasing lead is applied to the base of T_2). Note that in this example, applying the bias lead to the collector of T_1 would be incorrect—this would simulate a state which does not exist for the original circuit. Data for the normal state, with loads connected to ground, are given without parentheses. Data in parentheses ( ) show true variations between the beginning and the end of an actual metastable period; such variations are neglected in dc tests. Data in ⟨ ⟩ and [ ] brackets apply to the normal and the metastable state, respectively, for external loads connected to the +5.2 V supply rather than ground.

Example. Both states of the multivibrator shown in Fig. 5 can be dc checked with no essential departure from the true operational regimes, except for the base of T_2 which is arbitrarily kept near ground potential during measurement of the normally unstable state (this is done most simply by short-circuiting it via a flexible lead to ground, with a series resistor for protection against "illegal" shorts).

In this way, stationary voltages, currents, and impedances in metastable circuits can be studied in detail for either of the two states, using dc measuring equipment only.

9.9.3.3. Practical Voltage-Measuring Procedures. In order to make results truly representative of the circuit under test—rather than of the disturbance introduced by the meter—a quick evaluation of the unavoidable disturbance should be included as a standard routine with every meter reading.

Since the mathematically exact method [computing the driving-point resistance of the particular node from the given circuit diagram, dividing this by the voltmeter impedance to obtain the relative correction ε, and multiplying the actual reading by $(1 + \varepsilon)$] would be too lengthy for practical use on complex circuitry, we recommend a purely experimental procedure:

(a) In case the disturbance is large, reduce it by putting a known resistance in series with the voltmeter (in other words, use higher voltage range), and evaluate the ratio of the two respective meter deflections.

(b) In case the disturbance is <10%, increase it by shunting the voltmeter with a resistance equal to its internal impedance, and evaluate the difference between the two deflections.

It is evident that accurate results can be expected in case (b) only.

> *Example.* A high-resolution 10,000 Ω/V voltmeter gives a reading (after correction for any calibration errors) of 90.0 ± 0.5 V using the 100 V range. The circuit driving-point impedance is unknown.
>
> (a) Use of the 300 V range (or adding 2 MΩ in series) increases the voltmeter impedance three fold, but decreases the deflection only by a factor 2.47, corresponding to 108 ± 1.5 V on a 300 V scale. Assuming that the second reading includes one-third of the original disturbance, we add $(1/2) \times (108 - 90) = 9 \pm 0.8$ V to obtain the corrected result 117.0 ± 1.7 V. Use of the next higher range, e.g., 1000 V, gives a 116 ± 5 V reading to which $(3/7) \times (116 - 108) = 3.4$ V should be added; however, the accuracy of this result is affected by the decreased deflection sensitivity.
>
> (b) Shunting the voltmeter at the 300 V range by 3 MΩ (equal to its internal impedance) drops the reading from 108 to 98.2 V. Adding the difference of 9.8 ± 2.2 V gives the

corrected result 117.8 ± 2.7 V. When using the 100 V range, shunting with 1 MΩ causes a drop from 90 to 72 V, or by 20%, which prohibits the application of a simple linear formula.

In both approaches (a) and (b), the linear correction underestimates the actual disturbance: the true source voltage was $U_x = 120$ V (via $R_x = 333$ kΩ) in our example.

Modern digital voltmeters usually have such a high input impedance that corrections tend to be quite small. Method (b) provides a rapid check as to whether or not this is true. It must be kept in mind that the input impedance of these instruments is not necessarily proportional to the measuring range—on the contrary, it is usually constant for all ranges except the lowest one. The differential input facility is very useful when measuring base-emitter voltage of transistors whose emitter is not at ground potential.

9.9.3.4. Practical Current-Measuring Procedures.

Classical method. The circuit must be shut down, a connection unsoldered, the current meter inserted, and the power switched on again for a warm-up period, prior to every current measurement. Due to these inconveniences, until recently it has not been customary to check currents when testing electronic circuits.

Moreover, traditional multimeters generally have an internal impedance too high for straightforward current readings to be correct. Rapid correction procedure similar to those explained in Section 9.9.3.3 are suggested by duality considerations.

Example. The heater current of a tube powered from a 6.3 V supply is measured with a multimeter having a 0.8 V drop at full scale of its current ranges. Readings are:

264 mA on the 300 mA range
288 mA on the 1 A range.

Assuming that the drop of the second measurement is about 30% of that of the first one, we divide the difference of 24 mA by $3.33 - 1 = 2.33$, to obtain the correction:

24 mA/2.33 $\approx$ 10 mA
288 + 10 = 298 mA

is very close to the true current which flows when the ammeter is short-circuited.

Use of permanently installed dropping resistors. Computing a current from observed potentials and given circuit parameters is subject to error due to the inaccuracy of resistors or the presence of nonlinear circuit elements. Such errors are avoided if small, stable (wire-wound) resistors are permanently inserted in all leads carrying currents involved in routine testing. Reasonably small voltage drops can be combined with a convenient scale on a particular multimeter if the permanent dropping resistors in the circuit are chosen equal to the meter internal impedance (one then multiplies readings by a factor 2).

This method cannot be applied to very small currents such as base currents in transistors, or grid currents in tubes. The latter are evaluated from the change in plate current observed when the grid leak resistor is momentarily short-circuited.

The differential input facility and high sensitivity of recent digital multimeters allow current measurements at greatly reduced voltage drops: permanently installed dropping resistors of 1 Ω, for example, would be sufficient for "current" readings with a 10 μA resolution on a good instrument.

Clip-on current probe measurements. Checking of currents in electronic circuits without desoldering circuit components, nor having to cut off power supplies, is nevertheless possible, with the aid of modern clip-on current meters (see Section 9.9.2.3). We feel that permanently installed wire loops large enough to accomodate the current probe should be considered nowadays by circuit designers as a regular maintenance facility. No ohmic corrections need to be made with this measuring method.

9.9.4. Review of Oscilloscope Techniques

A particular section on oscilloscopes appears appropriate here, as these instruments are the basic tool for all dynamic testing. Practical procedures to assure correct interpretation of observations will be stressed.

9.9.4.1. Finite Bandwidth Corrections.

Risetime–bandwidth specifications. If the high-frequency 3 dB limit of an oscilloscope's over-all deflection system is specified as f_{high}, its response to a step function input is given by the following approximate risetime formula:

$$t_r(10\% - 90\%) \approx 0.35/f_{high}.$$

If cutoff frequencies are specified individually for the cascaded components (probe, preamplifier, main frame, CRT), the overall response is calculated conveniently in terms of the combined risetimes:

$$t_r = \sqrt{t_{r_1}^2 + t_{r_2}^2 + t_{r_3}^2 + \cdots} \; .$$

Example. An oscilloscope having 7 nsec risetime (main frame + CRT) is to be used with a 25 MHz preamplifier. The combined risetime is approximately

$$t_r \approx \sqrt{7^2 + t_{r_1}^2} = 15.7 \text{ nsec},$$

where

$$t_{r_1} \approx 0.35/25 \text{ MHz} = 14 \text{ nsec}.$$

If a 35 MHz current probe is used with this equipment, over-all risetime increases to

$$t_r' \approx \sqrt{(15.7)^2 + t_{r_2}^2} = 18.6 \text{ nsec},$$

where

$$t_{r_2} \approx 0.35/35 \text{ MHz} = 10 \text{ nsec}.$$

Risetime measurements. The risetime-squared addition must be taken into account for a correct interpretation of oscilloscope readings (apart from loading corrections, discussed in Section 9.9.5.3): the true risetime is calculated as

$$t_{\text{true}} \approx \sqrt{(t_{\text{obs}})^2 - (t_{\text{osc}})^2} \; .$$

The examples in Table I illustrate the critical effect of an error in oscilloscope risetime on the evaluation of an observed wavefront, assuming a nominal $t_{\text{osc}} = 10$ nsec (which may actually be a little shorter, say 9 nsec). It is evident from the table that t_{osc} can be determined experimentally by observing a wavefront whose risetime at the input is $\ll t_{\text{osc}}$.

TABLE I

Observed 10%–90% rise	100	30	20	14	12	11	10.3	10	nsec
True risetime									
(a) $t_{\text{osc}} = 10$ nsec	99.5	28.3	17.3	9.8	6.6	4.6	2.5	0	nsec
(b) $t_{\text{osc}} = 9$ nsec	99.6	28.6	17.9	10.8	7.9	6.3	5.0	4.4	nsec

Low-frequency cutoff. Modern oscilloscopes are dc coupled, in order to avoid loss of low-frequency information. However, ac coupling may prove necessary in certain applications, namely:

— when we wish to measure very small ac signals superimposed on a large dc component
— when the circuit to be tested does not permit any dc loading.

In such cases, a series capacitor is inserted into the oscilloscope input connections, usually with a value chosen to fix the 3 dB cutoff frequency, f_{low}, at a few hertz. Noise problems may necessitate in some cases more severe passband restrictions, while current probes by themselves introduce a relatively high cutoff (typically >50 Hz). The over-all performance, in terms of the frequency f_{low}, can be conveniently checked with a constant-amplitude sine wave generator connected to the input.

Sag corrections. Correct interpretation of slow phenomena observed in an ac coupled oscilloscope must take into account that a flat portion of the original signal will appear inclined. If most of this "sag" is caused by one single differentiating circuit, with time constant $\tau_{sag} \gg$ signal duration, the over-all distortion is a small change of signal slopes, proportional to U, the momentary signal amplitude at a given time. True signal slope can thus be evaluated as

$$\left(\frac{dU}{dt}\right)_{true} \sim \left(\frac{dU}{dt}\right)_{obs} + (U - U_0)/\tau_{sag}$$

where $U_0 =$ asymptotic level. The sag time constant can be determined experimentally from the response to a step function. Alternatively, an f_{low} specification may be used to calculate

$$\tau_{sag} = 1/(2\pi f_{low})$$

for estimating slope corrections.

Example. A 500 μsec wide, low duty cycle pulse observed with a current probe ($f_{low} = 50$ Hz) appears to sag from 100 to 84 mA. From f_{low}, we obtain $\tau_{sag} = 3.2$ msec. The correction formula yields

$$\left(\frac{dI}{dt}\right)_{true} \approx \frac{-16 \text{ mA}}{0.5 \text{ msec}} + \frac{100 \text{ mA}}{3.2 \text{ msec}} \approx 0;$$

i.e., the actual input has a flat top.

9.9.4.2. Deflection Accuracy and Resolution. *Definition of some terms*:

Sensitivity is the deflection as measured (in cm, or in divisions) on the CRT screen, divided by the applied input voltage; its reciprocal value is called *deflection factor*.

Sensibility is deflection expressed in spot widths, divided by input voltage; its reciprocal represents the smallest signal that can be resolved.

Relative resolution denotes spot width divided by the useful screen diameter; its reciprocal may be called *resolution factor*.

Differential nonlinearity is the relative deviation of a deflection factor from an ideally constant value.

Sensitivity limits. Thermal noise in a high-gain deflection amplifier increases the apparent spot width, thus reducing the meaning of the term "sensibility" effectively to a peak-to-peak noise specification.

In typical high-sensitivity oscilloscopes, the bandwidth is cut down to such an extent that trace widening due to noise remains within a few millimeters, even on the most sensitive ranges. Actual noise levels are around 50 μV_{pp} at 500 kHz bandwidth (or 10 μV in a 20 kHz band); thus, a deflection factor of 100 μV/division represents the practical sensitivity limit for a general-purpose instrument (1 mV/division for a 50 MHz oscilloscope).

Resolution. Spot widths of a few tenths of a millimeter (typical for a beam accelerated to 10 kV) on a screen of 8 cm × 10 cm of useful area offer a resolution of a few parts in 10^3; this is quite comparable to other analog instruments.

Many deflection amplifiers have an undistorted signal range of several times the screen diameter, so that improved resolution is available for certain applications.

Example. The top of a pulse may be examined with ∼0.1% amplitude resolution providing that its base line can be offset without affecting amplifier linearity nor bandwidth, to about 20 screen divisions.

Use of a general-purpose oscilloscope under such extreme conditions would necessitate careful performance checks with an accurately calibrated step generator. Special plug-in amplifiers are available which allow millivolt signal resolution at dc offset voltages as high as ±100 V.

Accuracy. Oscilloscopes have in general such a large number of components and of measuring ranges that over-all accuracy specifications (which have to cover worst case combinations) on the order of $\pm 3\%$ must be considered as satisfactory. This is roughly 10 times worse than full-scale resolution. The latter, however, sets the ultimate precision limit to which two similar objects (voltage steps, time intervals, etc.) can be compared on an oscilloscope. As with ordinary multimeters, measurements with a precision close to the resolution limit (say, $\pm 0.5\%$) are possible when the measuring range is carefully calibrated for the particular application.

Built-in digitized position markers have been suggested for this purpose. Excellent dc stability, a prerequisite for highly accurate measurements, has been achieved in modern oscilloscopes through the use of low-drift semiconductor devices.

9.9.4.3. Input Characteristics.

Standard input impedance. Modern general-purpose oscilloscopes have an accurate 1 MΩ input resistance when used on dc coupled ranges. The unavoidable parallel input capacitances are by no means standardized: 15 pF seems to be about the minimum, although values up to 100 pF may be encountered depending on the complexity of the circuits. The typical (1 MΩ||50 pF) combination is satisfactory for low-frequency testing but may load high-frequency and fast pulse circuits too heavily (see Section 9.9.5.3). The input impedance of any given instrument should be constant on all operation ranges: under this condition only will the effect of an attenuating probe not depend on range settings.

Standard voltage probes. Long leads from a test point to the oscilloscope input cannot always be avoided. As a protection against picking up stray signals, they should be shielded; this, however, increases the undesirable input capacitance still more, typically by ≥ 30 pF/m.

An attenuating voltage probe at the end of the shielded cable permits a substantial reduction of both capacitive and resistive loading. Typically, a 10:1 attenuator would have an input impedance of (10 MΩ||10 pF); in tests where capacitive loading is critical, 100:1 probes are preferable since their input capacitance may be as low as 2 pF. In using high-voltage probes (1000:1), additional bandwidth limitations must be taken into consideration.

Thanks to standardized 1 MΩ inputs, the same voltage probes may be used on oscilloscopes of different makes. It is evident that the probe capacitor must be readjusted after each transfer, by minimizing the distortion of a test square wave having a sufficiently long period ($T \gg$ input time constant, which is typically ~50 μsec, given by the input shunt capacitance multiplied by the input resistance).

Current probes. The advantage of a clip-on probe is evident (see Sections 9.9.2.3 and 9.9.3.4), although the models recommended for oscilloscopes cut off the lower audio frequencies. The range can, in principle, be extended towards lower frequencies if sensitivity is not the limiting factor.

Example. The secondary winding of a current probe, which has 125 turns, must be loaded with 125 Ω if it is desired to convert 1 mA (flowing in the single-turn primary) into a 1 mV signal on the secondary. A suitable ferrite core geometry gives $L_{\text{sec}} = 15$ mH, so that the low cutoff frequency becomes

$$f_{\text{low}} = \frac{R}{2\pi L} = \frac{125}{2\pi \cdot 15 \text{ mH}} = 1.3 \text{ kHz}.$$

If, however, conversion sensitivity may be reduced to 10 μV/mA, the secondary can be loaded with 1.25 Ω, resulting in

$$f'_{\text{low}} = 13 \text{ Hz}.$$

In the first example considered (125 Ω load), the primary series impedance would be calculated as (1 μH||4 mΩ) neglecting stray inductances, which is justified at low frequencies. High-frequency cutoff is usually quite satisfactory if the magnetic circuit is completely closed. From actual risetime values (18 nsec at 62 Ω load), we can conclude that secondary stray inductance was kept extremely small ($<10^{-3}$ of L_{sec}). High-frequency series impedance on the primary conductor is increased by stray inductances of a few nanohenries (typically, $\omega L < 1 \Omega$ even at 20 MHz).

Use of current probes in the presence of a strong dc bias requires special care. At 2 A biasing current, probe inductance may be down by a factor 2 so that f_{low} is increased correspondingly.

High-frequency input configurations. Beyond about 500 MHz, parasitic capacitances become such low impedances (e.g., $<300 \Omega$ for $C = 1$ pF) that a very high input resistance is no longer a desirable feature of test instruments. For testing the internal operation of a fast circuit, a 100 kΩ probe resistance shunted by less than 1 pF is in general satisfactory.

However, direct inputs of very fast oscilloscopes invariably operate on a much lower impedance level, typically 50 to about 200 Ω. Active probes (cathode followers and samplers) allow the necessary impedance transformation without attenuation. Passive probes allow increased signal swing, with input impedances up to 5000 Ω (for a 100:1 attenuation ratio), which is, however, not always sufficiently high for undisturbed internal circuit testing (see Section 9.9.5.3).

Multiple signal input. The need for "simultaneous" observation of waveforms at different points (e.g., (1) = input voltage, (2) = collector current, (3) = output voltage of a transistor amplifier stage) is generally recognized. Modern testing oscilloscopes have therefore at least two input connectors (A and B), with switching facilities for displaying either the individual signals, their sum or difference, or a succession of samples taken from signals A and B. The A/B switching is electronically controlled, either by an internal oscillator (chopped mode), or by an "end-of-sweep" pulse from the main frame (alternate sweep mode); in the simplest oscilloscopes, it must be done manually.

The main drawback of the switching method is that fast nonrecurrent phenomena cannot be observed simultaneously, because of the finite chop frequency. A double-beam CRT allows truly simultaneous observation of two signals, with full light intensity for each channel. If a double-beam oscilloscope is used with two dual-trace amplifiers, four traces can be displayed altogether.

Modern double-beam CRTs cover the full graticule area with either beam, with barely detectable alignment errors between the two deflection systems.

9.9.4.4. Trigger and Sweep Facilities. *Minimum requirements for general testing applications*:

(a) Single sweep, externally triggered by a passage of the trigger signal in the desired direction (selection of "up" or "down") through an adjustable voltage level.

(b) Single sweep, manually triggered (pushbutton).

(c) Recurrent sweep which can be phase-locked to the ac power line frequency (this feature permits a separation of random noise from power-line-correlated noise).

An output signal marking the start of every sweep should be available in operation modes (b) and (c), for controlling the equipment under test. In this way, unavoidable propagation delays act in the direction which facilitates observation of the early portions of the equipment's response.

Internal synchronization facilities, found in all oscilloscopes, are of little value in quantitative testing involving time and phase differences between signals. The preferred procedure uses a separate probe connected to the "external trigger" input; this probe is hooked to a suitable reference point so that the sweep is triggered under fixed conditions while the measuring probes can be moved freely through different sections of the circuit under test (including bypass points where one should not normally expect any visible signal).

Delayed sweep. Two methods are available for increasing the time resolution of oscillograms:

— inserting additional gain ("magnifier," e.g., $10\times$) in the horizontal deflection system; this solution is inexpensive but introduces a blurring background illumination, unless the electron beam is blanked out during the sweep portions falling outside the screen;

— use of a delay generator, either as an external accessory, or (in the more expensive oscilloscope models) as a built-in facility allowing easy switching between various modes of operation, e.g., display of a long-duration signal on the "delaying sweep" with an intensified zone indicating the portion which can be seen in detail on the "delayed sweep."

Ultimate time resolution is limited by:

— jitter of sweep start, due to variations of trigger input pulse
— fluctuations of sweep speed
— true signal jitter.

Jitter can be eliminated from the display by means of an "arming" feature which activates the delayed sweep trigger input at the end of the chosen delay period so that it may respond directly to the next occurring signal wavefront. Otherwise, a jitter $<10^{-4}$ of the delay period would be considered satisfactory.

Signal delay versus sweep delay. To observe the early parts of pulses which are not preceded by a trigger signal (e.g., random pulses from a nuclear particle detector), the unavoidable unblanking and sweep starting delay must be compensated by a larger delay in the signal channel. The necessary length of high-quality delay-line, about 100 nsec even for a system having fast sweep circuits, is normally built-in between amplifier stages and CRT deflection plates, because any external delay-line would be incompatible with the desirable high input impedance (this does not apply to gigahertz oscilloscopes having low-impedance inputs).

9.9.4.5. Sampling Techniques.

Principle.[18,19] While "realtime" oscilloscopes are limited in practice to the 100 MHz region, a stroboscopic time dilation technique allows an extension of effective bandwidth to 10 GHz and beyond. Any repetitive event can be reconstructed from samples taken at different relative times, $t_n = t_n^0 + n\,\Delta t$, during many recurrences; here, t_n^0 is a reference time related to the nth event $(n = 0, 1, 2, \ldots, N; 0, 1, 2, \ldots, N;$ etc.) and Δt the sampling delay increment. The samples, stretched to a duration of many microseconds, are displayed vertically in a conventional high-sensitivity oscilloscope whose horizontal deflection is made proportional to n (staircase with increments ΔX). To obtain a quasicontinuous presentation, the repetition rate should be at least 1000 events/sec, and the maximum number of horizontal steps, N, must be quite large. For example, choosing $\Delta t = 0.1$ nsec and $N = 100$ (or $\Delta X = 0.1$ screen divisions) would result, at an event rate of 1000/sec, in a flickering display of 10 sweeps/sec with an apparent 1 nsec/division time scale.

Triggering of sampling oscilloscopes. The events to be analyzed need not occur in a periodic succession, as long as it is possible to obtain a correct timing signal (t_n^0) for each event. A trigger signal is picked off the signal transmission line some 20 nsec ahead of the sampler, so that there is sufficient time for the trigger to start a fast ramp generator before the early parts of the event appear at the sampler (Fig. 6). Crossover of the fast ramp with a staircase voltage generated for the slow horizontal deflection defines the sampling instant, t_n.

To reduce jitter in the display of random signals having variable amplitude, only those pulses whose amplitude falls into a narrow window should be sampled. This has been done successfully using known fast-slow coincidence techniques.[20]

Time domain reflectometry (TDR).[21] A quick evaluation of 2- or 3-component complex impedances from step function response displays is already feasible with conventional oscilloscopes in the megahertz range. Short pieces of connecting cable between the oscilloscope and the unknown impedance would in this case not be recognized as transmission

[18] J. M. L. Janssen and A. J. Michels, *Philips Tech. Rev.* **12**, No. 2, 52; **12**, No. 3, 73 (1950).

[19] J. G. McQueen, *Electron. Eng.* **24**, 436 (1952).

[20] R. Sugarman, *Rev. Sci. Instrum.* **28**, 933 (1957).

[21] Time Domain Reflectometry, Programmed Instruction. 2 vols. Tektronics, Inc., Beaverton, Oregon (1966).

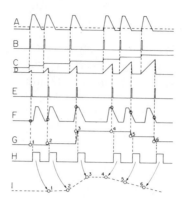

Fig. 6. Sequence of events in a sampling oscilloscope designed for correct internal triggering on randomly occurring pulses: (A) Input signals (trigger threshold indicated as a dashed line), (B) trigger pulses, (C) staircase voltage (each trigger adds a fixed increment), (D) fast ramp signals (started by B, stopped and reset by a comparator looking at C and D), (E) sampling pulse (timed by the comparator), (F) delayed input signal (at sampling gate input), (G) sampler-stretcher output, (H) intensifier pulses, (I) display on oscilloscope screen: G (vertical) versus C (horizontal), intensified by H.

lines, but rather be interpreted in terms of lumped constants. For example, a 100 MHz oscilloscope would permit a correct evaluation of L and R from the response curve A in Fig. 7 (neglecting the short piece of 50 Ω line) but would not "see" the difference between the cases B and C.

When sampling oscilloscopes are used for this method of impedance

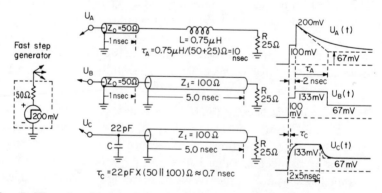

Fig. 7. Time domain reflectometry: Voltage waveforms observed with an ideally fast oscilloscope inserted in a 50 Ω transmission system at the input to one of three different complex loads. Their circuit components can be determined easily from the $U = U(t)$ curves. A conventional fast oscilloscope (100 MHz bandwidth) would smear the curves with its finite risetime (about 3.5 nsec), but would be sufficient to distinguish example A from B or C; it would, of course, not "see" the difference between B and C.

measurements, very small amounts of lumped capacitance and inductance show up as reflections along the signal path and can thus be localized from observed propagation times. Any sampling oscilloscope can be used in this manner, providing it has a high-impedance probe. The most advanced instruments of this category are specially designed for high-resolution time domain reflectometry (TDR), with direct calibrations in terms of reflection coefficient (vertical scale) and distance (horizontal scale).

Recorder output. The sampling method is ideally suited for recording fast or slow waveforms on a slow mechanical read-out system. Sampling oscilloscopes therefore are provided with analog X and Y outputs which can directly produce a permanent record on a mechanical plotter. Alternatively, digital records may be obtained by feeding the Y output to a digital voltmeter.

Single-event sampling. Since presently available sampling oscilloscopes can analyze only repetitive waveforms, a fast analog storage system is needed if one wants to apply this powerful method to single-event observations. Several solutions based on the memory capability of transmission lines have been suggested[22-24]; at present, actual prototypes are still in an experimental stage.

9.9.4.6. Summary on the Choice of an Oscilloscope.

Small oscilloscopes. Even the least expensive portable instrument should meet the "minimum requirements" listed above. Some small models can be battery-operated, a feature which allows the elimination of undesirable ground loops. Such instruments, covering typically dc to about 10 MHz with a maximum sensitivity of a few millivolts/division, are in general too slow for testing fast digital circuits. They are satisfactory for testing many analog systems, including operational amplifiers, and the slower types of digital circuitry.

The useful range of typical low-frequency oscilloscopes ($<$500 kHz) appears too limited for a test laboratory with widely varying needs.

Universal, general-purpose oscilloscopes. At least 100 MHz bandwidth is required if the equipment is to allow significant tests on the majority

[22] D. Maeder, *Proc. EANDC Conf. Automatic Acquisition Reduction Nucl. Data, Karlsruhe, July 13-16,* p. 379. Gesellschaft für Kernforschung m.b.H., Karlsruhe (1964).
[23] D. Maeder, *ZAMP* **15**, 426 (1964).
[24] D. Zenatti, Réalisation d'un dispositif d'échantillonnage d'un signal bref unique, Rep. CEA-R-3812, CEA-Grenoble (1967).

of present-day digital circuit families. For sufficient trace visibility in the single-sweep mode, the cathode ray tube needs about 12 kV of acceleration voltage.

Some saving can be gained from a modular design which allows a basic instrument of relatively simple construction (the "main frame") to cover all conceivable applications by means of special plug-in units. These must be interchangeable between similar main frames without noticeable effects on performance and calibration. Input impedances should be standardized (1.0 MΩ shunted by <100 pF) so that voltage probes etc. can be interchanged as well.

The dual-trace, wideband type of plug-in amplifier appears to cover the majority of applications. Sweep ranges from a few nanoseconds/division up to several seconds/division are standard.

Very fast oscilloscopes. Plug-in units for converting the general-purpose instrument into a sampling system are so expensive that a specialized, self-contained gigahertz oscilloscope is nearly always justified whenever subnanosecond applications occur occasionally. This instrument should cover a wide range of sweep speeds so that it can also be used for mechanical plotting and/or digital read-out of medium-frequency phenomena (see Section 9.9.4.5). It should be readily adaptable to TDR.

There is actually no convenient solution available for very fast single-transient observation. Oscilloscopes having 1 GHz bandwidth in real time exist, but their versatility is impaired by unsatisfactory sensitivity.

Special oscilloscope tubes. Many of the foregoing instruments can be equipped with special CRTs in order to add convenience:

— in viewing and comparing single events (variable persistence tubes, storage tubes)
— in photographing faint traces (blue luminescence).

Standard oscilloscope cameras using type 410 Polaroid film give satisfactory single transient pictures up to about 0.3 cm/nsec writing speed from a conventional CRT (12 kV, green luminescence). With specially designed CRTs (24 kV, type P-11 purplish-blue luminescence), 7 cm/nsec writing speed can be attained using an $f/1.2$ lens aperture.

9.9.5. Dynamic Testing

9.9.5.1. Significance of Quadrupole Parameters.
Any circuit can be sectioned into quadrupoles, by choosing arbitrarily one pair of nodes as input, another pair as output terminals. For example, to study the

power supply sensitivity of a multistage amplifier, the ± 15 V supply lines might be considered as input terminals ad hoc (while the normal input would be grounded), the collector and emitter of the nth transistor as the output pair of a quadrupole describing one particular aspect of the system.

In linear circuit theory it is shown that four parameters are sufficient to describe the dynamic behavior of a quadrupole completely (see Chapter 1.3), for example: input/output impedances, and forward/reverse gains.

These are, in general, complex functions of frequency, but should not depend on amplitudes (linear approximation). In a real circuit, however, they may vary considerably as a function of signal amplitude, and the measurement of such variations is practically important as they are sensitive indicators of nonlinearity: it will be recalled that integral nonlinearity of a response curve can remain very small when the slope change is already noticeable (e.g., a parabola deviates from a straight line by a mere $\pm 1\%$ in a range over which its slope varies by $\pm 4\%$).

Unless a frequency dependence of the quadrupole parameters is specified, the values given must be understood as low-frequency limits (midband frequency values in the case of an ac coupled system); these always have zero imaginary parts. Practical specifications should include at least the 3 dB cutoff frequencies for the magnitudes of the four parameters (measured near zero signal level), plus their effective magnitudes at a finite signal level (measured at low or midband frequency). A complete test in the complex frequency plane is quite laborious.

As a more practical alternative, step function responses associated with the four parameters (possibly measured at a number of different signal levels) give the full information in a convenient form for direct application in the time domain.

9.9.5.2. Circuit Requirements for Significant Dynamic Tests.

For *over-all circuit performance* measurements, the normally connected input and output devices must be replaced by some measuring equipment of similar impedances. With modern signal generators and measuring instruments, this is not difficult since for almost any test object (amplifiers, etc.):

— input impedance $\gg$ test generator output impedance
— output impedance $\ll$ oscilloscope (or voltmeter) input impedance.

In the special case of vhf apparatus, impedances are generally low, but are standardized to be identical with those of typical test instruments.

Thus, perfect matching is achieved—as far as necessary—by connecting additional loading elements across input and/or output terminals in parallel with the test equipment; no loading corrections are required under these conditions.

On the other hand, high-impedance coupling networks are often used in the interior of a circuit in order to push gain or bandwidth to the ultimate limits. Therefore, *internal circuit probing* cannot in general be done without modifying circuit performance: the results are subject to *loading corrections*, as explained in Section 9.9.5.3. In extreme cases, a normally stable circuit may even become oscillating when the capacity of one of its nodes to ground is increased by connecting a measuring probe to it (e.g., an internal node of a high-gain amplifier in a feedback loop). It is therefore advisable to keep an eye on a monitoring oscilloscope channel permanently displaying the normal output of the test object whenever a probe is connected to an internal test point.

9.9.5.3. Loading Corrections.

General remarks. The typical "high-impedance" parallel voltage measuring probe adds essentially a capacitance C_{probe} to the driving-point admittance of the circuit node under test. Its effect is expressed either by a voltage drop at high frequencies, or by an increase in voltage risetime.

A current-measuring clip-on probe inserts essentially a series inductance, L_{probe}, in the circuit branch under test—typically 3 nH. Its effects on current waveforms are usually too small to require any correction within the limited bandwidth of such probes (e.g., $\omega L_{\text{probe}} \approx 0.5\ \Omega$ at 20 MHz).

In principle, a step-down transformer could serve as a "high-input impedance voltage probe" in cases where the oscilloscope (or other measuring instrument) has a low input impedance. For example, a 10:1 transformer would convert a 50 Ω sampling oscilloscope input into 5000 Ω. However, bandpass limitations make this method unsuitable for a general-purpose wideband voltage probe, whose advantage would otherwise be the near-100% power efficiency (as opposed to resistive voltage dividers, with or without capacitance compensation). In practice, we have found that 4:1 transformers are good enough for fast pulse measurements, if one accepts the fact that frequencies below a few megahertz are cut off. Their main circuit loading effect is a voltage drop at low frequencies, or "sag" in time domain measurements. As the primary and secondary windings must be wound exactly on top of each other for good coupling, the parasitic parallel capacitance is not negligible; its effect is similar to C_{probe} of a conventional voltage probe.

In the following, we shall describe correction procedures for voltage measurements, as required by nonzero C_{probe} (typically 10 pF). For the effects of noninfinite R_{probe}, refer to Section 9.9.3.3.

Ground connections of probes should be made directly from the circuit under test to the probe shield with a very short lead. Even so, current spikes in the ground loops may cause spurious signals; when several probes are simultaneously used, one should always check to what extent the signal in a given channel appears to change when other probes are disconnected.

Frequency domain measurements. For a given current excitation, i_0, at frequency ω, the voltage *observed* at the circuit node under test is

$$u_1(\omega) = i_0(\omega) \cdot Z_0(\omega)/(1 + j\omega C_{\text{probe}} \cdot Z_0(\omega)).$$

The voltage is now *computed* which would exist at the same node if it were not loaded by the probe, namely,

$$u_0(\omega) = i_0(\omega) \cdot Z_0(\omega)$$
$$= u_1(\omega) \cdot (1 + j\omega C_{\text{probe}} \cdot Z_0(\omega)) = \ldots ?$$

An exact calculation would require a detailed analysis of the driving-point impedance Z_0 of the circuit node under test. As long as the correction is small, the following quick experimental procedure is applicable:

Double the loading effects, by connecting an additional, similar probe (or simply a suitable capacitor–resistor combination to ground) to the node under test, and observe:

$$u_2(\omega) = i_0(\omega) \cdot Z_0(\omega)/(1 + 2j\omega C_{\text{probe}} \cdot Z_0(\omega)).$$

The "true" (= unloaded) voltage is then extrapolated from the u_1 and u_2 magnitudes and their phases φ_1, φ_2 as:

$$u_0 \cong u_1 \cdot \frac{|u_1|}{|u_2|} e^{j(\varphi_1 - \varphi_2)}$$

or very roughly:

$$\text{Magnitude} \quad |u_0| \sim 2|u_1| - |u_2|$$
$$\text{Phase} \qquad \varphi_0 \sim 2\varphi_1 - \varphi_2.$$

Example. At 1.6 MHz, a node rms voltage was measured as $u_1 = 4.90$ V in a certain phase position, φ_1. When the probe (assumed to have $C_{\text{probe}} = 10$ pF) was shunted by an additional 10 pF, voltage dropped to $u_2 = 4.78$ V, and the waveform appeared delayed by $\varphi_2 - \varphi_1 = 5.4°$.

The approximate formula gives

$$u_0 = 9.80 - 4.78 = 5.02 \text{ V}$$

$$\varphi_0 = \varphi_1 - 5.4° \text{ (advanced with respect to } u_1\text{)}.$$

The true unloaded signal is 4.98 V, advanced by 5.6° instead of 5.4°, as shown by an exact calculation assuming $Z_0 = 1$ kΩ shunted by 10 pF.

Time domain measurements. For a given current excitation with a risetime t_{in}, the voltage wavefront *observed*[†] at the circuit node has a risetime given approximately by

$$t_{\text{obs}}^2 \sim t_{\text{in}}^2 + k^2(C_0 + C_{\text{probe}})^2 = t_{\text{true}}^2 + 2k^2 C_0 C_{\text{probe}} + \cdots$$

where k is a coefficient proportional to the resistive component of the driving-point impedance and C_0 is its shunt capacitance. For $C_{\text{probe}} \ll C_0$, the slowing-down caused by the probe loading is seen to be approximately proportional to C_{probe} and it can be estimated by a quick experimental procedure:

t_{obs} = result obtained with the smallest possible C_{probe}

t_{obs}' = result obtained with C_{probe} doubled.

Linear extrapolation yields for the risetime of the signal at the same circuit node if there were no probe connected to it:

$$t_{\text{true}} \approx 2t_{\text{obs}} - t_{\text{obs}}'.$$

When C_{probe} becomes comparable to C_0, this formula tends to exaggerate the corrections somewhat.

Example. $t_{\text{in}} = 20$ nsec, $R_0 = 450$ Ω, $C_0 = 20$ pF (use $k \simeq 2.2R_0$).

Unloaded risetime, calculated: $t_{\text{true}} = \sqrt{20^2 + 20^2} = 28$ nsec

Observed risetime with $C_{\text{probe}} = 10$ pF: $t_{\text{obs}} \approx 36$ nsec

with $C_{\text{probe}} = 20$ pF: $t_{\text{obs}}' \approx 45$ nsec.

[†] All observations are assumed to have been already corrected for finite oscilloscope risetime (see Section 9.9.4.1).

If R_0, C_0 were not known, linear extrapolation would have given:

$$t_{\text{true}} \approx 72 - 45 = 27 \text{ nsec.}$$

9.9.5.4. Small-Signal Response. The four parameters of interest (see Section 9.9.5.1) can be determined either in the *frequency or time domain*. In both cases, signal amplitudes must be chosen such that the output signal is well below the rated maximum. In case the corresponding signal level is not known for a particular section of the circuit, one should determine the parameters for at least two widely different signal amplitudes. If the parameters differ, the measurement is repeated with amplitudes reduced by an additional factor of two, and so forth.

Frequency domain testing. Input/output immittances (denotes *imped-ance* or ad*mittance*) can in principle be measured with conventional bridges; for transimmittances and gain ratios, special bridge configura-tions are required. This measuring method would, however, be impract-ical for a complete test since bridges are usually designed and calibrated for fixed-frequency operation, and the classical two-parameter adjust-ment procedures tend to be lengthy. Alternatively, a sine wave constant-voltage generator can be applied to either terminal, and parameter values calculated from measured current amplitudes and phase angles. Very elaborate equipment would be required for a complete test of this kind, which should yield real and imaginary parts of all four parameters over a wide frequency range.

Time domain testing. A step function response curve for each of the four parameters is sufficient to describe completely the small-signal per-formance of a circuit. As one single oscillogram covers about a 10:1 frequency range with satisfactory resolution, complete testing of any linear device is extremely rapid with this method.

Example. An amplifier with upper and lower cutoff frequencies of 100 MHz and 1 MHz is characterized, in time domain, by a 3.5 nsec risetime and by a 160 nsec sag time constant (see Section 9.9.4.1). Three oscillograms taken at 1, 10, and 100 nsec/division sweep speeds give complete coverage.

Gain (or transfer) functions are measured from a simultaneous display of input and output waveforms; with a suitable attenuation factor in the output channel, the two waveforms can be equalized at certain points of the time scale. This property is conserved in a linear system as one changes the input amplitude, thus providing an easy check on possible nonlinearities.

Input/output immittances are measured by TDR (see Section 9.9.4.5) with great versatility. Since sampling techniques are well-suited for digital recording, results can always be converted into the frequency domain by a computer, either from a punched-tape record or directly on-line with the oscilloscope.

9.9.5.5. Large-Signal Response. When the linear range of a system is widely exceeded, at least one element is either cut off or saturated. Under this so-called overload condition, the output signal tends to become stationary for a certain time interval, which is given by:

(a) the time until the input signal is brought back to its normal level,

(b) the time (a), lengthened by an internally created interval, usually called *dead time*.

At the end of such a period, the output needs additional time to reach its normal operating level to a given fraction of its range: this is called *recovery time*, and is tested by applying pulses of normal shape and duration but amplitudes much larger than normal (typically, 10 or 100 times larger).

The speed at which the output approaches its limiting levels during overload periods is called the *slewing rate*. Most operational amplifiers have a guaranteed minimum slewing rate. This is tested by applying a low-frequency square wave of sufficiently large amplitude to the input. When slewing occurs in the sine wave response of a quasilinear device, so that $|dU/dt| \leq S = \text{const}$, the output amplitude at frequency ω is limited to $U_0 \leq S/\omega$.

> *Example.* $S = 6.28$ V/μsec. The device operates up to 1 MHz in an approximately linear manner if output amplitudes are at most ± 1 V; up to 10 MHz if output is limited to ± 100 mV; but only to 100 kHz for a ± 10 V output.

The breakdown of gain-bandwidth terminology in this example indicates why overload phenomena are more naturally studied in time rather than frequency domain.

9.9.5.6. Testing Negative Feedback Systems. From the numerical example quoted in Section 9.9.3.1, the high information value of an open loop gain check in a feedback-stabilized system is evident. Such an operation needs a sound knowledge of circuit design, in order to avoid its pitfalls:

— Opening of the loop must be done without changing input or output loads nor operating biases: e.g., when a feedback resistor is taken off the output terminal, it should be connected instead to a suitable bias supply so that the input operating point is not changed. Its normal loading effect on the output terminal should be reconstituted by placing a similar load between output and another suitable bias supply.

— A stable system may become oscillating when its feedback is removed. Small internal parasitic and coupling capacitances may manifest their disturbing effect when the gain becomes very large. A series *RC* combination, applied as an additional load towards ground at a well-chosen circuit node, may in some cases re-establish stability against high-frequency oscillation by reducing the open loop gain as well as an undesirable phase shift at the critical frequencies.

The open loop gain, measured by the methods described earlier (Section 9.9.5.4), is the parameter which can be checked with the highest sensitivity for possible variations as a function of environmental conditions, signal overload, etc. In the closed loop system, undesirable variations are normally reduced to an extent which makes the residual effects difficult to detect, except at the high-frequency limit where feedback becomes ineffective. Therefore, risetime and overshoot characteristics of a fast pulse amplifier are not a priori stabilized by the use of negative feedback; rather, these have to be checked in closed loop running conditions, and adjusted for an acceptable compromise over the desired dynamic range.

9.9.5.7. Tests of Switching Circuits. Pure *digital* circuits must respond in a prescribed manner to a given set of "ON/OFF" input conditions. Worst case testing is done by applying somewhat deteriorated signals to the inputs, characterized by:

— a lower-than-normal "ON" level
— a higher-than-normal "OFF" level

according to well-established test specifications (e.g., $+2$ V instead of $> +3$ V nominal; $+0.8$ V instead of $< +0.4$ V nominal for the TTL integrated circuit family). These circuits therefore need not be tested with input voltages at any intermediate level, i.e., in the "forbidden zone." Output signals must reach at least the nominal levels even with the deteriorated input signals; output rise and delay times should be checked separately for transitions in either direction, both unloaded and at maximum permissible load. Depending on the circuit configuration, these times may vary in a very different manner as a function of loading.

For example, the *rise*time of an unloaded *NPN* emitter follower is generally short, but increases slightly as ohmic loads are added. Its *fall* time may, on the contrary, be longer in the absence of an external ohmic load, in cases where the internal pull-down current has been chosen very small.

Mixed circuits involving analog as well as digital-type signals require more detailed testing: their response generally depends in a critical manner on the analog signal levels, shapes, and/or timings; it should therefore be checked for a large variety of input conditions. This is true even in the case of a D → A converter, where the analog signal is capable of a finite number of stationary levels only—the conversion characteristic may have irregularities if the circuit is not properly adjusted, and the settling time depends strongly on the actual analog level. A discussion of specific procedures for a representative selection of instruments of this category (coincidence gates; threshold gates; discriminators; delay generators; pulse stretchers; time → voltage converters; A → D and D → A converters, etc.) would exceed the scope of this chapter. We shall only outline the basic steps recommended by the CERN Test and Instrumentation Group for a particular class of circuits[25] and refer to similar reports concerning other instruments[14] and from other laboratories.[26]

> *Example. Fast pulse discriminator.*[25] The unit under test is a modular NIM circuit with a calibrated front panel potentiometer threshold adjustment (-100 to -550 mV), internally terminated input (50Ω), and complementary logic outputs ("ZERO" $= 0 \pm 100$ mV; "ONE" $= -700 \pm 100$ mV, into a 50Ω load). When the input level crosses the threshold, the A output goes from logic 0 to 1 ($\bar{A}$ goes from 1 to 0), and is supposed to return to 0 after a fixed period has elapsed (4 nsec plus delay of an external cable). In the DT-mode of operation, the circuit has a dead time which is approximately equal to the output pulse width; when the input exceeds the threshold level

[25] G. Bianchetti and B. Righini, Rep. No. 14—Test of a Commercial Discriminator. CERN, NP Division, Internal Rep. 68-35 (Dec. 15, 1968).

[26] L. H. Reamond and W. E. Huebsch, BNL, EPS Div., Internal Rep. No. 12 (5 April 1968).

permanently, a symmetric wave train of frequency $1/(2 \times$ output pulse width) is produced. The following characteristics are evaluated in the tests (condensed typical results are quoted in parentheses):

(a) *Input reflections*, at nominal and higher than nominal amplitude of a 10 nsec wide, quasirectangular pulse (indicates some 20 pF parasitic capacitance and a slightly high ohmic termination, $\approx 53\ \Omega$ at 600 mV, 60 Ω at 1 V and above).

(b) *Output pulse shape* as a function of input pulse width, in normal mode (rise and fall times ≈ 1.6 nsec; minimum width ≈ 4.5 nsec independent of input width, except that when the two widths are similar the output width is found to be widened by 20%).

(c) *Threshold calibration* with 10 nsec wide pulses of different risetimes, and also with a dc input (correct within ± 10 mV for $T_r = 5$ nsec; idem for $T_r = 0.7$ nsec at low amplitudes, 10% sensitivity loss at 550 mV setting; idem for dc up to 400 mV, 10% sensitivity increase at 550 mV).

(d) *Second pulse sensitivity*, with 2 nsec wide input pulses, 4 nsec output pulse width, at 100 mV threshold setting (sensitivity is independent of the timing of a pulse following a triggering pulse for intervals >15 nsec, increases to a maximum at 10 nsec and drops to zero at <9 nsec).

(e) *Maximum frequency* (varies from 110 to 116 MHz for 10% $\rightarrow$ 100% threshold overdrive).

(f) *Propagation delay and time slewing* (or "walk"), at 100 mV threshold setting, with 2 nsec wide input pulses of different amplitudes <110 mV. The time from input threshold crossing to half-amplitude output is measured (approximately 9 nsec for large overdrive, increased by 0.8 nsec for 110 mV pulses).

(g) *Temperature test*, at 100 mV threshold setting and 10 nsec output pulse width, in the range $0 \rightarrow 65°C$ (effective threshold increases by about 25 mV, risetime to ≈ 2.4 nsec, fall time to ≈ 1.8 nsec).

(h) *Supply voltage sensitivity*, at 200 mV threshold setting with 10 nsec wide input pulses (effective threshold increases by about 20 mV for a 2% reduction of the ± 24 V dc supplies).

(i) *Vibration test*, see Section 9.9.1.8 (no electrical measurements during test; check for broken connections after test).

9.9.5.8. Amplifier Noise Testing. Tests similar to those outlined under Section 9.9.5.7 are indispensable for all kinds of pulse measuring equipment. In the case of linear amplifiers, the term "threshold" is of course replaced by "gain" for this purpose. In addition, high-gain amplifiers need to be checked for internal noise: for example, by measuring noise levels at the output as a function of ohmic resistance (acting as a noise source) connected across the input. For details, see Chapter 13.4.

Ac hum (including high harmonics due to current spikes in rectifiers) is distinguished from thermal and flicker noise by observing the output in an oscilloscope synchronized with the power line voltage. Remembering that, in a Gaussian distribution with rms value X_0, deviations in excess of $\pm 2X_0$ are relatively frequent (about 4%) while those exceeding $\pm 3X_0$ are quite rare ($<0.3\%$), one can also use the oscilloscope as a crude random noise meter by observing the trace widening. If we estimate an apparent peak-to-peak width X_{pp} so as to include 99% of the total distribution, the rms value is approximately obtained as one-fifth of X_{pp}.

9.9.6. Trouble-Shooting Procedures

9.9.6.1. Desirable Equipment Specifications. Repair of electronic equipment, if the nature of the troubles is not revealed by visual inspection, requires a knowledge of its supposed normal performance. A *complete* set of specifications is most conveniently arranged as a list showing:

(a) component and circuit node identification symbols corresponding to the circuit diagram;
(b) component parameters including tolerances and ratings; inductance values and ratings rather than type number for inductors, transformers and relays; type numbers for semiconductor devices and tubes;
(c) source and load impedances to be connected to inputs and outputs;
(d) supply voltages and currents, including dc potentials of heater supplies in tube circuits;
(e) dc potentials of all circuit nodes, for each stable (or metastable) circuit state;
(f) dc currents flowing in all circuit branches, for each stable (metastable) circuit state;
(g) waveforms (with amplitude and time scales marked) to be applied to inputs;
(h) waveforms expected at other circuit nodes.

If any of the circuit variables ((e) to (h)) depend on some switch or other control positions, these should be listed as well with each set of variables. Numerical data should be freed of instrumental influences (see Sections 9.9.3.3 and 9.9.3.4), or else a specification of the measuring instrument used is needed to make the data meaningful. Tolerances should be given, at least for the specifications (b), (c), (d).

For circuits of moderate complexity, it is possible to assemble most of this information on the circuit schematic diagram. To make such diagrams less crowded, it is customary to omit dimensional units (V, A, Ω, F, H) and sometimes even abbreviations for powers of ten (M, k, m, μ, n, p) in cases where no confusion is possible. Data concerning currents are identified by accompanying arrows, while voltage values are marked with the proper sign ($+$ or $-$) to distinguish them further from currents (whose sign is expressed by the arrow).

Completeness of this body of information helps to speed up diagnosis of circuit failure (see Sections 9.9.6.4 and 9.9.6.5), although it is understood that about half of it would be sufficient to calculate the rest.

9.9.6.2. Interpretation of Incomplete Specifications. Availability of a complete technical description is not a minor point in selecting the right equipment for a research laboratory, assuming that urgent repair jobs need to be done by laboratory personnel. In fact, technical manuals accompanying well-established standard-type instruments usually comply with the desiderata enumerated in Section 9.9.6.1. This is not always the case for special, custom-tailored or homemade apparatus: typically, only item (b) is found explicitly on circuit diagrams since this is all that is needed to wire the equipment, while additional information is dispersed in notebooks. On the other hand, once the circuit is assembled, component parameters are no longer accessible to direct checking: in general, some internal leads would have to be disconnected first. In reality, the critical user is more interested in items (e) to (h) measured under actual operating conditions rather than in individual component values as measured on the cold equipment.

If one had to chose between two extremes of abridged specifications for the labeling of circuit diagrams, the version of Fig. 8 would definitely by preferred over Fig. 9. The example is taken from a magnetostriction-type spark chamber read-out system developed at the author's laboratory. Essential characteristics such as gain, risetime, overload behavior, etc., are given by the user-oriented labeling (Fig. 8), from which the function of each component is immediately evident. Calculation of the missing

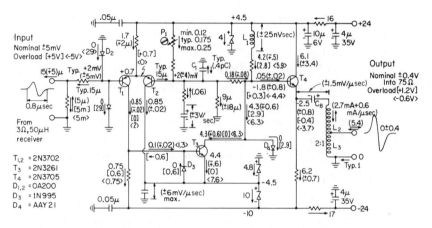

FIG. 8. Example of circuit diagram with minimum labelling: User-oriented, functional labels describe the operation of the circuit in detail and can all be checked in operation. The quiescent state is specified without parentheses, nominal signal swing in parentheses. Two overload conditions are given in brackets as far as necessary to show the function of the protection diodes. Potentials are in volts, currents in milliamperes, unless otherwise indicated. Only those components whose characteristics are not easily calculated from measurable voltages and currents are to be specified explicitly (e.g., bypass capacitors).

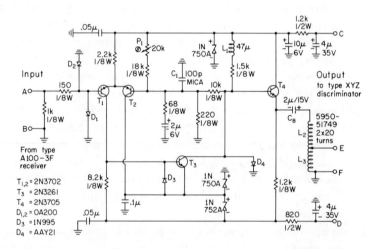

FIG. 9. Same circuit as in Fig. 8, showing another extreme of minimum labeling: fabrication-oriented, component parameter labels neglect functional information concerning the finished circuit. In these labels, Ω is omitted, k stands for $k\Omega$, μ for μF or μH, p for pF. Electrical checking of these data would require disassembly of circuit components. Explicit potential or current indications should be given at least for the input/output and power supply leads.

R, L, C values using the indicated voltages, currents, and their time derivatives is a very simple matter, and could be done for any suspected component without calculating all the others.

The inverse process of inferring the detailed functioning of the circuit from the fabrication-oriented labels alone, Fig. 9, is a rather lengthy one, as the reader may try for himself (assuming that Fig. 8 had been lost). The indications found in Fig. 9 are thus of little help for the troubleshooter: first, he would have to use them in a theoretical circuit analysis. As opposed to this, all the data of Fig. 8 can be directly checked with an oscilloscope, using suitable voltage and current probes.

As neither way of minimizing numerical indications in circuit diagrams would be completely satisfactory in practice, a combined presentation is to be preferred. Redundancy is, in addition, a necessary protection against errors of all kinds encountered in schematic diagrams.

9.9.6.3. General Strategy. Two different classes of repairable circuit troubles can be distinguished:

(a) the beginnings of self-destruction have been noticed (overheating, sparking, fuse blowout, noises, smoking, odors);

(b) the apparatus can be maintained in operation without danger, but its output signals deviate considerably from the normal response (incorrect gain, wrong timing, etc.).

In case (a), a visual inspection may reveal some defective components immediately, thereby indicating which circuit sections should be disconnected before attempting to switch the power supply on again. The suspected sections are checked for gross defects (short and open circuits); when these are eliminated, supply voltages are reconnected tentatively until the complete instrument is in order or has at least been brought to status (b).

Case (b) allows a completely systematic procedure: providing that supply voltages have their nominal values, any anomaly in a resistive element (i.e., resistors, semiconductors, and tubes) is revealed by a deviation of one or several measured circuit node dc potentials from their normal values. Alternatively, a test signal is applied to the input and the response checked along the signal path until a significant deviation from normal behavior is found (see Section 9.9.6.5).

9.9.6.4. Interpretation of Dc Measurements. In circuits where the wiring permits access for current clip-on probes, each element can be checked individually by combining its current reading with a conventional voltage drop measurement: any deviation from its expected I/U-characteristic is immediately revealed without the necessity of circuit calculations.

Printed circuits do not in general allow such direct checks, except for a few particular locations where rigid wire loops are provided for the sole purpose of giving access for current probing (this is standard practice in the author's laboratory). The remainder of the circuit branch currents must be inferred from potential measurements and component specifications. Using these data, all branch currents can at least be estimated, and the balance of currents is checked at each circuit node. In this balance, a result significantly different from zero indicates that one of the adjacent branches deviates from the anticipated behavior. Since such a condition shows up on both ends of the respective circuit branch, a compilation of the current unbalance results leads to a straightforward identification of a faulty resistive element.

The largest uncertainty in these evaluations is due to the individual spread and temperature dependence of diode and base-emitter forward characteristics. For example, from the measured forward voltage drops across a Si signal diode one can, in practice, conclude only that the diode is:

carrying negligible current ($\ll 1$ mA) if (1 mV $<$) $U_f < 0.5$ V

strongly conducting (>1 mA)   if 0.6 V $\lesssim U_f \lesssim 0.8$ V

probably $\begin{cases} \text{(short-circuited)} \\ \text{(open-circuited)} \end{cases}$ defective

if $U_f < 1$ mV

if $U_f > 1$ V

Apart from these restrictions, resistive elements deviating from specifications can thus be located, in principle, without any unsoldering. If their replacement does not remove the observed anomalies, the dc measurements were incorrectly made (see Section 9.9.3.3), or involved some ac component across a nonlinear element (hum, pulse, or high-frequency pick-up; this can be checked by capacitive short-circuiting to ground).

Reactive circuit elements cannot, of course, be diagnosed from dc measurements, except for short-circuited capacitors and open-circuited inductors.

9.9.6.5. Fast Strategies Based on Signal Flow. A rough check of signal flow is useful at the start of trouble-shooting, in order to rule out areas of the equipment which are not affected. If waveforms observed in the malfunctioning section do not immediately point out the defective component, the straightforward dc procedure (Section 9.9.6.4) may be applied next.

Operating potentials of the active elements (transistors, ICs, tubes) are checked first; other circuit nodes are checked in case no trouble is revealed otherwise, or if complementary information is needed. If the completed dc checks fail to identify the defective element(s), further investigation is necessarily based on signal flow observation and usually requires some circuit analysis.

The remark at the beginning of Section 9.9.6.4 applies to dynamic as well as to dc tests: thus, reactive elements too could, in principle, be diagnosed in situ and without any circuit calculations if simultaneous signal current and voltage drop observations were made, using a wideband current probe on one oscilloscope channel, and two high-impedance voltage probes on two more oscilloscope channels operating in an accurately balanced differential mode.

> *Example.* In the amplifier circuit shown in Figs. 8 and 9, consider a current probe and two voltage probes connected to (a) the coupling capacitor, C_8, and (b) the series-peaking inductor, L_1. The amplifier input receives 5 mV rectangular test pulses of 2 μsec duration at a low repetition rate.
>
> (a) The current pulse through C_8 is found to rise from 2.7 to 3.8 mA (averaging 3.3 mA); the initial potential difference changes by 3 mV during the 2 μsec interval. The capacitance value is thus obtained as $C_8 \approx 3.3$ mA $\times 2$ μsec$/3$ mV ≈ 2.2 μF.
>
> (b) After a relatively fast rise to a 0.4 V peak, the voltage across L_1 decays with an approximate 30 nsec time constant while the current rises by 0.5 mA. The voltage average over the observed current risetime of about 80 nsec is estimated as 0.3 V, so that $L_1 \approx 80$ nsec $\times 0.3$ V$/0.5$ mA $= 48$ μH.

In crowded circuitry where current probing is impossible, the value of a given reactive component could also be inferred from observed signal voltages only, making more or less complicated calculations assuming that all other components have their nominal values. However, the following experimental procedure, applicable in all cases where signal shapes resemble a step function, is much faster:

With respect to signal flow, a given circuit element is classified either as a series coupling link or as a parallel load. Qualitatively, a

> series capacitor causes sag with excessive slope if $C_8 <$ nominal
> parallel capacitor causes excessive risetime if $C_p >$ nominal.

Similarly, a

> series inductor causes excessive risetime if $L_s >$ nominal
> parallel inductor causes sag with excessive slope if $L_p <$ nominal.

Without having to shut down the equipment (for unsoldering some circuit connections) we can always double a given capacitance, or halve a given inductance temporarily, simply by holding, with an insulating tool, a similar element in parallel against the given circuit component. Thus, any suspected capacitor can be checked in situ by observing what value of parallel capacitance has to be added in order to increase the respective time constant by a factor 2. Similarly, inductors are measured in situ by observing at what value of momentarily connected parallel inductance the respective time constant is reduced by a factor 2. When this method is applied to the estimation of wiring capacitances, it must be kept in mind that risetime components do not add linearly (see Section 9.9.4.1). On the other hand, small slopes due to different sag components may be added linearly. An oscilloscope with a storage screen is extremely helpful in this kind of manipulation.

> *Example* (same configuration as in preceding example, except that no current probe is available).
>
> (a) The voltage across C_8 is seen to sag by 1.5 mV only (instead of 3 mV originally) when 2.2 μF are placed in parallel to C_8. Therefore, $C_8 \approx 2.2 \, \mu$F.
>
> (b) The voltage across L_1 is seen to drop to a 0.2 V average over a slightly reduced current risetime, 65 nsec, by holding a 47 μH choke in parallel. Since 65 nsec $\times$ 0.2 V = 13 nVsec is about one-half of the originally observed voltage-time integral, we conclude that $L_1 \approx 47 \, \mu$H.

Precautions are required when this method is to be applied to large capacitors, especially in high-voltage circuits. To avoid destructive transients in shunting a large capacitor by an equally large external capacitor, the latter should first be charged to the correct dc potential difference by a temporary high-resistance connection, using insulating tools.

9.9.6.6. Causes and Prevention of Circuit Trouble. If a fault is developing slowly, one day, one or several internal adjustments may be found to arrive at the limit of their range. This is a good moment to turn them all back to their center positions and start trouble-shooting while the instrument is still in a near-operational condition. When some anomaly or defect has been located, it is desirable to also determine its causes, in order to prevent—if possible—such conditions from developing again after the repair.

The various categories of components may be classified tentatively as follows (in order of decreasing likelihood of failure):

(a) vacuum tubes (cathode life; dissipation; mechanical);
(b) transistors and other semiconductor devices when used in hybrid circuits (temperature; voltage);
(c) unsealed wire-wound resistors (mechanical);
(d) carbon resistors (dissipation);
(e) electrolytic capacitors (temperature; voltage);
(f) potentiometers (dissipation; mechanical);
(g) ceramic capacitors (voltage);
(h) power transformers (dissipation; voltage);
(i) switches (mechanical);
(j) semiconductor devices used in low-voltage systems.

In this list, the kinds of stresses which are most likely to affect the various elements are indicated in parentheses. Apart from insufficient component derating (permissible dissipation or voltage being a function of ambient temperature), any coincidence of several unfavorable operating conditions (warm-up period; certain combinations of switch and control settings; extreme signal size or duration; input pulses of wrong polarity; excessive duty cycle; output short or open-circuited instead of nominal load; external voltage or current applied to output; repetitive power supply switching) may be the origin of a chain reaction leading to damage.

9.10. Telemetering*

9.10.1. Introduction

Telemetry technology is concerned with the transmission of data from one point to another. Frequently, many channels of data are multiplexed onto one telemeter link. The principal methods of multiplexing are time division (commutation) and frequency division (subcarriers). In the former, time is divided into slots, one or more for each data channel; and in the latter, the frequency domain is divided into slots, one or more for each channel. In the latter case, one or more frequency slots are occupied by a subcarrier modulated by the data or by a time division multiplex of data. The telemeter link may consist of a leased telephone line, land line, pneumatic coupling, radio link, acoustic link, etc. In the case of wire lines, the multiplex (baseband) can be impressed directly on the line, or it can modulate a carrier for transmission down the line.

In case the frequency-division multiplex using frequency-modulated subcarriers is frequency modulated onto a carrier, the telemeter is labeled FM/FM. In case a pulse code modulation time-division multiplex is phase modulated onto a carrier, it is labeled PCM/PM. In case a pulse amplitude time-division multiplex is modulated onto a frequency-modulated subcarrier which amplitude modulates the carrier, a PAM/FM/AM system results.

Generally, the instruments which transform the physical stimulus into voltages or currents are not included as part of a telemetry system so that the telemetry system accepts an input signal voltage representing the physical quantity. This is not always the most economical design. Frequently, magnetic tape storage is used in which case the baseband multiplex at the output of the carrier demodulator is recorded, or the modulated carrier itself, suitably transformed in frequency, is recorded. In the former case, upon playback, the tape recorded output feeds the demultiplexer directly; in the latter case, the tape recorded output feeds the carrier demodulator.

---

* Chapter 9.10 is by Myron H. Nichols.

Degradation of the output data can occur for many reasons. Among the important are distortion of waveforms in the various parts of the system, including distortion due to propagation effects in the transmission medium, additive noise such as thermal noise in the low-level parts of the system, interfering signals, atmospheric noise, time base error in the tape, etc. The properties of the various methods of modulation and multiplexing in the presence of such degradation, as well as matters of inventory and cost, are considerations in the choice of methods.

Because of the breadth and depth of the field, and the space limitation, this chapter is intended to be of a survey nature with a bibliography for detailed information.[†]

9.10.2. Requirements

The basic requirement in the choice (design) of a telemeter system is the rate at which the data must be transmitted. In analog systems, this can be conveniently expressed in terms of bandwidth required and output signal-to-noise ratio, the latter being interpreted to include the effects of waveform distortion in the transmission.

In digital systems it can be expressed in terms of the number of data samples (words) per second, the permissible quantization error, and the permissible error probability. Quantization error is the error incurred in rounding off the data to the digital quantum levels. The error probability may be defined in terms of word error probability, probability of error in symbols used to form the words, etc.

[†] General references are:

[1] C. E. Shannon and W. Weaver, "Mathematical Theory of Communication." Univ. of Illinois Press, Urbana, Illinois, 1949.

[2] M. H. Nichols and L. L. Rauch, "Radio Telemetry." Wiley, New York, 1956.

[3] H. L. Stiltz (ed.), "Aerospace Telemetry." Prentice-Hall, Englewood Cliffs, New Jersey, Vol. 1, 1961; Vol. 2, 1966.

[4] S. W. Golomb et al., "Digital Communications." Prentice-Hall, Englewood Cliffs, New Jersey, 1964.

[5] J. M. Wozencraft and I. M. Jacobs, "Principles of Communication Engineering." Wiley, New York, 1965.

[6] A. J. Viterbi, "Principles of Coherent Communication." McGraw-Hill, New York, 1966.

[7] R. G. Gallager, "Information Theory and Reliable Communication." Wiley, New York, 1968.

[8] H. L. Van Trees, "Detection, Estimation, and Modulation Theory," Part 1. Wiley, New York, 1968.

In terms of information theory, the information rate of digital data source is[1]

$$H = - \sum_{i=1}^{M} p_i \log_2 p_i \qquad \text{bits per word} \qquad (9.10.1)$$

where M is the number of digital quantum levels (words) and p_i is the probability that the ith quantum level (word) occurs. The quantity H is called the entropy of the data source. In the process of transmission, errors occur which result in loss of information. Let x_i represent the ith transmitted word, and y_j the jth received word (in which errors may have occurred). Then the loss of information, called equivocation,[1] is

$$L = - \sum_{i=1}^{M} \sum_{j=1}^{M} p(x_i, y_j) \log_2 p(x_i \mid y_j) \qquad \text{bits per word} \qquad (9.10.2)$$

where $p(x_i, y_j)$ is the joint probability of x_i and y_j and $p(x_i \mid y_j)$ is the conditional probability of x_i under the hypothesis that y_j has been received. The conditional probability $p(y_j \mid x_i)$ depends on the properties of the transmission channel only. But $p(x_i \mid y_j)$ and $p(x_i, y_j)$ depend on the channel and $p(x_i)$.

The rate of data transmission is $H - L$ bits per word. Alternatively, if there are N words transmitted per second, the source entropy can be expressed in bits per second by multiplying Eq. (9.10.1) by N and the equivocation of Eq. (9.10.2) by N. The corresponding rate of transmission is R bits per second.

Depending only on the properties of the transmission system, there exists a maximum average rate at which information can be transmitted. This is called channel capacity, C. For example, the capacity of a transmission system of bandwidth W and average power S, perturbed by white additive Gaussian noise of power N, is[1]

$$C = W \log_2(1 + S/N) \qquad \text{bits per second.} \qquad (9.10.3)$$

If the information rate fed into a system exceeds C, then the equivocation increases so that the rate of transmission does not exceed C. It is evident that by use of Eq. (9.10.3), the data rate requirement for analog systems, as stated above, can be interpreted in terms of data channel capacity.

9.10.3. PCM Telemetry

9.10.3.1. Coding. Figure 1 is a conceptual block diagram of a PCM telemeter. The commutator samples the data channels in cyclic-serial sequence. If the sampling rate of a data channel is greater than twice the

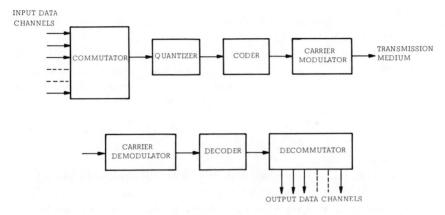

INPUT DATA CHANNELS

COMMUTATOR → QUANTIZER → CODER → CARRIER MODULATOR → TRANSMISSION MEDIUM

CARRIER DEMODULATOR → DECODER → DECOMMUTATOR

OUTPUT DATA CHANNELS

FIG. 1. Block diagram of a multichannel PCM telemetry system.

highest frequency component in the data, then it is possible, in principle, by interpolation techniques (filtering) to recover the original data signal without error.[2,9] If the condition is not satisfied, an error called "aliasing" occurs. Passing the samples through a low-pass filter is a frequently used method of analog interpolation.

The quantizer rounds the samples off to prescribed quantum levels. The quantum levels are selected in terms of requirements. For example, linear quantizing divides the range of the data variable into equal steps. Logarithmic quantizing divides the logarithm of the variable into equal steps and so on. The number of quantum levels is determined by the allowed quantization error.

The coder assigns a distinct waveform to each quantum level. Frequently, this waveform is constructed from pulses each of constant value. For example, the waveform can be constructed of binary pulses—i.e., from pulses having two possible values, usually $\pm a$ where a is constant (volts, amperes, etc.). These are often called bit pulses. Bit pulses belong to the class called binary symbols or binary digits. Perhaps the simplest code is the natural one shown in Fig. 2. If all quantum levels are equally likely, the information of this source is 3 bits per word—see Eq. (9.10.1). The information per bit pulse is one bit. If there are M quantum levels (words) per sample and N samples per second, the minimum number of bit pulses f_b per second is

$$f_b = N \log_2 M \qquad \text{bit pulses per second.} \qquad (9.10.4)$$

[9] D. D. McRae and R. C. Davis, *Int. Telem. Conf. Proc.* **4**, 530 (1968).

QUANTUM LEVEL	CODE
——	1 1 1
——	1 1 0
——	1 0 1
——	1 0 0
——	0 1 1
——	0 1 0
——	0 0 1
——	0 0 0

FIG. 2. Example of an eight quantum level system using binary symbols in a 3-bit natural code. Each group of symbols corresponding to a quantum level is called a word.

For example, if $N = 1000$/sec and $M = 128$ different quantum levels, then $f_b = 7000$ bit pulses (binary symbols) per second. If the waveform is constructed of pulses having eight distinct values, then the number of pulses (symbols) required is $7000/\log_2 8 = 7000/3$ and so on.

The choice of the code depends on a number of telemetry system considerations. For example, if it is important to conserve carrier power and if bandwidth is available,[†] then a code utilizing a bandwidth greater than the minimum required, under the particular circumstances, can be constructed. This is a form of redundancy. Examples are special waveforms such as simplex, orthogonal, and bi-orthogonal codes.[4-6] These codes are defined in terms of the cross correlation (with zero time shift) between pairs of the words of the code. In the simplex code, the normalized zero-time-shift cross correlation between all pairs of words is $-1/M$, where M is the number of words of the code. In the orthogonal codes, the cross correlation is zero between all of the word pairs. A bi-orthogonal code can be obtained from an orthogonal code by adding an equal number of new words such that each of the new words has a normalized zero-time-shift cross correlation of -1 with one each of the original orthogonal words.

Simplex, orthogonal, and bi-orthogonal waveforms constructed of binary pulses can be generated rather simply by shift-register techniques.[4] Let $n =$ number of data bits per word. In a simplex code the number of words is $M = 2^n$. The number of binary pulses per word is $M - 1$. An orthogonal code can be obtained from a simplex code by adding a binary zero to each word. A bi-orthogonal code can be obtained by adding the complement of each word in the orthogonal code. Assuming additive white Gaussian noise, the optimum decoding process requires cross correlation of each received word with all possible code words.

[†] Rf bandwidth occupancy is regulated by the agencies of the government. Leased line (telephone, for example) bandwidth is limited by cost.

For example, a 32 code word simplex system requires 31 binary pulses per word and requires 32 cross-correlation calculations for each word decoded. The optimum decoder selects the code word having the highest cross correlation with the received word, assuming all words are equally likely to occur. Figure 3 is a block diagram of such a decoder. In this decoder, the signal is decoded a word at a time. Assuming perfect synchronization, additive white Gaussian noise, $n(t)$, and all words equally likely, Fig. 4 shows the word error probability of simplex codes. For $n = 5$ or more, there is very little difference in word error probability among simplex, orthogonal, and bi-orthogonal codes.[4,6]

Note that for very large n, the system approaches the Shannon limit for error-free transmission obtained from Eq. (9.10.3) under the condition $S/N \ll 1$. If the data bit rate is R, then the data are encoded in blocks of $n = RT$ data bits, where T is the time per word. Thus, error-free transmission implies large coded blocks. Since $M = 2^n$, the number of words in the code grows exponentially with block length. Since the

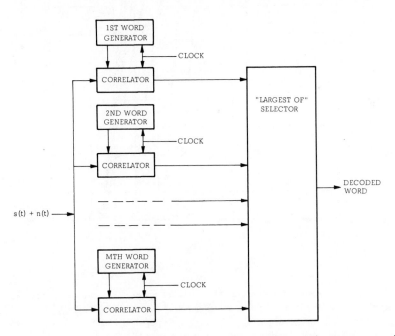

FIG. 3. Block diagram of a correlation PCM receiver. The correlator computes the time average of the product of $s(t) + n(t)$ and each of the locally generated words of the code. The average is over a word period. A "matched filter" can be used for the correlator in which case $w_j(-t)$ is the waveform of the jth word, where $w_j(t)$ is the weighting function of the jth matched filter.

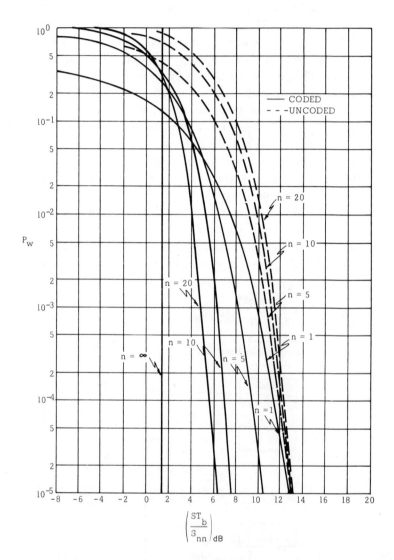

FIG. 4. Word error probability P_w for simplex codes. The abscissa is the ratio of energy per *data bit* E_b to the noise power spectral density S_{nn} (2-sided). If S = signal power, then $E_b = ST_b$, where T_b = word length per data bit. The solid lines are for the simplex code and the dotted lines are for uncoded PCM—see Fig. 2—detected a bit at a time. The parameter n is the number of data bits per word. The curve for $n = \infty$ is the Shannon limit for error-free transmission.

number of binary impulses required per word is $M - 1$, the if band-width required is approximately $2^n/T$ so that the required bandwidth becomes exponentially infinite as the Shannon limit is approached. Also the decoder grows exponentially with T as shown by Fig. 3.

With other codes providing more efficient use of bandwidth, it is possible to achieve the limiting probability-of-error properties of the simplex codes while avoiding the unlimited growth in bandwidth.[5,7] Binary parity-check codes satisfying these criteria can be generated by shift registers and modulo 2 adders. Convolution codes are a subclass of parity-check codes and are rather simply generated. However, decoding techniques to exploit the properties of these codes are more difficult to realize mainly because of the computing capacity implied. Various methods are under development. For a comparison of different types of coding, see Muller.[10] Such coding and decoding schemes can operate with binary-symbol or multiple-level-symbol signal channels with symbol-by-symbol detection.

The simplest error-detecting code using binary pulses is obtained by adding a parity bit[10] to each data word—for example, the data words of Fig. 2. The parity bit can be assigned on the basis of an even or odd number of binary "ones" in the data word. At the output of the bit detector, the number of "ones" is counted, and the count compared to the value of the parity bit. If there is a disagreement, the word is flagged as an error. It is evident that all odd numbers of errors in a word, includ-ing the parity bit, are detected; even numbers of errors are not detected. Thus, if the bit-error probability is small—say 10^{-3}, or less—then the probability of undetected error goes roughly as $n^2 P_e^2/2$, where P_e is the bit pulse error probability and n is the number of bit pulses per word. If the data waveforms are somewhat oversampled, then the loss of the few words in which the errors are detected is not serious.

On the other hand, if bandwidth is at a premium but sufficient carrier power is available, then symbols of more states than binary can be used. For example, the bandwidth required is roughly proportional to the number of symbols sent per second. Thus, if the code is constructed of 8-level pulses, it takes roughly one-third the bandwidth required for binary pulses. The above are illustrations of Eq. (9.10.3). In this formula, holding C constant, W and S/N can be traded off. If, for example, S/N is $\gg 1$,

$$S/N \cong 2^{C/W}. \tag{9.10.5}$$

[10] R. W. Muller, *Int. Telem. Conf. Proc.* 5, 123 (1969).

Thus, if the bandwidth W is halved, the S/N ratio must be squared to hold C constant. The reduction of bandwidth can be at a considerable cost in power, but in many cases this is justifiable.

9.10.3.2. PCM Pulse Waveforms. Figure 5 defines various pulse waveforms used in binary PCM. The nonreturn-to-zero level (NRZ-L) is frequently used in PCM/FM radio telemetry because it occupies minimum bandwidth and is amenable to efficient bit detection. NRZ-Mark and NRZ-Space are convenient for magnetic tape recording because the change in flux produces a voltage upon playback. The spectrum of a

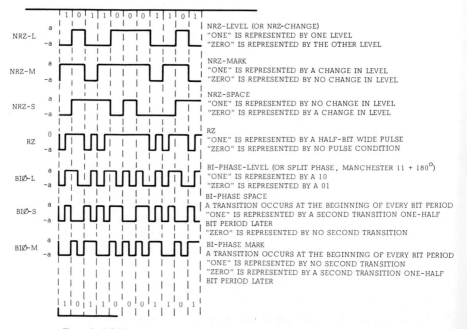

FIG. 5. PCM waveforms as defined by the IRIG telemetry standards.

random NRZ process is shown in Fig. 6. The spectrum of the split-phase process is also given in Fig. 6. Note that the split-phase process has no power density at zero frequency. This makes post-carrier detection magnetic recoding easier because the recorder is not required to have low-frequency response. However, approximately twice the bandwidth is required relative to the NRZ format. It should be understood that the spectra of Fig. 6 represent an average over all possible realizations. A particular realization such as 101010—in NRZ-L—would have peak power at half the bit rate, for example.

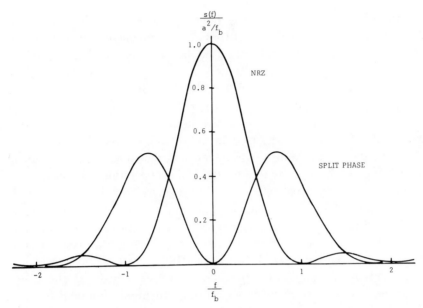

FIG. 6. Power spectral densities of random NRZ and split-phase PCM processes with pulse height $\pm a$.

Another waveform, referred to as the Miller code, proposed recently to remove the requirement for low-frequency response from the tape recorder, is defined as follows: a binary one is represented by a transition in the middle of the bit period and a binary zero followed by a binary zero is represented by a transition at the end of the first zero's bit period. This implies a coding delay of one bit. This waveform has a power spectrum which peaks at approximately half the bit rate. Performance and optimization data on this code when applied to PCM telemetry have not been published.

9.10.3.3. Bit and Frame Synchronization. For bit-by-bit detection, it is necessary to obtain bit clock. This can be done by transmitting a clock over a second channel, or clock can be obtained from the bit stream itself. It is evident that bit timing can be obtained from zero crossings, i.e., level changes. However, in a data bit stream the direction of the zero crossings is random. For example, if the bit stream waveform is differentiated, pulses will occur at the zero crossings, but their polarity and occurrence at the transition time is random. It is therefore necessary to rectify these pulses in order to generate discrete sinusoidal frequency components in the spectrum. The first component is at the bit rate.

Clock can then be recovered by filtering out a harmonic component, or by locking a phase-lock loop (PLL) to the component. The component at the bit rate is most commonly used by bit synchronizers to produce bit clock.

There are several methods of detecting zero crossings in use, but they all involve a nonlinear operation. Generally, a PLL is used to filter out the component at the bit rate. The PLL itself has a threshold. In addition, the nonlinear operation referred to above has a threshold. Also, with NRZ-L, during the time there are no bit transitions as, for example, in a string of binary ones or zeros, the PLL is operating open loop—i.e., there are no error signals.

With split-phase there is at least one transition per bit period, so the PLL operation is better in this respect. It is evident that narrowing the closed loop bandwidth of the PLL improves the performance against both the random nature of the bit stream and against the noise, which is enhanced by the nonlinear operation. However, narrowing the bandwidth increases the difficulty in acquiring synchronization and tracking time base error.

If the bit clock at the transmitter end is not steady, or if time base error is introduced, as by tape recorders, then the PLL must have sufficient bandwidth to track the time base error. The allowable untracked time base error—i.e., phase jitter on the bit clock relative to the PCM waveform—places a lower limit on the loop bandwidth. Thus, in case time base error is present, there is a trade-off between this effect and the two mentioned in the previous paragraphs.

The loop bandwidth and signal-to-noise ratio together with the uncertainty in bit rate and acquisition procedure determine the length of time required to acquire bit clock.

In spite of the fact that bit synchronizers have been in use for several decades, there is still a lack of good test data available to better understand their performance. Note that it was not until 1966 that definitive threshold data on the ordinary PLL were published. Coded PCM systems require improved synchronizers. See, for example, Mallory,[11] Lindsey and Anderson,[12] McBride and Sage,[13] Hurd and Anderson,[14] and Simon.[15]

[11] P. Mallory, *Int. Telem. Conf. Proc.* **4**, 1 (1968).

[12] W. C. Lindsey and T. O. Anderson, *Int. Telem. Conf. Proc.* **4**, 259 (1968).

[13] A. L. McBride and A. P. Sage, *IEEE Trans.* **COM-18**, 48 (1970).

[14] W. J. Hurd and T. O. Anderson, *IEEE Trans.* **COM-18**, 141 (1970).

[15] M. K. Simon, *IEEE Trans.* **COM-18**, 589, 686 (1970).

In addition to bit synchronization, it is necessary to obtain frame synchronization. This is accomplished by using a frame synchronization word sufficiently long that the probability of its accidental occurrence in random data bits is adequately small. For example, if the number of bit pulses in the word is Q, the probability of accidental occurrence is 2^{-Q}. Special words have been found, which, when imbedded in random data bits, have sharply peaked autocorrelation properties. Many frame synchronizers operate in several modes,[16,17] for example, search mode, check mode, and lock mode. In search mode, the word recognizer is open to the bit stream. When a frame synchronization word appears, it shifts to the check mode in which the word recognizer is gated out until a full frame bit count is reached at which time the recognizer is opened for one frame synchronization word length. If the word appears in this time slot a programed number of times, the synchronizer is shifted to the lock mode. In this mode, if the word fails to appear a programed number of times, the synchronizer shifts to the search mode, and so on. Having frame synchronization, word synchronization can be obtained by counting bit slots.

9.10.3.4. Carrier Modulation for PCM. When a carrier is used for PCM, it can be phase, frequency, or amplitude modulated. Because of the theoretical and technical advantages of angle modulation, amplitude modulation for PCM is rarely used. In binary PCM/PM, the carrier is deviated $\pm\beta$ degrees, where generally $\beta \leq 90°$. In this case, the system threshold can be reduced by using a phase-lock loop as a carrier demodulator.[6] Two methods of PLL carrier demodulation can be mechanized. If sufficient power is left in the carrier, by keeping $\beta < 90°$, the loop can track the carrier component directly by keeping the PLL tracking bandwidth much less than the bit rate. The instantaneous phase error is the demodulated PCM wave plus noise. The PLL tracking is sometimes facilitated by using the split-phase waveform which removes the modulation power from the neighborhood of the carrier—see Fig. 6.

The other method is used when the phase modulation is $\pm90°$. In this case, the waveform is passed through a nonlinear element, such as a rectifier, which generates a component at twice the carrier frequency. This is tracked by a PLL and divided by 2 to provide a demodulation reference frequency which is multiplied into the PCM/PM wave thereby

[16] M. W. Williard, *Int. Telem. Conf. Proc.* **2**, 483 (1966).

[17] R. S. Van de Houten, *Int. Telem. Conf. Proc.* **3**, 595 (1967).

translating the PCM waveform plus noise into the baseband where it is bit-detected. Another loop, known as the Costas loop, can be used in this case.[6] The performance after acquisition is identical but it requires only half the PLL loop bandwidth. If the loop is in lock and the loop error is small, essentially coherent carrier demodulation is accomplished in either case.

Figure 7 is a block diagram of PLL demodulation of PCM/PM for the first method. Best performance in PCM/PM with PLL carrier demodulation is obtained when the bandwidth of the carrier channel is sufficiently wide to reproduce the shape of the binary symbols reasonably accurately and when the bit detection is done by correlators as shown in Fig. 3. When the bit pulse shape is NRZ and rectangular, the familiar integrate-and-dump bit detector performs as a correlator (matched filter) since its $w(-t)$ has the same shape as a rectangular binary symbol. Here $w(t)$ is the weighting function defined as the response to a unit impulse input. An integrate-and-dump detector integrates the received waveform over one bit period. The output of the integrator is sampled at the end of the bit period. If the output is greater than zero, it is read as a binary one. If it is less than zero, it is read as a binary zero. The content of the integrator is then dumped and the process repeated.

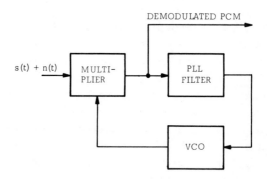

Fig. 7. Block diagram of a coherent carrier detector for PCM/PM.

Carrier-frequency modulation for PCM is usually mechanized in two ways. One is frequency shift keying, PCM/FSK,[6] which amounts to switching between two oscillators which are independent—i.e., not phase coherent. This causes a change in frequency and, in general, a change in phase. If the switching is abrupt, which is usually the case, then the changes in frequency and phase are abrupt. PCM/FSK is generally detected by use of bandpass filters, one at each frequency. The filter which puts out the larger amplitude at the end of each bit period is selected as the best estimate of the binary state of the transmitter. This is noncoherent detection because phase information is not utilized.

The other way is to modulate a voltage-controlled oscillator, VCO, with the PCM waveform. In this case there is no phase discontinuity. This method is called PCM/FM and is detected by an ordinary frequency discriminator. In order to reduce spectrum occupancy, the waveform is frequently low-pass filtered before carrier modulation. It has been determined by experiment that for NRZ-PCM/FM, the optimum peak-to-peak deviation is $0.7f_b$ and the optimum if bandwidth is f_b, where f_b is the bit pulse rate.[18,19]

A comparison of the three methods is given in Fig. 8. The solid lines are theoretical and are accurately confirmed by experiment. There is some spread in the NRZ–PCM/FM data points because of variations in the test configurations. In all cases, ideal synchronization is assumed. No data have been published for split-phase waveforms for PCM/FM and PCM/FSK. For split-phase, the filter bandwidths must be about twice as large so some degradation relative to NRZ must be expected. Near receiver threshold, the degradation may amount to several dB. In PCM/PM, as long as correlation detection is used, the results are independent of waveform provided the bandwidths are sufficient to reproduce the waveform reasonably accurately and the wave forms have a normalized zero-time-shift cross correlation of -1.

9.10.3.5. Bit Detection. Generally, two types of bit detection are used. One is the previously mentioned integrate-and-dump which is matched to the rectangular binary pulse shape. The other consists of video filtering and sampling at the appropriate time. The video filtering is usually accomplished with a low-pass filter with the 3 dB point at about half the bit pulse rate and with a roll off of -36 dB/octave or more. The sampling time is adjusted to coincide with about the peak output time of a single bit pulse. When the bandwidth is restricted as it usually is in PCM/FM, the integrate-and-dump detector is not matched because the pulse shape is no longer rectangular. Depending on the filtering and adjustment of the sampling time, the filter-and-sample detector can perform a little better than the integrate-and-dump in the narrow-band

[18] E. F. Smith, *IRE Trans.* **SET-8**, 290 (1962).
[19] G. M. Pelchat, *IEEE Trans.* **SET-10**, 39 (1964).

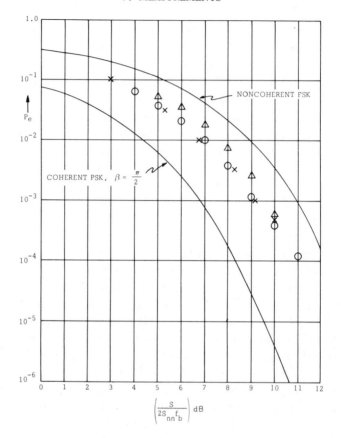

FIG. 8. Bit error probabilities for NRZ-PCM/FM, NRZ-PCM/FSK, and PCM/PM with $\beta = \pm 90°$ (coherent PSK). The NRZ-PCM/FM data sources are:

$\triangle$ = Experimental Determination of Signal-to-Noise Relationships in PCM/FM and PCM/PM Transmission. NASA Contract NAS5-505, 20 Oct. 1961. Electro-Mechanical Res., Inc., Sarasota, Florida.

$\bigcirc$ = Telemetry Systems Study-Final Report. Aeronutronic Publication U-743. Aeronutronic—a subsidiary of Ford Motor Co., Newport Beach, California.

$\times$ = J. J. Hayes, C. H. Chen, and W. J. Kubichi, *Int. Telem. Conf. Proc.* **4**, 233 (1968).

systems.[20] Also, the integrate-and-dump detector tends to be more sensitive to time base error in the bit clock.[20] On the other hand, unless the sampling time is kept in adjustment, the filter-and-sample detector appears to be a little more sensitive to bit pulse distortion due to restricted bandwidths.

[20] W. C. McClellan and M. H. Nichols, *Int. Telem. Conf. Proc.* **6**, 155 (1970).

Roughly, Peavey[21] and Roche and Mallory[22] indicate some bit synchronizers cause very little degradation in performance with P_e values as high as 0.2 provided that the PLL bandwidth is small, say less than 0.1% of the bit rate. However, in the presence of time base error characteristic of longitudinal tape recorders, the performance is degraded by as much as several decibels and slippage (momentary loss of lock) may prevent the use of the higher P_e range.[20] When error correcting codes are used, performance in the higher P_e range can be important.[11–15] It should be noted that these data are for Gaussian noise added to the bit stream. In PCM/FM, at the lower carrier S/N, the baseband noise departs from Gaussian because of the "pop" noise effect.[23]

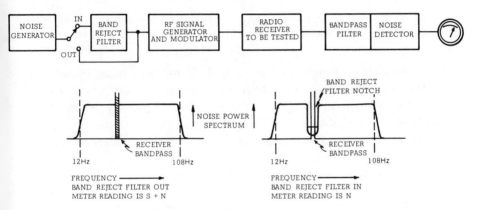

FIG. 9. Notch-noise test of a radio link.

9.10.4. Frequency-Division Multiplexed Telemetry

The functioning of frequency-division multiplexed telemetry can be understood easily by examining notch-noise test data. A block diagram of the tester is given in Fig. 9. The baseband of a frequency-division multiplex is simulated by a corresponding band of noise, as shown. At the input end, notch filters can be switched in. These filters remove the noise ("signal") in a narrow-band region without appreciably affecting the level of modulation. A corresponding bandpass filter at the receiving end selects the power in a narrow band of frequencies included in the notch filter. When the notch filter is out, the output power of the

[21] B. Peavey, *Int. Telem. Conf. Proc.* **3**, 407 (1967).
[22] A. O. Roche and P. Mallory, *Int. Telem. Conf. Proc.* **4**, 420 (1968).
[23] S. O. Rice, *Proc. Symp. Time Ser. Anal.* Wiley, New York, 1963.

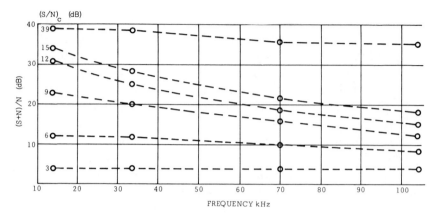

FIG. 10. Notch-noise test results. Flat input spectrum between 12 and 108 kHz. Rms carrier deviation 78 kHz [F. J. Schmitt, *Int. Telem. Conf. Proc.* **3**, 347 (1967)].

bandpass filter is "signal" plus noise, $S + N$, the noise being introduced by the radio link. When the notch is in, the output power is noise, N, since the attenuation of the notch is very large—typically 70–80 dB in commercially available test equipment. By inserting shaping filters, the spectrum of the noise generator can be altered to simulate subcarrier pre-emphasis schedules.

Figure 10 gives notch-noise test data on a typical commercial FM telemetry transmitter and receiver combination using a 500 kHz final if bandwidth.[24] In this figure the parameter is carrier $(S/N)_c$. The input noise spectrum is flat and the rms carrier deviation is 78 kHz. At low $(S/N)_c$ the noise in the receiver output is dominated by the thermal noise in the rf stages of the receiver, and the output noise spectrum is flat across the baseband. As the receiver $(S/N)_c$ is increased, the thermal noise power density in the output decreases more rapidly at the low-frequency end, until at about 12 dB $(S/N)_c$ the noise power decreases uniformly across the baseband shown and, since the $(S + N)/N$ curve drops at 6 dB/octave, the power spectrum is nearly parabolic, which is according to FM theory when above threshold. As the carrier power is increased further, the thermal noise in the output decreases to the point where intermodulation power dominates, so that no matter how large $(S/N)_c$ becomes, there is no further improvement in the output $(S + N)/N$.

Intermodulation is the result of frequency translation due to non-

---

[24] F. J. Schmitt, *Int. Telem. Conf. Proc.* **3**, 347 (1967).

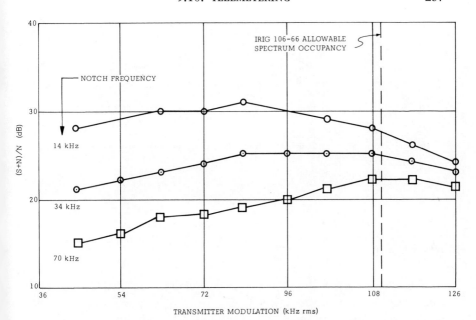

FIG. 11. Effect of transmitter deviation on notch-noise data with flat input spectrum between 12 and 108 kHz. Carrier signal-to-noise ratio, $(S/N)_c = 12$ dB [Frank Schmitt, Double Sideband (DBS) Telemetry Study. Lockheed Electron. Co. Tech. Memo 68-1, WAO 1327. White Sands Missile Range, New Mexico (Jan. 1968)].

linearities in the radio link. These are of two basic types; static[25] and dynamic.[26] Static, or simple nonlinearity, occurs mostly in carrier modulator and demodulator, in which the input/output relation can be expressed as a simple power series. Frequently the dominant term is the cube which produces third harmonic distortion. Dynamic nonlinearity is the result of phase distortion, mostly in the final if of the receiver. When the phase characteristic of the if departs from a linear function of frequency, the various signal-frequency components suffer a nonuniform time delay. Since the frequency demodulator responds to the rate of change of phase, intermodulation power density due to dynamic nonlinearity is zero at zero frequencies, and, depending on the input spectrum, increases with increasing frequency.

Static nonlinearity generally is at maximum at low frequencies, and decreases with increasing frequency. The effect of the two is evident in Fig. 11, which shows notch-noise data with the same transmitter/receiver

[25] W. R. Hedeman and M. H. Nichols, *Int. Telem. Conf. Proc.* **6**, 390 (1970).
[26] A. F. Ghais, E. J. Ferrari, and C. J. Boardman, *Int. Telem. Conf. Proc.* **3**, 26 (1967).

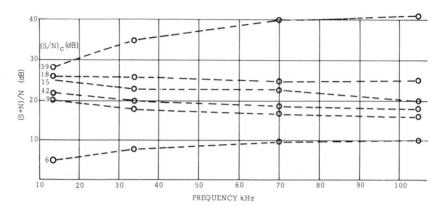

FIG. 12. Notch-noise test results with 6 dB/octave pre-emphasis of input spectrum from 23 to 108 kHz. Rms carrier deviation 78 kHz [F. J. Schmitt, *Int. Telem. Conf. Proc.* **3**, 347 (1967)].

combination as a function of total carrier modulation in kHz rms. The carrier $(S/N)_c$ for these data is 12 dB. Note that at low deviation the output $(S + N)/N$ improves with increasing deviation until intermodulation becomes a controlling effect. Thus, at any particular carrier $(S/N)_c$ there is an optimum modulation level. In missile test-range telemetry, there are restrictions on the rf spectrum utilized, and hence a limit on the deviation. For the 250 MHz frequency band, the limit is shown in Fig. 11.

It is sometimes desired to adjust the shape of the input spectrum to maintain the $(S + N)/N$ in all slots constant at some carrier $(S/N)_c$ figure. Figure 12 is an example for $(S/N)_c$ of 18 dB.[24] The rms carrier deviation is 78 kHz. Note here that for larger $(S/N)_c$ the intermodulation due to static nonlinearity has a serious effect at low baseband frequencies.

When FM subcarriers are used, the subcarrier FM improvement is applied to the baseband noise. The first FM/FM system standardized by IRIG[†] consisted of 18 channels, the lowest being centered at 400 Hz, and the highest at 70 kHz, with a ratio of adjacent center frequencies approximately 1.3. This is called the proportional baseband. The maximum deviation was 7.5% of subcarrier frequency. Later, provisions were made for ±15% for five channels at the high-frequency end of the

---

[†] The Inter-Range Instrumentation Group, IRIG, of the Range Commanders' Council establishes and publishes telemetry standards via the Telemetry Working Group. These standards are intended to define an equipment interface between the military test ranges and the range users.

band. The FM baseband was later expanded to include twenty-one
±7.5% channels, with maximum frequency 165 kHz, and provision for
a total of eight ±15% channels.

Figure 13 gives individual channel output S/N measured and plotted

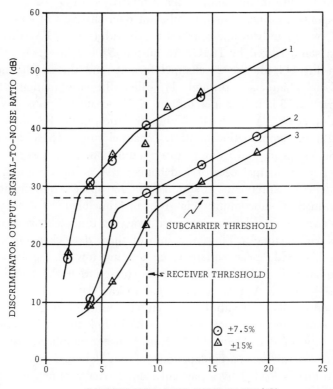

FIG. 13. Fully modulated signal-to-noise ratio performance for IRIG FM/FM multiplex. The ratio is the mean square value of the channel output when fully modulated
by a sinusoid divided by the mean square value of the output noise. Curve #1 with
⊙ points is the IRIG proportional channels 1 through 18 all with ±7.5% deviation
and with · points is for proportional channels 1 through 16 at ±7.5% and channel
E with ±15%. Curve #2 is for proportional channels 1 through 21 all with ±7.5%
deviation. Curve #3 is for proportional channels 1 through 19 at ±7.5% and H at
±15%. In all cases, the discriminator output signal-to-noise ratio is for the highest
frequency subcarrier. The subcarrier deviation ratio is 5. For curve #1 the carrier
deviation is 75 kHz rms and for curves #2 and #3 it is 56 kHz rms. In all cases, pre-
emphasis of subcarrier amplitudes was set to provide equal output S/N from all channels
with a carrier-to-noise ratio of 9 dB [Telemetry FM/FM Baseband Structure Study.
Final Rep., Contract #DA-29-040-AMC-746(R), 14 June 1965. Electro-Mechanical
Res., Inc., Sarasota, Florida].

as a function of carrier $(S/N)_c$.[27] There are two thresholds clearly indicated; one is the carrier threshold, and the other is the subcarrier threshold. Because of the noise improvement of about 26 dB resulting from a subcarrier deviation ratio of 5, the maximum output S/N, which is limited by intermodulation, is about 60 dB—see Fig. 12.

In response to a need for a number of data channels of constant and relatively wide bandwidth, a constant bandwidth (CBW) FM/FM format has been standardized by IRIG. An output S/N ratio versus $(S/N)_c$ is given in Fig. 14 for the 21-channel format, and the 21-channel plus proportional channels 1 through 11, called the combinational format. Note that the 3 kHz channel output S/N becomes limited by inter-modulation at $(S/N)_c = 15$ dB. With a deviation ratio of 2, the constant bandwidth and combination multiplex channel output (S/N) are limited to a little less than 40 dB due to intermodulation.

One of the principal uses of wideband CBW is for vibration, shock, and acoustic measurements. A summary of aerospace requirements compiled by an SAE subcommittee is given in Himelblau.[28] In many cases, the data are of noise-like character, with varying time delay and coupling between data channels. Thus, the sum of these data channels tends to be noise-like with mean square value tending toward the sum of the mean squares. For this reason, and the reason that the amplitude of suppressed carrier AM is proportional to the data, it turns out that suppressed carrier AM subcarriers—either double or single sideband—provide more efficient modulation of a radio link than FM subcarriers for this type of data channel.[29,30] Note that the amplitude of an FM subcarrier remains constant independent of modulation. Thus, in FM/FM the radio link is loaded by a particular subcarrier channel, whether the data signal is zero or not.

In addition, with the suppressed carrier AM, the principle of automatic gain control can be utilized in some applications to maintain the carrier modulation at an efficient level. In suppressed carrier AM, double sideband (DSB), a phase coherent reference frequency for each data channel must be available for demodulation. This can be derived separately for each channel from its sideband power or from a separately transmitted pilot tone after suitable frequency translation. Obviously, if

[27] Telemetry FM/FM Baseband Structure Study. Final Rep., Contract #DA-29-040-AMC-746(R), 14 June 1965. Electro-Mechanical Research, Inc., Sarasota, Florida.

[28] H. Himelblau, *Int. Telem. Conf. Proc.* **2**, 232 (1966).

[29] W. O. Frost, *IRE Trans.* **SET-8**, 238 (1962).

[30] M. H. Nichols, *Int. Telem. Conf. Proc.* **3**, 361 (1967).

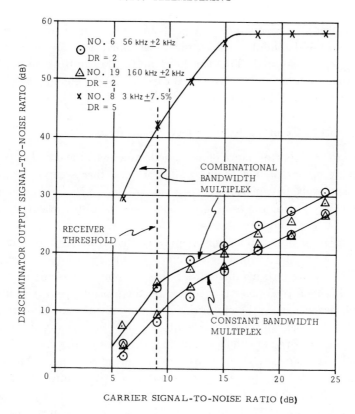

FIG. 14. Fully modulated signal-to-noise ratio performance of constant and combinational basebands. The ratio is the mean square value of the channel output when fully modulated by a sinusoid divided by the mean square value of the output noise. The carrier modulation for the constant bandwidth multiplex is 27 kHz rms and for the combinational the carrier modulation is 49 kHz rms. In all cases, pre-emphasis of subcarrier amplitudes was set to provide equal output S/N from all channels with a carrier-to-noise ratio of 9 dB [Telemetry FM/FM Baseband Structure Study. Final Rep., Contract #DA-29-040-AMC-746(R), 14 June 1965. Electro-Mechanical Res., Inc. Sarasota, Florida].

a common pilot tone is used for a number of subcarriers, the frequencies of the subcarriers must be harmonically related and coherent in phase.

In single sideband suppressed carrier AM (SSB) a pilot tone must be used to generate coherent reference frequencies if high fidelity is required; or, if less than perfect frequency translation can be tolerated, a local frequency source can be used as in SSB radio telephony. Since the operations required for the generation of the reference frequencies required for demodulation, either from the separately transmitted pilot tone or from the modulation sidebands in DSB, lead to amplitude ambiguities in the demodulated data, provisions must be made to remove the ambiguity, if required. This can be done by transmitting an additional pilot tone or tones. The suppressed carrier AM baseband systems have been standardized by IRIG in document 107-71.

9.10.5. Effects Resulting from Magnetic Tape Recording

The over-all performance of a magnetic tape machine is rather complex involving effects of the magnetization process, bias setting, record and playback levels, irregularities in the oxide coating, variation in tape speed, etc.[31] The following is a brief discussion of several of the important effects related to recording of data signals.

Modulation noise appearing in playback is an amplitude modulation of the output signal which is explained by the aggregation of magnetic particles during tape manufacture. Signal dropout is extreme modulation noise. If signal-plus-noise is recorded on the tape, as from the output of a radio receiver, then the amplitude modulation is the same for the noise as for the signal, and, to this extent, the S/N is unchanged. However, if the output of the tape falls so low that the noise in the playback output amplifier is important, then the S/N in the output is reduced. Modulation noise can be reduced by recording an automatic gain control (AGC) pilot along with the data signal and using an AGC loop to degenerate the modulation noise.[24] If the data signal frequency modulates a carrier or carriers recorded on the tape, then the modulation noise is leveled by the limiters so that as long as the FM demodulator is above threshold, the effect of modulation noise is greatly reduced.

Additive noise in the output arises mostly from intermodulation due to nonlinearity in the record and playback process[25] and thermal noise in the electronics—mostly in the playback amplifier which is often heavily compensated to extend the frequency range at the high frequencies because of the finite gap widths.

Variation in tape speed as it crosses the record and playback heads produces a time base error in the output since time is associated with the distance along the tape.[32] Let the average tape speed be s_0 and the variation from the average be $\Delta s(t)$. The average speed is maintained by a synchronous drive motor or by means of a tape drive servo using a

---

[31] S. W. Athey, Magnetic Tape Recording. NASA Technol. Utilization Div., Technol. Surv., NASA Publ. SP-5038 (1966.)

[32] S. C. Chao, *IEEE Trans.* **AES-2,** 214 (1966).

recorded pilot tone as a frequency reference. Suppose that a sinuisoid of frequency f_i Hz is recorded on the tape. Let $\Delta f(t)$ be the variation in frequency in the output due to variation in tape speed in the record and playback process. Then

$$\Delta s(t)/s_0 = \Delta f(t)/f_i = q(t). \tag{9.10.6}$$

Suppose a frequency-modulated carrier is recorded on the tape. Let its instantaneous frequency be $f_i(t)$. Then the instantaneous frequency $f_0(t)$ in the output, aside from a possible constant frequency offset, is

$$f_0(t) = f_i(t) + f_i(t)q(t). \tag{9.10.7}$$

The term $f_i(t)q(t)$ produces an error in the demodulated output. When FM subcarriers are recorded on tape, a method of correcting for the tape speed is to measure $q(t)$ by use of an unmodulated pilot tone recorded along with the FM subcarriers. Then $q(t)$ is multiplied by the unmodulated subcarrier frequency, suitably delayed to correct for propagation time through the filters, and scaled and subtracted from the demodulated output of the subcarrier. Obviously, this correction is an approximation because the unmodulated frequency is used for scaling whereas from Eq. (9.10.7) for the exact correction, the instantaneous frequency is required. However, if the degree of modulation is small—say a few percent—and $q(t)$ is small—say the order of one-tenth percent—the approximation is not bad. More precise correction can be made by computer scaling based on the instantaneous frequency. Note that this does not correct for the error in time base which appears in the data.

Since time is essentially given by the distance along the tape which is the integral of the instantaneous speed, the instantaneous time base $\gamma(t)$ in the tape output is given by

$$\gamma(t) = t + \eta(t) \tag{9.10.8}$$

where

$$\eta(t) = \int_0^t q(\varepsilon) \, d\varepsilon. \tag{9.10.9}$$

Thus, for example, if $\cos \omega t$ is recorded, the output will be

$$\cos \omega[t + \eta(t)].$$

In coherently demodulated systems, such as DSB and SSB amplitude modulation, phase modulation, etc., it is necessary to synthesize from the modulated carriers, or from a pilot tone, a demodulation reference tone. It is necessary for the recovery filter or PLL to have sufficient bandwidth to track the instantaneous time base to within a tolerable error. The bandwidth required can be estimated from the spectrum of $\eta(t)$.[33] The spectrum of $\eta(t)$ can be obtained from the spectrum of $q(t)$ by multiplying by $1/(2\pi f)^2$ as seen from Eq. (9.10.9).

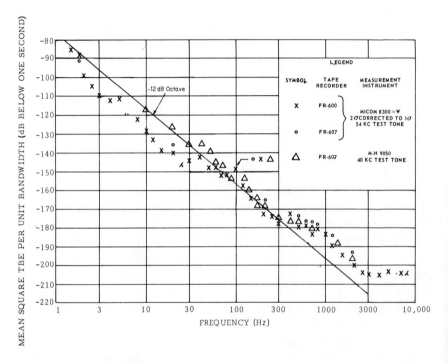

FIG. 15. Typical measured time base error spectrum. Data from two different spectrum analyzers and two different tape machines are compared [M. H. Nichols and F. J. Schmitt, *Int. Telem. Conf. Proc.* **4**, 399 (1968)].

Figure 15[33] is a typical time base error spectrum. The untracked time base error spectrum is obtained by multiplying the spectrum of $\eta(t)$ by the square of the magnitude of the phase error transfer function of the tracking loop. The spectrum of $q(t)$ can be measured by recording an unmodulated sinusoid and playing back through a frequency discriminator followed by a spectrum analyzer. Commercial equipment for this purpose is available. The spectrum can also be obtained, often with

[33] M. H. Nichols and F. J. Schmitt, *Int. Telem. Conf. Proc.* **4**, 399 (1968).

greater difficulty, by measuring the autocorrelation of $\eta(t)$ followed by a Fourier transform.[34]

In certain advanced and more expensive tape machines, the time base error has been reduced to the point where a variable electronic delay line with feedback can be used to reduce it to very small values.

9.10.6. Data Quality Assurance

In a telemeter, degradation of equipment performance and/or environmental noise and interference can seriously reduce the quality of the data. Thus, it is generally desirable to monitor the performance of the entire system either periodically or continuously.[35] In order to do this, it is desirable to simulate the input signal in some practical way and to make a minimum number of observations on the output which are definitive in assessing the performance of the telemeter.

In frequency-division systems, a method developed by the commercial communication industry is the notch-noise loading test already discussed in Section 9.10.4. This test can be applied periodically or continuously (in unused frequency slots) to assess the additive noise and intermodulation in telemeter links or parts thereof such as summing amplifiers, tape recorders, etc. Figure 16 is a block diagram showing the application of the notch-noise test to a complete frequency-division receiving station. Note that this test does not include the baseband processing since the analysis is performed at the output of the carrier demodulator. Thus, it is necessary to calibrate the subcarrier demodulators. This can be performed by the use of a baseband simulator or calibrator independently of the rf and storage parts of the system.

In PCM, a test developed for commercial use consists of a shift register generated pseudonoise (PN) sequence which has an autocorrelation function suitable for closed loop synchronization. Figure 17 is a block diagram showing the application of the test to a complete PCM ground station. The output of the shift register modulates a test signal generator which feeds the test signal into a directional coupler. An identically connected shift register is locked to the output of the telemeter by means of the autocorrelation properties of the sequence. The output sequence of the locally synchronized[20] shift register is then identical to the input sequence and locked to it in phase. Thus, the output of the bit detector can be

[34] J. Y. Sos, W. B. Poland, J. M. Cole, and G. Weiss, *Int. Telem. Conf. Proc.* **3**, 611 (1967).

[35] M. H. Nichols, *Int. Telem. Conf. Proc.* **2**, 535 (1966).

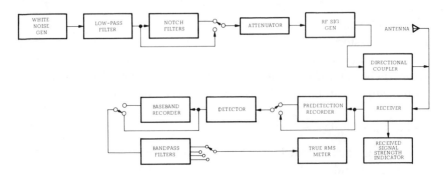

FIG. 16. Block diagram of frequency-division telemetry receiving system test utilizing the notch-noise method.

compared to the output of the local shift register and the errors counted. Also, slippage in the bit synchronizer can be detected and logged. Testing, if required, of the purely digital portion of the ground station which is connected to the output of the bit detector can be accomplished with a digital simulator independently of the testing of the analog parts of the system shown in Fig. 17.

For other time-division systems, such as PAM/FM, a test pattern having PN autocorrelation properties can be obtained by digital-to-analog conversion of the words in the shift register. The output of the PAM sample detector, for example, can be compared to the D/A output of the local shift register and the rms value of the difference logged.

It should be noted at this point that the notch-noise test is not effective in detecting impulse noise unless it is very severe. However, the PCM bit-error test is effective in detecting impulse noise. Thus, if this test is run in conjunction with the notch-noise test, a complete test results.

In the tests of Figs. 16 and 17, the directional coupler, feeds, and antenna aperture can be calibrated in terms of power flux incident on

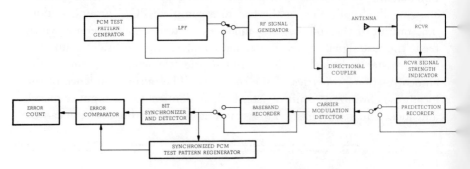

FIG. 17. Block diagram of PCM telemetry receiving system test.

the antenna by use of a calibrated far-field source. Then for every power setting P_s of the signal generator, an equivalent value of the flux density incident on the antenna can be computed by the relation

$$J_a = \alpha P_s \qquad (9.10.10)$$

where J_a is the equivalent incident flux density in W/m^2 of specified polarization and α is the constant determined by the calibration. Under certain conditions, the sun and/or radio stars can be used for the far-field calibrated source.[36] By the use of Eq. (9.10.10), test data such as in Figs. 10, 12, and 8 can be expressed directly in terms of J_a. This defines a performance interface between the telemetry transmitting system and the receiving system which removes the guesswork in telemetry mission planning.

The above described tests are part of the telemetry ground station test procedure standards adopted by the Telemetering Working Group of IRIG in document 118-71.

9.10.7. Choice of Modulation and Multiplexing

The considerations involved in the choice of carrier modulation and multiplexing are frequently complicated, involving data requirements, economic and engineering considerations, spectrum occupancy, data reduction, inventory, personal sophistication and preference, signal propagation, etc.[37] The following is a brief discussion of some of the important considerations.

9.10.7.1. Digital versus Analog.
The errors incurred in an analog telemetry system depend upon the nature of the communication link design, maintenance, calibration, and other factors. In aerospace telemetry, a rule-of-thumb indicates digital telemetry if less than about 1% error relative to full modulation is required. In current PCM aerospace telemetry practice, feedback encoders are used which utilize the null principle of measurement. Twelve-bit accuracy at 1-megabit rates has been achieved. Twelve-bit accuracy implies peak quantization error of one part in 8000.

In addition, PCM frequently offers greater flexibility in programing, adaptive telemetry, etc. Also, PCM signals are more easily relayed because

[36] W. R. Hedeman, *Int. Telem. Conf. Proc.* **4**, 330 (1968).
[37] K. M. Uglow, *IEEE Trans.* **COM-16**, 133 (1968).

the signal can be bit detected at repeater stations without appreciable degradation, whereas in the analog systems the distortion in frequency translators and amplifiers accumulates. Frequently the output of tele-meters must interface with digital computers and computer-controlled data handling equipment and it is frequently helpful to have the tele-metered data already in digital form. Of course, if the input data are already in digital form such as the output of digital counters, bank balances, inventory, etc., then PCM telemetry, often via telephone facilities,[38] is generally required.

Digital telemetry implies time-division multiplexing because the data must be sampled. The PCM waveform can modulate a subcarrier of a frequency-division system.

9.10.7.2. Time Division versus Frequency Division. When a large number of slowly varying data channels must be transmitted, time division is obviously more practical than frequency division because of the channel filtering required in the latter. When a number of rapidly varying data channels is required, the choice becomes more subtle. It used to be that high-speed time division was more expensive, but modern technology has narrowed the cost difference. Already in some cases, time division (PCM) is actually less expensive or will be in the near future.

Spectrum occupancy is in some cases very important. For noise-like data, such as vibration and acoustic data, suppressed carrier AM is utilized to make effective use of spectrum and rf power. Development with the objective of removing data redundancy is under way and may be effective in reducing spectrum occupancy in PCM.

9.10.7.3. Carrier Modulation. Angle-modulated carriers are generally used in radio telemetry because there is no requirement for amplitude linearity in the power stages of the transmitter. In angle modulation, either FM or PM, the carrier can be modulated at low-power-level stages and then multiplied and/or amplified without regard to amplitude fidelity. If coherent carrier detection is required either for system threshold reduction, where applicable, or in order to regenerate a phase coherent carrier for purposes such as carrier doppler shift measurements, then phase modulation is required. Otherwise, because of the simplicity of generation and demodulation, frequency modulation is widely used. Telemetry systems generally use a low FM carrier modulation index in

[38] P. Hersch, *IEEE Spectrum* **8**, No. 2, 47 (1971).

order to conserve rf bandwidth. However, if bandwidth is available, higher carrier modulation indices can be utilized to obtain more improvement over noise and interference but with an increase in carrier threshold. The increase in threshold can be reduced by use of PLL or FMFB (FM feedback) carrier demodulation.[39,40]

[39] J. A. Camp, *IEEE Trans.* **COM-18**, 191 (1970).
[40] M. M. Gerber, *IEEE Trans.* **COM-18**, 276 (1970).

10. MICROWAVES

10.1. Definition of Microwaves*

Microwaves are electromagnetic waves of relatively short wavelengths; however, the microwave frequency spectrum is not very clearly or closely defined. Usually the region over the range of 30 to 1 cm in wavelength or 1000 MHz to 30 GHz in frequency is termed the microwave (centimeter wave) portion of the electromagnetic spectrum. Less specifically, one generally thinks of the microwave region of the spectrum as being that part where one must use in analysis the full wave equations and not a quasistationary approximation as is used in low-frequency or lumped constant analysis. Also in contradistinction to the optical portion of the spectrum where, of course, the wave solution is also necessary, microwave apparatus is of a size such that the wavelength-to-size ratio is of the order of unity. It is really this last fact which gives to microwave equipment its particular character.

One of the original primary applications of the microwave spectrum was radar during World War II. However, in addition to the sophisticated radar systems in use today, microwaves are utilized in a large variety of ways, some of which are communications (both on earth and in space), industrial control, aircraft guidance, heating, medical application, and physical research.

10.2. Microwave Circuits

These are not circuits in the usual low-frequency sense, but are interconnections of transmission lines, cavities, and antennas. A transmission line is generally thought of as a device for transferring electromagnetic energy from one point to another. In addition, one assumes that the equations in the direction of transmission take the form given in Section 1.4.1; that is, one can write the propagation equations in a trans-

---

* Chapters 10.1 and 10.2 are by **Robert D. Wanselow.**

mission line as

$$E = f_1(x, y)e^{\gamma z} + f_2(x, y)e^{-\gamma z}$$
$$H = f_3(x, y)e^{\gamma z} + f_4(x, y)e^{-\gamma z}. \tag{10.2.1}$$

In these equations z is the longitudinal or propagation direction along the transmission line, and x and y are transverse coordinates. The functions f_1, f_2, f_3, and f_4 give field variations in the transverse directions, and γ is the propagation constant which in general is complex.

The field equations in certain types of transmission lines cannot, of course, be written in simple form. For instance, a spherical wave may be considered as propagation along a spherical transmission line. The fields, expressed in asymptotic forms at large radius, are of the type of Eq. (10.2.1). However, near the origin, these simple forms cannot be used. One can cite numerous examples of these more complex cases; however, when one thinks of a transmission line one ordinarily is considering a propagating wave in a space such that equations similar to (10.2.1) govern.

10.2.1. Transmission Lines

Several common forms of transmission lines used in the microwave portion of the spectrum are shown in Fig. 1. The first, Fig. 1a, is the simplest form which is the two-wire line consisting of two parallel conductors properly supported and insulated from each other. This line is described more fully in Section 1.4.1. The lines shown in Figs. 1b through 1d are useful variations of the two-wire line and all are used over the lower portion of the microwave frequency spectrum, that is, below 10 GHz. Above 3 GHz, more typical of the microwave transmission-line apparatus is that indicated in Figs. 1e through 1g, i.e., hollow pipe transmission lines, or, as they are more commonly called, waveguides. It should be noted that any transmission line is, strictly speaking, a waveguide. However, the hollow pipe form of transmission line has become most commonly referred to in this way and the name is ordinarily reserved for this type. Figure 1h is a conical transmission line most commonly used either as a transmission section between two sizes of hollow pipe lines or as an antenna. Figure 1i represents a spherical transmission line consisting of two metallic cones placed point-to-point; this is most commonly used as an antenna.

Lines shown in Figs. 1a through 1g can be described with perfect generality by equations in the form of (10.2.1). The field equations are

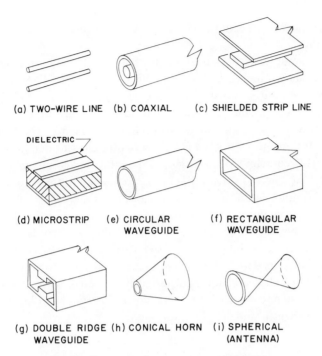

FIG. 1. Common microwave transmission lines.

amenable to solution in terms of an orthogonal function expansion valid for the given geometry. The theory on these orthogonal-mode solutions is beyond the scope here and reference is made to the literature.[1-3] A general description of the modes, however, is very useful in thinking of waveguide circuits.

Each orthogonal mode has a given electric and magnetic field configuration. Some modes have only the electric field transverse to the direction of propagation, some have only the magnetic field transverse to the direction of propagation, and certain fields have both electric and magnetic fields transverse to the direction of propagation. Modes for which only the electric field is transverse to the direction of propagation are called TE. Similarly, TM, or transverse magnetic modes, consist of those modes for which only a magnetic field exists transverse to the direction of propagation, and TEM or transverse electromagnetic modes

[1] C. G. Montgomery, R. H. Dicke, and E. M. Purcell, Principles of Microwave Circuits, Massachusetts Inst. Technol. Radiat. Lab. Ser., Vol. 8. McGraw-Hill, New York, 1948.

[2] S. A. Schelkunoff, "Electromagnetic Waves." Van Nostrand, New York, 1943.

[3] R. E. Colin, "Field Theory of Guided Waves." McGraw-Hill, New York, 1960.

have electric and magnetic fields both transverse to the direction of propagation.

The commonly used coaxial mode is a TEM mode and has a field pattern as shown in Fig. 2a. The TEM mode has the characteristic that it will propagate in a transmission line with any size-to-wavelength ratio from 0 to ∞. All other modes will propagate freely in a transmission line only in a certain size-to-wavelength range. Hence, the term wavelength cutoff or cutoff frequency is frequently used to denote the lowest theoretical frequency of propagation possible for a given guide size. However, the lowest practical operating frequency is normally more than 10% above the cutoff frequency because of the reactive transmission effects in the vicinity of the cutoff region. If a waveguide has a cross section as shown in Fig. 1g, it is called ridge waveguide. This guide geometry not only exhibits the characteristics of wider frequency bandwidth, lower impedance, higher attenuation, and lower power handling capability, but also, and most importantly, operates with a lower cutoff frequency when compared to rectangular waveguide (Fig. 1f) of the same inside width and height dimensions.

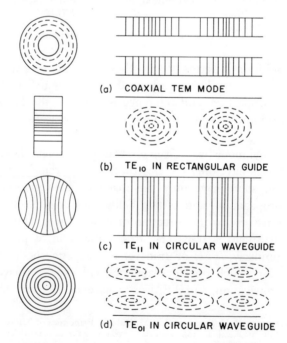

(a) COAXIAL TEM MODE

(b) TE$_{10}$ IN RECTANGULAR GUIDE

(c) TE$_{11}$ IN CIRCULAR WAVEGUIDE

(d) TE$_{01}$ IN CIRCULAR WAVEGUIDE

FIG. 2. Dominant-mode field patterns for some common transmission lines: H – – – –, E ———.

A word is necessary concerning the nomenclature used for identification of waveguide modes. In transmission lines which are circular cylinders, the first subscript is the number of full wavelengths of field variation around the circumference, while the second subscript refers to the number of half-wavelength variations in the field in the radial direction. In transmission lines of rectangular cross section, the first subscript refers to the number of half-wavelength variations in the direction of the wide dimension, and the second subscript refers to the number of half-wavelength variations in the narrow dimension.

Figure 2b shows the TE_{10} mode in a rectangular hollow pipe waveguide, while Fig. 2c shows the TE_{11} mode in a circular hollow pipe waveguide. These are the modes having the lowest cutoff frequency for given transverse dimensions. They are sometimes called the dominant modes, and are the modes most commonly used.

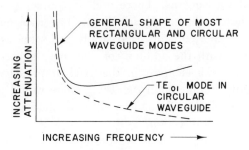

FIG. 3. Waveguide attenuation characteristics for various propagation modes.

Most rectangular and circular waveguide modes exhibit a decrease in attenuation with an increase in the propagating frequency above cutoff. However, when one takes into account the wall losses of a transmission line, the real part of the propagation constant γ exhibits a loss characteristic as indicated by the solid line curve in Fig. 3, i.e., at very high frequencies above cutoff the attenuation increases. However, the TE_{01} mode (see Fig. 2d) in circular waveguide is one important exception. As the dashed curve in Fig. 3 shows, the attenuation of this mode decreases continually with frequency whereas all others rise after reaching a minimum. Hence the use of circular waveguide for very long distance transmission or for millimeter wave transmission is apparent.

The stripline shown in Fig. 1c is of particular importance for many applications because of its particular ease of application and its small size. One method of making such a line is to use a flat metal plate as bottom surface with a thin sheet of dielectric on top of this, then a thin flat strip of metal for the center conductor, another piece of dielectric, and then a top metal plate to complete the sandwich. Such a line can be made in a very compact form. A slight variation on this type of TEM line is to leave off both the top plate and top dielectric sheet, having a stripline which might be termed "unshielded" but is better known as microstrip as indicated in Fig. 1d. When a low-loss, high dielectric constant material is utilized, a very compact microstrip line is realized which has become the transmission medium of present-day high-frequency integrated circuit technology.

10.2.2. Resonators and Filters

The resonator can be considered as a transmission line shorted at each end. It resonates at a frequency such that the line length is an integral multiple of half-wavelengths. There are, of course, as many kinds of resonators as there are transmission lines. The nomenclature used in identifying modes in resonators is basically the same as that used for transmission lines with the addition of one more subscript to identify the number of half-wavelengths in the axial or propagation direction. Thus a rectangular resonator in the TE_{101} mode would have the transverse field pattern of Fig. 2b, and would have one-half wavelength field variation in the length direction.

Cavity resonators are used as wavemeters or as elements of filters because of their possible high Q—ranging from 1000 to 100,000. Use as a wavemeter is illustrated in Fig. 4a, where a cylindrical cavity resonator is shown with an end plate made movable by means of a micrometer

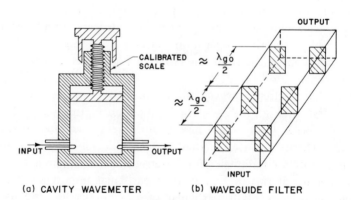

(a) CAVITY WAVEMETER   (b) WAVEGUIDE FILTER

FIG. 4. Resonant circuits.

for calibration. Accuracy of calibration ranges from 0.003% up to 1%, depending upon the mode used in the cavity, the Q, and the care used in arranging the tuning mechanism.

In microwave systems, filter networks may be employed to perform any one of a variety of functions such as to channel energy, separate the frequency spectrum, integrate coherent signals, extract frequency modulation information, or to maintain accuracy of frequency separation between two transmitters. Hence the filters required may be low-pass, high-pass, bandpass, or stopband and, depending on the application, may be constructed in any one of the transmission lines shown in Figs. 1a through 1g.

For example, a straight section of waveguide (see Figs. 1e through 1g) displays all the characteristics of a high-pass filter since it passes all frequencies above its dominant-mode cutoff frequency and reflects or rejects all frequencies below. Filters are generally iterative obstacles deriving their characteristic properties by virtue of a rapid phase change with frequency. Figure 4b denotes a typical two-section or two-cavity direct-coupled waveguide bandpass filter where the coupled obstacles shown are pairs of inductive vertical vane irises[†] spaced approximately one-half guide wavelength ($\lambda_{g0}/2$) apart at the desired bandcenter frequency.

10.2.3. Transmission Line Junctions and Obstacles

Any of the common transmission lines can be joined in a number of different ways to form tees, Y's, corners, etc. Almost always some provision must be made for matching the impedances at the junction. There will always be set up at the junction higher order modes to satisfy the boundary conditions. Stated another way, one must, in order to satisfy the modified boundary conditions at the junction, invoke the presence of an infinite series of orthogonal terms such that the expansion in this series produces a field which just matches the geometric configuration at the junction. If the waveguide is operated in a frequency range such that only one mode will propagate, that is, the dominant mode, then the higher order modes which are set up at the junction must be considered as localized fields. They store energy but they cannot transmit it from one point to another. Thus, one can consider the energy stored in these

---

[†] If the vanes were horizontal, across the larger width of the waveguide, they would constitute a capacitive obstacle—see Section 10.2.3.

higher order fields as being a reactive term in the impedance of a junction.[4]

The basic idea that a change in geometric configuration induces a field variation which is tantamount to reactive energy storage at localized spots along the transmission line can be used as an impedance matching technique. For instance, a small perturbation introduced in a hollow pipe waveguide will produce a series of higher order modes in the immediate vicinity of the perturbation. Since these modes cannot propagate, they only store energy in the immediate vicinity of the perturbation. This energy storage is equivalent in a low-frequency analogy to a reactive element in shunt or series with the transmission line at the point of perturbation. By adjusting the perturbing element properly, one can introduce reactive elements of known values. These can then be used to produce reflected waves such that they cancel reflected waves from some further point along the transmission line. Thus, the perturbing elements can be such as to produce equivalent inductive or capacitive reactances or combinations, which then become resonant circuits, as for example the vertical vane irises shown in Fig. 4b.

Aside from transmission from one type of transmission line to another, that is, coaxial line to waveguide, or coaxial line to strip line, etc., perhaps the more important types of junctions are those having the properties of bridges. The hybrid tee is shown in Fig. 5a. From symmetry alone it is evident that, if lines 2 and 3 are terminated in equal impedances, power incident upon line 1 cannot be transmitted to line 4 and vice versa. In fact, one can show that this junction is fully equivalent to the bridge circuit of low frequencies. It can, therefore, be used for the same purposes, for instance, as an impedance measuring device as discussed later in Section 10.5.1.

The hybrid ring or "rat-race" circuit shown in Fig. 5b is another useful form of waveguide bridge circuit which has an advantage over the hybrid tee in that it can be readily implemented in either a strip line or microstrip transmission medium since it is a planar network, i.e., all four ports lie in the same geometrical plane. The inductive post obstacle, shown in Fig. 5c, is sometimes used instead of the inductive vane iris (see Fig. 4b) in construction of waveguide filters because of the simplicity of fabrication. Figure 5d indicates an obstacle which produces a resonant circuit at one point or transverse plane in a wave-

[4] N. Marcuvitz, Waveguide Handbook, Massachusetts Inst. Technol. Radiat. Lab. Ser., Vol. 10. McGraw-Hill, New York, 1951.

guide. That is, the resonant window,[5] equivalent to a capacitive iris and an inductive iris (see footnote, page 257) located in the same plane normal to the transmission line, produces a parallel resonant circuit in shunt with the line. This type of obstacle can be employed as a circuit element in a bandpass network. If a low-loss dielectric is sealed across the window opening, a pressure window is formed which can stand a substantial pressure differential at this junction.

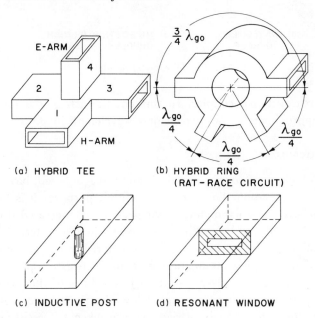

(a) HYBRID TEE

(b) HYBRID RING
(RAT-RACE CIRCUIT)

(c) INDUCTIVE POST

(d) RESONANT WINDOW

FIG. 5. Circuit junctions and obstacles.

Another element having many of the characteristics of a bridge is the directional coupler. A simple example is shown in Fig. 6a. Energy incident from the left in transmission line 1 is coupled through a series of holes to transmission line 2. The series of holes can, if one wishes, be considered as a directional antenna radiating from line 1 into line 2. It is excited in such a way that it radiates preferentially in the forward direction as shown. Essentially none of the wave energy transmits backwards in line 2, thus, the name directional coupler. Figure 6b indicates by arrow direction the forward direction of power flow of a branch-type

[5] L. D. Smullin and C. G. Montgomery, "Microwave Duplexers," Massachusetts Inst. Technol. Radiat. Lab. Ser., Vol. 14, pp. 67–114. McGraw-Hill, New York, 1948.

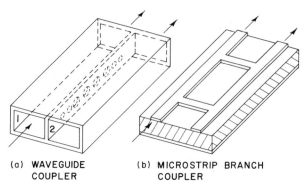

(a) WAVEGUIDE
COUPLER

(b) MICROSTRIP BRANCH
COUPLER

FIG. 6. Directional couplers.

microstrip directional coupler. In this circuit each of the four branch arms is approximately one-quarter of a wavelength long.

10.2.4. Attenuators

No conductors, of course, are ideal; any waveguide will have a certain amount of attenuation. For microwave circuit purposes it is often convenient to have transmission lines having very much increased attenuation. This can be produced by making the waveguide of a material which has high resistivity or by introducing such a material into a waveguide. One method of introducing attenuation is to use a resistive plate mounted in such a way that the plate is parallel to the lines of the electric field. By moving this plate from regions of low electric field to high electric field, the attenuation can be varied because when the resistive plate is in a position where the electric field is low, there will be small currents flowing in the resistance and, hence, small losses. As the plate is moved into the region of high electric field, the current flow, and therefore the loss, increases. By making the resistive plate of glass with a metalized film on it, for instance, it can be made to be very stable, and by producing a rigid and accurate mechanical support for the plate it can be constructed in such a way that the attenuation produced is accurately reproducible. Such a resistive plate attenuator is indicated in Fig. 7a. A simpler form of attenuator for laboratory use is the flap or top wall attenuator as shown in Fig. 7b. However, for greater accuracy and better repeatability of measurements the side wall attenuator is preferred.

Another form of microwave attenuator is based upon the fact that a hollow pipe waveguide will not transmit appreciable energy at frequencies below a certain critical or cutoff frequency. If one sets up a microwave field in the end of a hollow pipe guide which is too small to transmit energy freely at that frequency, there will still be a field set up within this space. It is not a field corresponding, however, to a traveling wave but is simply a field, all in-phase, whose magnitude varies exponentially with length along the axis of the hollow tube. By inserting a movable pickup means in the other end of the tube, one can pick up a portion of this exponentially decaying wave, and by moving the pickup, one has produced a variable attenuator. This waveguide-below-cutoff attenuator can be made very accurate and is capable of producing very high values of attenuation—of the order of a hundred or more decibels. For a round waveguide the attenuation is about 32 dB per diameter so long as the frequency is well below the cutoff point.

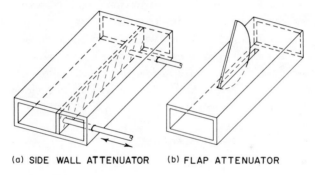

(a) SIDE WALL ATTENUATOR   (b) FLAP ATTENUATOR

FIG. 7. Resistive wave attenuators.

10.2.5. Ferrite Devices

Certain ferromagnetic substances have come into common use in microwave apparatus. These are ferrites which are materials similar to magnetite but with some of the iron atoms replaced by other elements. In the presence of a steady magnetic field, the electron spins which are responsible for the ferromagnetic behavior of the substances will tend to precess about the direction of the magnetic field. In the presence now of both the steady magnetic field and an alternating magnetic field, certain other effects become apparent.

Any linearly polarized wave can be considered as two oppositely rotating circularly polarized waves. One of these waves has a rotation in the same direction as the precession motion of the electron spins about the magnetic field. The other wave rotates in the opposite direction. It is at least qualitatively evident that the wave rotating in the opposite direction to the electron precession will have zero average effect, whereas the circularly polarized component rotating in the same direction will have a definite effect. Such a device is a Faraday rotator, that is, a plane wave incident on such a medium will suffer a rotation of its plane of polarization because the two oppositely rotating circular components are affected differently (have different propagation constants in the medium) and hence, when one attempts to combine them again at any point in the medium, they combine to give a linearly polarized wave at a different angle.

A variety of nonreciprocal waveguide elements has been constructed utilizing this principle. The principal microwave components are isolators, which have the property of transmitting with small loss in one direction and large loss in the opposite direction, nonreciprocal phase shifters, which are transmission lines having small attenuation but different phase constants in the different directions of propagation, and circulators. A circulator is a device whose properties can be understood from the symbol commonly used to represent it in a circuit diagram. This is shown in Fig. 8a. A wave incident on line 1 will exit through line 2 and will not be coupled to line 3. Similarly, a wave incident on line 2 will exit through line 3 and will not be coupled to line 1.

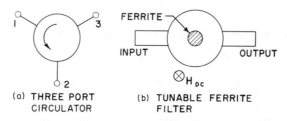

(a) THREE PORT CIRCULATOR   (b) TUNABLE FERRITE FILTER

FIG. 8. Ferrite devices.

The 3-port circulator of Fig. 8a can easily be made into a 2-port isolator by simply terminating port 3 in a matched load. Thus with port 1 as the input and port 2 the output, any reflected power back into port 2 will be terminated in port 3 and port 1 will remain isolated from the output load match.

To provide radio-frequency selectivity in receivers, tunable filters are required in some microwave systems. These networks may be mechanically tuned or they may utilize the resonance properties of ferrite materials which depend on an applied magnetic field. Thus, as shown in Fig. 8b, if a properly located piece of ferrite is placed in a resonant cavity, the application of an adjustable dc magnetic field to the ferrite will cause a shift of the resonant frequency of the cavity.[6] Thus, since the effective Q of the ferromagnetic resonance of the material may be quite high, an electronically tunable, highly selective filter can be realized.

10.2.6. Antennas

Because the size-to-wavelength ratio is near unity, one expects that radiation is relatively efficient from microwave apparatus of conventional transmission-line size. In fact, any crack or aperture in a microwave setup is apt to be a distressingly efficient radiator.

For intentional radiation, dipoles and dipole arrays as used at low frequencies are common, as are mirrors and lenses as used in the optical region. A characteristically microwave type of apparatus is the horn radiator, which consists simply of a waveguide flared out as in an acoustic horn. Another is a dielectric rod radiator which consists of a terminated dielectric cylindrical waveguide without external walls. Because of the termination, there will be radiated waves originating near the end of the structure. The parabolic mirror or reflector is a very common means of concentrating microwave energy in a given direction, and lenses are often used in place of the mirror. In addition to these general radiator types, electronically scanned antenna arrays have recently become an important addition to the microwave field since the beams of high-gain or large aperture antenna arrays can be shifted or rotated (nonmechanically) over relatively large conical angles in space through the use of either progressive phase shift of each radiating element or change in frequency to the array feed.

The subject of antennas, of course, is a very large one and we must refer to more extensive treatments for further details.[7-9]

[6] C. E. Nelson, *Proc. IEEE* **44**, 1449 (1956).

[7] S. Silver, "Microwave Antenna Theory and Design," Massachusetts Inst. Technol. Radiat. Lab. Ser., Vol. 12. McGraw-Hill, New York, 1949.

[8] H. Jasik, "Antenna Engineering Handbook." McGraw-Hill, New York, 1961.

[9] R. C. Hansen, "Microwave Scanning Antennas," Vols. 1–3. Academic Press, New York, 1966.

10.3. Microwave Power Sources*

10.3.1. Microwave Tubes

10.3.1.1. Grid-Control Tubes at Microwave Frequencies. There exists a series of specially designed grid-control tubes which, together with specially designed circuits, can be used at frequencies as high as 4 GHz as amplifiers or oscillators. The basic difficulty in designing grid-control tubes, i.e., triodes or tetrodes, for microwave frequencies has to do with the very close spacings required between electrodes in order to keep the transit times down to less than a cycle in the various interaction regions. In addition, there are other requirements imposed on the design to decrease lead inductance and other deficiencies of standard triodes. Such a typical grid-control tube used at microwave frequencies is usually similar to the schematic diagram shown in Fig. 8, Section 3.5.2.

The cathode, grid, and anode planes are very closely spaced, are usually in the form of disks, and the construction is such that the edges of the disks extend outside the vacuum envelope, as shown in the figure. Tubes of this kind can be used as oscillators or amplifiers, and can deliver reasonable amounts of continuous wave power, and powers of the order of kilowatts in pulsed operation. These tubes are characterized by low gain and are, therefore, not as suitable as amplifiers as klystrons. At some frequencies, however, say 2500 MHz or less, commercial availability makes it more convenient to get powers of the orders of tens of watts from these kinds of tubes than from klystrons. Also, since the cavities are not an integral part of the tube, it is possible to use the same tube over a wide range of frequencies by the use of different cavities or by plungers in one cavity.

The most common tube of this kind, 2C39, operates at frequencies as high as 3 GHz. Typical performance is of the order of 10 W at 2500 MHz and 25 W at 500 MHz with gains, respectively, of 5 and 10. The same tube can deliver 4 kW of pulsed power at 2500 MHz. Because this tube does use external circuits which can be fabricated without great

---

* Chapter 10.3 is by Marvin Chodorow and Ferdo Ivanek.

264

difficulty by the user, it is possible to use it over a range of frequencies where there may not be other types of tubes available.

10.3.1.2. Klystrons.

10.3.1.2.1. GENERAL CHARACTERISTICS. Probably the most commonly used type of microwave tube is the klystron. The simplest form of this is the two-cavity amplifier, and the description of its operation will serve to describe the general principles involved in all types of klystrons.[1-3] The two-cavity amplifier shown schematically in Fig. 1 consists of a unipotential cathode, two resonant cavities separated by a drift space, and a collector. Electrons are injected from the cathode into the first cavity in which there is an rf field. As electrons pass through the

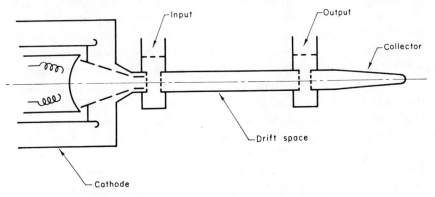

FIG. 1. Schematic of two-cavity klystron amplifier.

narrow gap of this cavity they will be either accelerated or decelerated depending on the phase of the field. The electrons which left the cavity during the decelerating portion of the cycle will tend to fall back towards the electron (denoted as O) which passed through earlier with zero acceleration, whereas electrons in the portion of cycle which left later will tend to overtake electron O.

The result is that, as the electrons drift toward the second cavity, there is a grouping with an increase in the density of electrons in the

[1] E. L. Ginzton, "Microwave Measurements," pp. 3–33. McGraw-Hill, New York, 1957.

[2] R. R. Warnecke, M. Chodorow, P. R. Guenard, and E. L. Ginzton, *Advan. Electron.* **3**, 43 (1951).

[3] A. H. W. Beck, "Velocity-Modulated Thermionic Tubes." Cambridge Univ. Press, London and New York, 1948.

neighborhood of electron O and a decrease in the neighborhood of the electron O' which left the cavity a half-cycle earlier or later than O. If an rf field exists in the second cavity, so phased that it exerts a decelerating force at the time that electron O is passing through and an accelerating force at a time when O' is passing through, there will be more electrons losing energy than gaining energy because of the greater electron density around O, and there is thus a net transfer of power to the output cavity. If the distances, current, and cavity parameters are chosen properly, one will get gain.

This idea can be extended to more than two cavities. In such a case, the power transferred to the second cavity by the beam results not in delivery of power to a useful load but merely in a much greater voltage occurring across the gap of the second cavity than that across the gap of the first. This voltage produced by the bunched beam arriving there will in turn remodulate the beam, thereby producing even greater variation of density at the third cavity, and this process can be continued. The net effect is to give appreciably greater gain than in a two-cavity device.

The parameters describing the cavities are the resonant frequency; the shunt impedance R, which is defined by $R = V^2/2P$ where V is the peak gap voltage and P the power dissipated in the cavity, in complete agreement with a similar definition for lumped constant circuits; and the quality factor Q, which is defined as $\omega W/P$ where ω is (angular) frequency and W is energy stored in the cavity. The electronic properties of this device are usually measured by the beam impedance, which is the ratio of the beam voltage to the beam current.

10.3.1.2.2. AMPLIFIERS. Klystron amplifiers have been built for a variety of powers and frequencies. They are most useful for cases where one requires extremely high peak power, high average powers, or high gains. At frequencies from about 200 MHz up to 10 GHz, peak powers up to 2 to 10 MW have been obtained; and for a special application, tubes of this kind have been built which deliver peak powers of 30 MW at 3 GHz. Average powers have been obtained in commercially available tubes ranging from 200–300 kW in ranges up to 9 GHz. Typical gain characteristics are 10 dB for a two-cavity tube, 30 dB for a three-cavity tube, and 50 to 60 dB for a four-cavity tube. The efficiency of these tubes varies from about 20 to over 50%. Since the cavities have fairly high Q's, these amplifiers have relatively narrow bandwidths, 0.5% being typical, although it has been possible with multicavity tubes and stagger tuning to get as much as 8–10%.

It should be apparent from the description of these amplifiers that one can convert them into oscillators by feeding a suitable amount of the output power back to the input cavity, with appropriate phase and amplitude control. Changes in phase shift will cause changes in oscillation frequency. Similarly, if one changes the beam voltage slightly, causing a change in the transit angle between cavities, the frequency of oscillation will change. The bandwidth over which this kind of oscillation occurs will, however, be limited to the order of 0.5%, since if the frequency is too far from the resonant frequency, the phase angle between the driving current in the electron beam and the voltage produced in the off-resonant circuit (the cosine of this angle is essentially the power factor) becomes too large to deliver sufficient power.

As amplifiers, the tubes have complete frequency stability since the output is extremely well-isolated from the input, and there is thus no tendency of the output load to change the frequency of the amplifier. If the acceleration voltage on the electron beam in an amplifier is changed, there will be a phase shift in the output, and if this acceleration voltage is changed periodically, as by power supply ripple, this will produce phase modulation of the output.

10.3.1.2.3. REFLEX OSCILLATORS. Probably the most common type of klystron used is the reflex klystron.[4,5] This consists of only a single cavity, and the electrons pass through this cavity twice. After passing through the first time and being velocity modulated, the electrons pass into a region where there is a retarding field produced by an electrode which is negative with respect to the resonator. The electrons which have been accelerated in passing through the cavity will penetrate further into the retarding region and, therefore, take longer to return to the cavity than electrons which have been decelerated in passing through the rf gap. The slower electrons overtake the fast ones and the net effect is that one has a nonuniform beam returning to the cavity. If the potential of the reflector is properly chosen, the regions of increased density in the returning beam pass through the cavity at a phase such that the rf field extracts energy from them and there is a net transfer of energy to the cavity.

Analysis shows that the optimum condition for such an energy transfer is for the transit time of the electrons in the retarding field to be $n + \frac{3}{4}$

[4] J. R. Pierce and W. G. Shepherd, *Bell Syst. Tech. J.* **26**, 460 (1947).
[5] Ginzton,[1] pp. 33–53.

cycles, where n is an integer. If, by changing the voltage of the reflecting electrode, this condition is violated slightly, so that the electrons arrive slightly earlier or later than this optimum, one has a resonant circuit being driven slightly out-of-phase by an alternating current, and the tube will oscillate at a frequency different from the resonant frequency of the cavity. This then is a method of electronic tuning, since by changing the voltage on the reflector electrode one can change the frequency. Typical electronic bandwidths are $\frac{1}{2}$ to 1% of the operating frequency of the klystron. The reflector electrode can be easily modulated since it draws no current.

Reflex klystrons of the kind described here are used mainly for local oscillators in receivers and as signal sources for bench measurements. The electronic tuning feature is particularly valuable for these purposes. The power from such tubes is usually small, 10 W being about the maximum available in commercial tubes, with more common values being of the order of ten to several hundred milliwatts. Commercial tubes exist for frequencies from 1 to 60 GHz. The efficiency of these tubes is relatively small, usually of the order of 1 or 2%, although in rare cases 6 or 7% has been obtained. For any type of work which simply requires some microwave power, these are the most convenient of all microwave tubes.

While the electronic tuning is usually an advantage, special voltage control methods have to be used in cases where extreme stability of frequency is required. These methods involve monitoring the output frequency and using this information in some sort of feedback loop to adjust the reflector voltage to maintain a constant frequency.[6] The methods devised are sufficiently good so that one can get very good stability for most purposes.

As has been indicated, it is possible to frequency modulate these tubes by modulating the reflector electrode. Changes in the frequency, however, are always accompanied by changes in the amplitude of the signal, the amount of the amplitude modulation depending on how much frequency deviation is imposed. Figure 2 shows the typical qualitative dependence of the output power and frequency on variations in the reflector voltage. The ease of modulation of the reflector electrode has made the reflex klystron very useful in many applications where the frequency has to be altered.

---

[6] C. G. Montgomery (ed.), "Techniques of Microwave Measurements," Massachusetts Inst. Technol. Radiat. Lab. Ser., Vol. 11. p. 58. McGraw-Hill, New York, 1947.

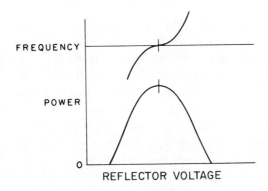

FIG. 2. Power and frequency variation of reflex klystron as a function of reflector voltage (linear scales).

10.3.1.3. Traveling-Wave Tubes.

10.3.1.3.1. GENERAL CHARACTERISTICS. The traveling-wave tube is a device in which an electromagnetic wave travels along a suitably designed circuit at a velocity much less than the velocity of light so that the fields of the wave can interact with an electron beam. Though other types of propagating circuits have been used, the helix is the most common one, and we shall confine our discussion to it, although with some obvious changes the statements made here can be applied to other types of propagating circuits.[7-10]

An electromagnetic wave is launched on the helix. The wave travels along the helical wire at approximately the velocity of light, which means that the axial velocity is considerably less, the exact value depending on the pitch of the helix and its diameter. A wave of this kind will have an axial electric field along the center of the helix, and an electron beam injected along the same axis of symmetry and having approximately the same velocity, as the wave will interact with the fields of the wave. At certain phase positions the electrons will be accelerated and at others decelerated, and one can see from Fig. 3 that there will be a tendency for the electrons to group around positions of zero field. If the electrons have an average velocity slightly greater than the average velocity of the wave, then the group which has been formed by the velocity modulation

[7] J. R. Pierce and L. M. Field, *Proc. IRE* **35**, 108 (1947).

[8] R. Kompfner, *Proc. IRE* **35**, 124 (1947).

[9] H. J. Reich, P. F. Ordung, H. L. Krauss, and J. G. Skalnik, "Microwave Theory and Techniques," Chapter 15. Van Nostrand, Princeton, New Jersey, 1953.

[10] J. R. Pierce, "Traveling-wave Tubes." Van Nostrand, Princeton, New Jersey, 1950.

will tend to move forward into a decelerating part of the wave and lose energy which is transferred to the circuit. The net effect is that the wave grows exponentially along the circuit, and the device acts as an amplifier. The power is fed into and out of the circuit by suitable transducers. Traveling-wave tubes with power output ranging from several tens of milliwatts up to several kilowatts exist over a range of frequencies from somewhat higher than 50 GHz down to about 400 MHz. There also exist pulsed tubes delivering from 50 kW to several megawatts. These use propagating circuits other than a helix.

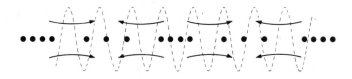

FIG. 3. Schematic of field distribution and electron velocity distributions in a traveling-wave tube. The dashed line indicates the helix along which the electromagnetic wave travels. The arrows show the direction of the electric field near the axis, leading to a bunching of the electrons.

10.3.1.3.2. APPLICATIONS. The most interesting property of this type of tube arises from the fact that the propagating circuits used are nonresonant. Therefore, such an amplifier will operate with constant gain over a large bandwidth, typical values being as high as 2 to 1 in the frequency range (for example, from 2 to 4 GHz) for a helix, and perhaps 10% for other circuits. The gains available are of the order of 30 to 40 dB. Greater gains than this are possible, in principle, but are not actually designed into most tubes because of a tendency to oscillate. It will be apparent that a slight reflection at the output and another reflection at the input can lead to an oscillation if the product of these reflections times the total gain is greater than unity. As a consequence, it is necessary to put some attenuation along the middle of the tube and to hold the gain at values such that this oscillation is not apt to occur. Traveling-wave tubes offer very interesting possibilities as wideband amplifiers, particularly for short pulses, since bandwidths of the order of several thousand megahertz are not at all unreasonable.

Other interesting possible applications arise from the noise figures which it has been possible to achieve with these amplifiers. At 3 GHz commercial tubes exist with noise figures less than 6 dB with comparable performance at 6 GHz and perhaps 10 dB at 11 GHz. With tubes of this kind, extremely weak signals can be amplified without the addition of undesirable tube noise.

10.3.1.3.3. WIDEBAND TRAVELING-WAVE OSCILLATOR (BACKWARD-WAVE OSCILLATOR).[11-14] Another interesting device which is related to the traveling-wave tube results from the use of certain types of propagating circuits in which the phase and group velocities are in opposite directions. In such a circuit an electron beam will interact with the wave in the same way as described above, namely that there will be velocity modulation with some portions of the beam being accelerated and some decelerated, depending on the phase positions. This will result in grouping and nonuniform density in the beam just as in the traveling-wave tube. However, in this case the energy delivered by the beam to the wave does not travel in the forward direction with the beam but travels back toward the cathode end of the tube.

One can describe the process qualitatively by saying that the beam adds increments of energy to the circuit as it moves along. All of these add up in-phase for the energy traveling toward the cathode end. In this case the electromagnetic wave does not grow exponentially, but there is a definite growth of the wave in the backward direction. Because of this internal feedback mechanism, such a device acts as an oscillator, with no reflections at the ends of the circuit being required.

For oscillation, however, one requires synchronism between the electron beam velocity and the phase velocity of the wave. As one changes the accelerating voltage, the frequency of oscillation changes as determined by the dispersive characteristics of the propagating circuit, so that at each voltage the frequency of oscillation is the one for which the phase velocity is the same as the electron velocity. This device has been called the backward-wave oscillator. Its principal characteristic is that it can be used to produce large frequency changes by changing the operating voltage; no mechanical tuning or circuit tuning is required. Frequency ranges of the order of 2 to 1 are not at all untypical. These usually require voltage changes of the order of 4 to 1. Some tubes have been built which tune continuously from about 3 to 10 GHz. The power output of such tubes as have been built is usually in the range of milliwatts to watts.

[11] R. Kompfner, *Bell Lab. Record* **31**, 281 (1953).
[12] R. Kompfner and N. T. Williams, *Proc. IRE* **41**, 1602 (1953).
[13] H. Heffner, *Electronics* **26**, No. 10, 135 (1953).
[14] H. Heffner, *Proc. IRE* **42**, 930 (1954).

10.3.1.4. Magnetrons.

10.3.1.4.1. GENERAL CHARACTERISTICS. A magnetron[15] is a device in which the electrons move perpendicularly to both a static electric and magnetic field which are mutually orthogonal. In crossed fields of this kind the average velocity of the electron perpendicular to the two fields is proportional to the ratio E/B. The most common type of magnetron consists of a cylindrical cathode which is concentric with a hollow cylindrical anode, and the electrons move in a circular or cycloidal path around the cathode. The electrostatic field is applied between the two electrodes and the magnetic field is parallel to the axis of the cylinders.

The anode of such devices consists of a propagating circuit similar in behavior to those described in Section 10.3.1.3.1 (not a helix, however) except that the circuit is arranged in a circle so that it closes on itself. Because of this, it is possible to have an electromagnetic wave traveling in either direction around such a closed propagating circuit, and the electron cloud which travels around the cathode in a particular direction, as determined by the polarity of the magnetic field, interacts with a rotating electromagnetic wave. The voltages and magnetic field are so adjusted that the electron cloud moves at roughly the same velocity as the rotating wave.

In this device, again, one gets an interaction similar to that in a traveling-wave tube. However, the dynamics of electrons in a crossed field are such that as the electrons lose energy to the rf field, they tend to drift toward the anode and thereby pick up additional energy from the dc field present. As a result, the electrons, starting from the cathode, continually transfer energy from the dc field to the alternating fields. Because of the closed structure of the propagating circuit these devices are oscillators with built-in feedback. They are extremely efficient, and as pulsed sources can produce peak powers up to several megawatts; continuous-wave (cw) versions, which are less commonly used, can produce powers of the order of one or two kilowatts at 3 GHz, considerably less at high frequencies. Pulsed magnetrons which are run at peak powers of one, two, or more megawatts have efficiencies of the order of 50% at frequencies from about 1 to 30 GHz.

10.3.1.4.2. CARCINOTRONS AND VOLTAGE-TUNED MAGNETRONS. There exist two varieties of devices related to the magnetron which are oscillators

---

[15] G. B. Collins, "Magnetrons." Massachusetts Inst. Technol. Radiat. Lab. Ser., Vol. 6. McGraw-Hill, New York, 1948.

and in which the frequency can be tuned by voltage change. In one kind, called a carcinotron, the circuit is not a closed loop, there is no recirculating electron beam, and instead, an electron beam is injected into the crossed field, passes along the circuit, and frequency is controlled by means of the applied voltage. This is closely related to the backward-wave oscillator. The other device, called the voltage-tuned magnetron, does have a rotating re-entrant beam as in a conventional magnetron, but by suitable design of the circuit, the applied voltage rather than the circuit resonance controls the oscillation frequency.

10.3.2. Solid-State Microwave Power Sources

While the above described microwave tubes, with the exception of the triode, are being designed and built as integral units incorporating their oscillator or amplifier circuit, microwave solid-state sources use active devices which are, in general, independently designed and separately manufactured, and can function in a variety of circuits. The solid-state devices for microwave power generation can be classified according to their principle of operation, as follows:

1. Microwave devices which have a counterpart at submicrowave frequencies where the development actually started. They include the transistor, the tunnel diode, and the varactor diodes. These have been covered in Part 2.

2. Devices whose principle of operation can be practically exploited only at frequencies in the microwave range, due to the characteristics of the readily available semiconductor materials and the realizability of device configurations. These comprise two new device categories most commonly known as avalanche diodes and Gunn-effect devices, which will be treated in the following sections.[16,17]

Table I gives an overview of the oscillator and amplifier applications of the aforementioned active microwave semiconductor devices. As can be seen, the avalanche diodes and Gunn-effect devices are assuming

[16] J. E. Carroll, "Hot Electron Microwave Generators." Amer. Elsevier, New York, 1970.

[17] Review articles on the physics and state of the art of avalanche diodes and Gunn-effect devices are contained in the following two special issues of IEEE periodicals: Special issue on microwave circuit aspects of avalanche-diode and transferred electron devices. *IEEE Trans.* **MTT-18**, No. 11 (1970); Special issue on microwave semiconductors. *Proc. IEEE* **59**, No. 8 (1971).

TABLE I. Summary of the Oscillator and Amplifier Applications of the Active Microwave Semiconductor Devices

Function	Type of circuit / Description	Microwave semiconductor device				
		Transistor	Tunnel diode	Avalanche diode	Gunn-effect diode	Varactor
Oscillator	Fundamental frequency oscillator	Major applications below ~5 GHz	Minor applications	Emerging major applications over entire microwave spectrum		
Oscillator	Multiplier source (Fundamental frequency oscillator followed by harmonic multiplier)	Major applications as fundamental oscillator. Some multiplier applications			Minor applications	Major applications as harmonic multiplier
Amplifier (Single-Port)	Injection-locked oscillator	Minor applications	Minor applications		Emerging major applications	
Amplifier (Single-Port)	Negative-impedance amplifier		Major applications for small-signal amplification		Emerging applications for large-signal amplification	
Amplifier (Single-Port)	Parametric amplifier					Major applications for small-signal amplification
Amplifier	Two-port	Major applications below ~5 GHz for small-signal and large-signal amplification	Negligible applications		Minor applications	
	Remarks	Applications at frequencies above ~5 GHz under development	Main disadvantage: low-power capability	Not suitable for small-signal amplification because of excessive noise		Other major applications as frequency up-converters and electronic tuning of oscillators

important roles in microwave power generation. The following are the most significant resulting trends in the solid-state area:

— Fundamental-frequency avalanche and Gunn oscillators are displacing multiplier sources in an increasing number of applications.
— Avalanche diodes and Gunn-effect devices find important applications for large-signal amplification at microwave frequencies where transistors are not available.

The two devices under discussion are, of course, serious competitors to microwave tubes, as well.[18] Avalanche and Gunn oscillators have displaced tubes in virtually all new low- and medium-power applications. Amplifier applications are lagging, especially where higher gain and larger bandwidths are needed. High-power microwave tubes, on the other hand, are still uncontested by commercially available solid-state devices.

10.3.2.1. Phenomenological Device Description.

10.3.2.1.1. AVALANCHE DIODES. The basic oscillation mode of avalanche diodes exploits a combination of the following two processes (Fig. 4):

1. Periodic carrier multiplication by impact ionization in a reverse-biased P-N junction or in a Schottky barrier.
2. Drifting of the carriers through a depletion zone at saturated velocities.

When the device is operated in a suitable circuit, the finite time of the avalanching process and the carrier transit time can result in approximately 180° phase shift between rf voltage and current. The device then delivers to the microwave circuit part of the energy it obtained from the dc bias field. There is a natural "avalanche frequency" at which the dc-to-rf energy conversion efficiency peaks, but operation over bandwidths up to approximately one octave is obtainable by mechanical and/or electronic circuit tuning.

The described form of avalanche diode operation is commonly known as the IMPATT mode (IMPact Avalanche and Transit Time). It has

---

[18] The most complete collection of performance data of both microwave tubes and active microwave solid-state devices is presented in T. S. Saad, "Microwave Engineers' Handbook," Vol. 2, pp. 149–178. Artec House, Dedham, Massachusetts, 1971. The pace of technological developments makes it necessary, however, to supplement the information referred to above with more recent data published in periodicals. *The Microwave Journal, Microwaves,* and *Microwave System News* are among the most informative periodicals for this purpose.

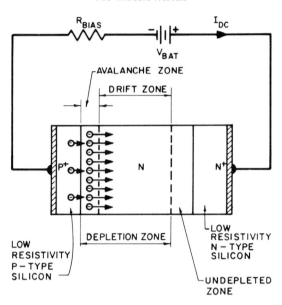

FIG. 4. Schematic representation of reverse-biased *p-n* junction avalanche diode (from application note 935, Hewlett-Packard Company, courtesy of Hewlett-Packard Company). Chip package and microwave circuit not shown (refer to Section 10.3.2.3.1 and Section 10.3.2.3.4).

been realized in germanium, silicon, and gallium arsenide. While important investigations had been carried out with germanium avalanche diodes, the latter two materials showed more promise for practical applications and are used in industrial production of IMPATT diodes. These are either of the *P-N* junction type, P^+NN^+ (Fig. 4), or of the Schottky-barrier type, also with an N^+ substrate.

Another important mode of avalanche diode operation can exist when the large-signal limits of IMPATT operation are substantially exceeded. The TRAPATT mode (TRApped Plasma Avalanche Triggered Transit) is triggered by an avalanche shock front which rapidly sweeps across the diode, filling it with an electron-hole plasma whose space charge produces an abrupt drop of the rf voltage. The essential advantage of the TRAPATT mode is a relatively high efficiency due to switching between a high-voltage low-current state and a low-voltage high-current state. However, the TRAPATT mode imposes more stringent requirements on both the device and the circuit, and is limited to frequencies which are much lower than the "avalanche frequency" because the dissipation of the plasma takes much more time than the avalanche transit. Efficient TRAPATT operation was obtained only in germanium and silicon.

Negative impedance can be produced in avalanche diodes by some other modes of operation as well. The only one of practical interest, so far, is the parametric mode which exploits variable reactance effects. Its operational requirements are also much more complicated than that of the IMPATT mode.

10.3.2.1.2. GUNN-EFFECT DEVICES.[†] The Gunn-effect devices basically differ from all other active solid-state devices in that their power generating mechanism does not exploit a P-N junction, but an intrinsic material property, namely, the existence of more than one minimum in the conduction band of some compound semiconductors, like N-type GaAs. For a sufficiently high (threshold) electric field applied across a uniformly doped chip of such material, electrons from the lowest conduction band

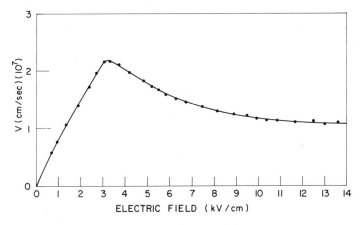

FIG. 5. Velocity-field characteristic in GaAs (from J. G. Ruch and G. S. Kino, *Phys. Rev.* **174**, 921 (1968); courtesy of *Physical Review*).

"valley" are transferred into the next higher one where they have a higher effective mass and lower mobility. The average electron drift velocity then varies as a function of the applied electric field as shown in Fig. 5. A variety of oscillation and amplification modes can be sustained above "threshold," in the region of negative differential mobility indicated by the negative slope of the velocity-field characteristic (Fig. 5). The properties of the various modes depend on the material doping, device configuration, circuit parameters and applied bias field. Other materials,

---

[†] Alternative names: transferred electron devices, bulk semiconductor devices.

like InP, CdTe, and GaInSb have qualitatively the same velocity-field characteristic and also exhibit the Gunn effect, but GaAs is, so far, the only material of technological importance in this field.

The most commonly used method of application of the Gunn effect involves so-called domain modes. In a semiconductor, an internal fluctuation of current density is accompanied by a fluctuation of electric field and the effect of the electrical field is to reduce the current fluctuation essentially by dielectric relaxation. With a negative differential mobility, a fluctuation of this kind grows with time and distance, and with the proper preparation of the device, this growth stabilizes into the form of a propagating domain. This domain forms very close to the cathode and travels across the whole device. It is a dipole layer which consists of a charge depletion region followed by a very high charge accumulation region with the whole domain being electrically neutral. Depending on the circuitry used, material constants, the length of the sample, etc., this domain can propagate entirely across the whole sample, in which case the operating frequency is determined by the transit time of the domain.

Under other conditions, the voltage (dc plus rf) falls sufficiently low during the transit that the total voltage is no longer in the negative differential mobility region and the domain is quenched or it may have a delayed nucleation. In these conditions, the operating mode is determined by the material properties plus the circuit properties, and the device can be operated over a much greater bandwidth by tuning the circuit. It is also possible under conditions of sufficiently low doping not to have domains formed, and the device then presents a stabilized negative resistance which peaks at the transit time frequency. It can produce either amplification or oscillation, depending upon the circuit parameters.

There is also an oscillation mode in which domains do not form, although the material doping is sufficiently high. By a proper and rather critical choice of circuit parameters, bias conditions, and material properties, one can obtain the LSA (Limited Space-charge Accumulation) mode in which the formation of domains is prevented. Instead, one has for a short portion of the cycle a spatially uniform electric field, in the negative differential mobility range, across the device. The LSA mode is particularly interesting for high-power, high-frequency operation.

The LSA device is unique in the area of microwave power generation in that it involves no transit time effect. A "hybrid" oscillation mode representing a transition between the LSA and the domain modes has also been identified.

10.3.2.2. Power Generation Capabilities of Avalanche Diodes and Gunn-Effect Devices.[17,18] As of late 1971, commercially significant production was established only for avalanche diodes operating in the IMPATT mode and for Gunn-effect devices operating in one of the domain modes. These devices were designed for low- and medium-power cw applications. Best performance characteristics were achieved in the 5–15 GHz frequency range where commercially available Gunn-effect devices can reliably deliver cw power outputs up to approximately 0.5 W with typical efficiencies in the 3–5% range, while IMPATT diodes are capable of generating output powers of the order of 1 W with dc-to-rf efficiencies of the order of 10%. Roughly twice as much power and efficiency can be obtained from GaAs IMPATT devices than from Si IMPATT devices. Typical efficiency figures at, or near, optimum operating conditions are in the 4–7% range for Si and in the 8–12% range for GaAs.

Considerably higher efficiencies are expected from cw-operated Si TRAPATT avalanche diodes. So far, these have been experimentally tested only at frequencies below 5 GHz (2.2 W with 24% efficiency at 4.3 GHz). Comparable efficiencies are also anticipated in the 5–15 GHz frequency range but the cw power capability of TRAPATT diodes is not expected to exceed that of IMPATT diodes, due to heat dissipation limitations.

Both IMPATT avalanche diodes and Gunn-effect devices are being manufactured in a variety of designs which provide continuous coverage of the 4–40 GHz frequency range. The power capability declines with increasing frequency. At 40 GHz, for example, commercially available GaAs IMPATT diodes produce up to 100 mW and Gunn-effect devices up to 50 mW of cw microwave power with up to half the efficiency obtainable at the lower microwave frequencies. However, a modified version of the IMPATT diode promises to substantially boost the power capability at higher microwave frequencies.

The "double drift" avalanche device, consisting essentially of two complementary Si IMPATT diodes in series, demonstrated the capability of doubling both the power and efficiency as compared to conventional IMPATT diodes. This is due to the use of both types of carriers from the same avalanche source; in the conventional "single drift" diode only the electrons are utilized in the power generating process. Experimental "double drift" Si IMPATT diodes have produced 1 W of cw power with 14% efficiency at 50 GHz and 0.18 W with 7% efficiency at 92 GHz.

Power outputs one or several orders of magnitude higher than the above quoted figures are obtainable from pulse operated avalanche and Gunn-effect devices. The best performance characteristics were obtained in the TRAPATT avalanche and LSA Gunn-effect modes whose commercial availability is lagging, especially that of the TRAPATT diodes. Readily repeatable power capabilities from LSA diodes are 500 W peak, 0.5 W average in the 1–3 GHz frequency range, 200 W peak, 0.2 W average at 5 GHz, and 100 W peak, 0.1 W average at 10 GHz. The dc-to-rf conversion efficiencies are typically in the 5–10% range. Comparative figures for TRAPATT diodes are 200 W peak, 1 W average power with 30% efficiency at 1 GHz and 100 W peak, 1 W average with 25% efficiency at 2 GHz. Both TRAPATT and LSA devices have been experimentally operated at higher power levels and efficiencies, as well as outside of the aforementioned frequency ranges.

Substantial improvements of all performance characteristics and frequency coverage are to be expected in the near future.

10.3.2.3. Application Considerations and Techniques. The following are some of the prevalent considerations and techniques used in the various commercial and laboratory applications of avalanche diodes and Gunn-effect devices.

10.3.2.3.1. PACKAGING OF ACTIVE DEVICE. Commercially available avalanche diodes and Gunn-effect devices come in ceramic packages most of which were developed for varactor applications. In some microstrip and millimeter-wave waveguide applications the actice device is used in chip form. Elimination of package parasitics generally results in better oscillator or amplifier performance, but the need for adequate chip protection can pose serious technological problems.

10.3.2.3.2. Dc BIAS. Gunn-effect devices may be chosen in some cases in preference to avalanche diodes solely because of their substantially lower dc bias voltage. Representative figures for low- and medium-power operation around 10 GHz, for example, are 10 V for Gunn-effect devices, 95 V for Si IMPATT diodes, and 60 V for GaAs IMPATT diodes. For these transit time devices, the dc bias voltage scales, to a first approximation, in inverse proportion to the operating frequency.

10.3.2.3.3. HEAT SINKING. Operation at all but the very low-power levels requires effective heat dissipation. A satisfactory solution for most applications consists of mounting the packaged device into a copper holder or block which is, in turn, mounted in or attached to a larger aluminum heat sink plate. Where necessary, heat dissipation can be enhanced by additional arrangements, like forced-air cooling.

10.3.2.3.4. CIRCUITS. Coaxial, waveguide, or microstrip circuits—or combinations thereof—are used depending on the specific requirements and the operating frequencies. In coaxial circuits the active device is usually mounted in series with the center conductor (analogously in microstrip circuits), whereas the most common method used in waveguide circuits consists of mounting the device under a post inserted across the narrow side of the rectangular waveguide. In general, coaxial and microstrip circuits are the natural choice for applications requiring low-Q values, whereas waveguide circuits are more suitable for high-Q applications. In terms of the operating frequency, the use of coaxial and microstrip circuits predominates in the lower microwave range. Waveguide circuits are almost exclusively used in the higher microwave range. The intermediate range, approximately between 5 and 15 GHz, offers the widest selection among alternative circuit configurations.

10.3.2.3.5. POWER COMBINING. Some applications of avalanche diodes and Gunn-effect devices may depend on overcoming the power limitations of available devices. This can be done in hybrid, periodic, or other composite circuits which combine the output powers of two or more devices. Capability to produce microwave powers up to an order of magnitude above the output power of a single packaged device has been demonstrated.

10.3.2.3.6. INJECTION LOCKING. While this technique has been known for some times and had previously been used with some microwave tubes, it represents a particularly valuable tool for large-signal amplification with avalanche diodes and Gunn-effect devices. It provides, in many cases, higher power gain per stage than negative-impedance amplifiers.

10.3.2.3.7. NOISE. Due to their respective electronic mechanisms, avalanche diodes are much noisier than Gunn diodes. Noise figures range typically around 40–50 dB for the former and around 20 dB for the latter. Thus, neither device qualifies for low-noise applications. However, for large signal amplification, one or the other device, or a combination of both, used in an appropriate circuit (negative-impedance amplifier or injection-locked oscillator) can satisfy all but the most demanding, special applications. Passive and active noise reduction techniques are available in the form of high-Q stabilization cavities and injection locking by low-noise, low-power sources.

10.3.2.3.8. FREQUENCY STABILIZATION. Proper choice of oscillator cavity materials can enhance the frequency stability of simple free-running oscillators by approximately an order of magnitude. High-Q external stabilizing cavities can bring about an additional order of magnitude of improvement. For most stringent frequency stability requirements, it becomes necessary to use locking to crystal-controlled, low-power multiplier sources or other highly stable signals.

10.3.2.3.9. MODULATION. A great variety of amplitude and frequency modulation schemes have been applied to cw oscillators using avalanche diodes or Gunn-effect devices, and several pulse-modulation schemes were realized, as well. Most of these were derived from modulation techniques previously used with other microwave semiconductor devices or microwave tubes. Some have already found commercial applications, for example, in amplitude- or frequency-modulated low-power communication transmitters, in pulsed radar transmitters, and in swept-frequency oscillators.

10.4. Detectors and Receivers for Microwaves*

A detector is essentially a nonlinear element which enables one to translate energy from a high-frequency region to a low-frequency, or direct current, region. Of the several kinds of nonlinear elements which have been tried in the microwave portion of the spectrum, many have received intensive development.

10.4.1. Crystal Detectors

The crystal detector is a semiconductor diode constructed in such a manner that the capacitance (C) shunting the nonlinear resistance (R) element is extremely small (see Fig. 1b). The resistance r is the spreading resistance in the semiconductor resulting from the constriction of current-

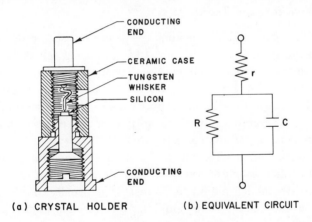

(a) CRYSTAL HOLDER   (b) EQUIVALENT CIRCUIT

FIG. 1. Microwave crystal rectifier.

flow lines in the semiconductor near the contact. These diodes are constructed by touching a very fine wire to the surface of a suitably prepared silicon or germanium crystal as shown in Fig. 1a.

A diode can be used in a number of different ways. The simplest way is to use it as a rectifier of the high-frequency energy. In the low-current

---

* Chapters 10.4 and 10.5 are by **Robert D. Wanselow.**

region, such a device acts as an approximately square law element. Thus upon applying an alternating current to the rectifier, one finds flowing in the output a direct current which is proportional to the square of the amplitude of the input sinusoid.

The crystal must be mounted in the transmission line in such a manner that it approximately matches the characteristic impedance of the line in order that the device may be used over a broad band of frequencies without retuning.

In attempting to detect very low powers, the noise introduced in the crystal element itself becomes of importance since, if the noise introduced in the crystal is greater than the incident signal that one is attempting to detect, it may be very difficult to pick out the required signal from the background. Used as simply a rectifying detector, germanium and silicon crystals are rather poor from the point of view of their noise performance; crystals are used in this manner in certain types of microwave apparatus simply as indicators, for instance, as an indicator in a wavemeter or in a standing wave instrument. For other uses see Section 10.4.3.

10.4.2. Bolometers and Thermistors

Both bolometers and thermistors are power measuring devices; they are discussed more fully in Section 10.5.2.2. Being relatively sensitive, however, they are often used simply as detectors of energy or for relative measurements. They are used in this form sometimes as substitutes for crystals. They have the advantage over crystals that their output indication is quite accurately proportional to power, and ratios of power over large ranges may be accurately computed.

Both crystals and bolometers can be used with amplitude-modulated sources. In this use one measures not the dc component of the output, but the low-frequency modulated signal. It is usually more convenient to use an ac-coupled electronic amplifier for detection than it is to use a dc-coupled amplifier or a sensitive galvanometer.

10.4.3. Frequency Conversion

The poor noise figure of crystals used as direct detectors can be obviated by making use of the superheterodyne principle. A local oscillator is mixed with the incoming signal in a nonlinear crystal element such that the difference frequency is some place in the megacycle region. Commonly, the beat frequency or the intermediate frequency is of the order of 1 to 100 MHz. In this frequency band the crystals produce very much less noise than in the low-frequency region. As a consequence, the minimum signal which can be detected is very much smaller.

It is also possible to produce sum and difference frequencies, as are required in the superheterodyne principle, by making use of nonlinear reactances as well as by using nonlinear resistances as discussed above. An ordinary microwave crystal biased in the back direction so that very little current flows will form a nonlinear condenser. The microwave electric field applied across the crystal produces a variation in capacitance which is proportional to field strength. Thus the device is nonlinear and will produce mixing or the generation of harmonics and sum and difference frequencies. It can be shown that such a device can not only produce mixing with very low noise, but in addition can produce gain.[1,2] These so-called parametric amplifiers produce very little noise; a noise figure of 1 to 2 dB is achievable with relatively little complication. By cooling to liquid nitrogen or liquid helium temperatures, even lower noise input circuits can be constructed. Low noise is inherent in these devices since one is making use of a nonlinear reactance which produces no thermal noise of its own.

10.4.4. Duplexers and Diplexers

It is sometimes convienent in microwave apparatus to be able to transmit and receive on the same antenna circuit or via the same transmission line at least. In order that the transmitter and receiver may be connected simultaneously to the same line, one must have some variety of duplexer to separate the transmitted and received signals. In a pulse system, where the transmission and reception are at different times, one can make use of so-called TR and ATR tubes. These are gas discharge tubes which will be unaffected by the received low-power signal (see also Chapter 4.6). Connection of these devices in a transmission line to separate pulsed transmission and reception is indicated in Fig. 2a. On transmission, both tubes fire. The ATR tube is mounted in a stub line, one-quarter wavelength long, in shunt with the main transmission line. Upon firing, this one-quarter wavelength line presents an open circuit at the shunt point.

The TR tube is similarly mounted a quarter-wavelength away from

[1] L. A. Blackwell and K. L. Kotzebue, "Semiconductor-Diode Parametric Amplifiers." Prentice-Hall, Englewood Cliffs, New Jersey, 1961.

[2] K. K. N. Chang, "Parametric and Tunnel Diodes." Prentice-Hall, Englewood Cliffs, New Jersey, 1964.

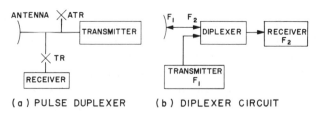

FIG. 2. Transmit–receive circuits.

the main line. Upon firing, it shorts and prevents any energy from the transmitter from entering the receiver. It also presents to the main transmission line an open circuit in shunt with the line so that there is no disturbance of the outgoing wave. On reception both tubes are unfired and are open circuits in shunt with the line. An open circuit a quarter-wavelength away from the line at the ATR point becomes a short circuit across the line, whereas an open circuit in the receiver line allows energy to flow freely from the antenna to the receiver.

When it is necessary to receive and transmit simultaneously, some form of bridge circuit is required. The hybrid junction of Fig. 5a of Chapter 10.2, for instance, can be utilized. By placing a matched load on line 3, the antenna on line 2 which is also matched, the receiver on line 4, and the transmitter on line 1, one has a situation where transmitted power through 1 splits equally between the dummy load on 3 and the antenna on 2, but none enters the receiver. On reception, energy from 2 splits equally between lines 3 and 4 with none of it entering line 1. Similarly, the hybrid ring or rat-race circuit shown in Fig. 5b may also be employed for simultaneously transmitting and receiving rf energy.

The ferrite circulator is another device which could obviously be used for such an application. In Fig. 8a of Chapter 10.2 one could attach a transmitter to line 1, an antenna to line 2, and a receiver to line 3. The transmitted energy would proceed from line 1 through the circulator and out 2. Received energy on the antenna would come in 2 and proceed around the circulator and out line 3 to the receiver.

Frequency diplexer circuits operate on the principles of filter networks. That is, if the diplexer network in Fig. 2b is a bandpass filter tuned to the frequency F_2 and F_1 is sufficiently displaced from F_2 that the skirt isolation of the filter adequately rejects F_1 from the receiver, then the transmitter and receiver circuits can simultaneously operate on a cw basis with minimum interaction. This circuit philosophy is fully utilized in microwave repeater-relay communication networks where, in some cases, two antennas are employed rather than the one as shown in Fig. 2b.

10.5. Microwave Measurements[1-3]

The field nature of microwave apparatus argues against the use of voltage and current as they are used in low-frequency electrical measurements. However, one is not restricted as in optics to measuring simply the intensity, that is, the energy. Since voltage and current are, in microwave apparatus, ambiguous concepts, one measures instead the field strength or sometimes simply relative amplitudes as when one measures impedance or the reflection (or scattering) coefficient.

The mathematical characterizations of impedance relations on transmission lines is given in Section 1.4.1.1. On a transmission line which will support a TEM mode, one can write the relation between voltage and current, which on such a line are not ambiguous, as

$$V = IZ = I/Y \qquad (10.5.1)$$

where V is voltage, I is current, Z is impedance, and Y is reciprocal impedance or admittance. The power flow is given by the product of voltage and the complex conjugate of current,

$$P = VI^*. \qquad (10.5.2)$$

In a similar description applicable to modes other than TEM on transmission lines, one takes V not as a unique quantity but simply as a proportionality constant in the field description. Suppose that one is describing, for instance, the dominant mode in a waveguide. The electric and magnetic fields can then be written as follows:

$$\begin{aligned} \mathbf{E}(x, y) &= V\mathbf{e}(x, y) \\ \mathbf{H}(x, y) &= I\mathbf{h}(x, y) \end{aligned} \qquad (10.5.3)$$

where x and y are assumed to be transverse coordinates on the waveguide, with z, which does not appear in Eq. (10.5.3), as the longitudinal coordinate. One assumes that both sides of Eq. (10.5.3) are proportional to $e^{\gamma z}$.

From Poynting's theorem, one has the magnitude of the energy flow as

$$|\mathbf{P}| = |\mathbf{E} \times \mathbf{H}^*| = |(VI^*)\mathbf{e} \times \mathbf{h}| = VI^*. \qquad (10.5.4)$$

[1] E. L. Ginzton, "Microwave Measurements." McGraw-Hill, New York, 1957.

[2] M. Wind and H. Rapaport, "Handbook of Microwave Measurements," Vols. 1–3, 3rd ed. Polytechnic Inst. of Brooklyn, New York, 1963.

[3] A. L. Lance, "Introduction to Microwave Theory and Measurements." McGraw-Hill, New York, 1964.

Equations (10.5.3) and (10.5.4) then define what the quantities **e** and **h** must be. One has characterized by **e** and **h** the variations in fields in the transverse directions, but they are normalized to have a cross product of unity. Characteristic impedance for the mode can then be defined as the ratio V/I.

It is evident that there is still a certain degree of arbitrariness in the definition of V and I. V is sometimes taken as the line integral of the electric field between some two definite points in the waveguide, as for instance the line integral along the line of maximum field strength. This definition, along with those above, would then define I as being some particular current flow along some portion of the waveguide walls. The definitions are highly artificial in any event but they are a very great aid to thought. It is relatively easy to show that impedance as defined above is equal to any other consistently defined impedance to within a constant multiplying factor.[4]

In place of an impedance description for a transmission line, it is sometimes very convenient to make use of a somewhat different, but rather closely related concept, the scattering characterization of the line behavior. One writes

$$(t + r)\mathbf{e}(x, y) = \mathbf{E}(x, y, z)$$
$$(t - r)\mathbf{h}(x, y) = \mathbf{H}(x, y, z)$$

$$(10.5.5)$$

with t and r identified with the transmitted and reflected waves, respectively, that is,

$$t = | t | \, e^{-\alpha z - i\beta z}$$
$$r = | r | \, e^{+\alpha z + i\beta z}$$

$$(10.5.6)$$

with α equal to the attenuation constant and β the phase constant appropriate to the particular mode in the particular line. Using the definitions of Eq. (10.5.6), one has from Poynting's theorem again,

$$| \mathbf{P} | = | \mathbf{E} \times \mathbf{H}^* | = | t |^2 - | r |^2$$

$$(10.5.7)$$

which states that the power flow is equal to that in the transmitted wave minus that in the reflected wave.

---

[4] C. G. Montgomery, R. H. Dicke, and E. M. Purcell, "Principles of Microwave Circuits," Massachusetts Inst. Technol. Radiat. Lab. Ser., Vol. 8, pp. 10–59. McGraw-Hill, New York, 1948.

Each method, the impedance characterization or the scattering characterization, of describing field relations of a transmission line or in a microwave circuit element or junction has its own merits and its own field of usefulness and application. The two are, of course, closely related to one another.

10.5.1. Impedance Measurements

Impedance measurements, or as they might equally properly be termed, scattered wave measurements, are performed by two general methods with microwave apparatus. These two methods, the standing wave ratio method and the bridge method, will be described in this order.

The standing wavemeter is a fundamental and important item of microwave test equipment. It consists of a section of waveguide or transmission line so arranged that one can sample the field strength in the line as a function of distance along the line. A useful arrangement in a hollow tube waveguide, for instance, is to cut a slot along one surface of the guide in such a position that it does not interrupt any appreciable current flow in that wall surface; for the TE_{10} mode, the slot would be in the center of the broad side of the waveguide. Into this slot a very small probe or loop is inserted which will act essentially as an antenna to pick up a small fraction of the energy inside the tube and conduct this energy to a detector.

The probe structure is mounted on a sliding carriage which is arranged to slide on ways machined on the waveguide surface rather similar to the ways on a lathe. Thus, the field may be sampled along the transmission line as a function of distance in the propagation direction. One should be careful that the probe which is inserted through the slot to measure the field does not set up reflected waves of its own and thus disturb the very field which one is attempting to measure. The device must be designed carefully so that there is no reflected wave generated where the standing wavemeter is attached to a line or device to be measured. It is also necessary that the inside dimensions be aligned with the outside ways upon which the probe is moved. In general, a good standing wavemeter is built with the same accuracy as a good machine lathe.

The principle of the standing wave measurement is outlined in Barlow and Cullen.[5] In the microwave case one measures a quantity proportional to field strength and hence proportional to our artifically defined voltage

---

[5] H. M. Barlow and A. L. Cullen, "Microwave Measurements," pp. 118–164. Constable, London, 1950.

and current; see Eq. (10.5.3). The standing wave ratio is given by

$$\text{VSWR} = \frac{|V_{\max}|}{|V_{\min}|} = \frac{|t| + |r|}{|t| - |r|} = \frac{1 + |\varrho|}{1 - |\varrho|} \qquad (10.5.8)$$

where

$$\varrho = (r/t) = \text{reflection (or scattering) coefficient.} \qquad (10.5.9)$$

VSWR stands for voltage standing wave ratio.

One assumes that the detecting instrument connected to the probe produces an output proportional to field strength and hence proportional to the quantity V above. Standing wave ratio is defined as the maximum voltage along the transmission line divided by the minimum voltage, and it is quite evidently related in a simple way to the amplitudes of the transmitted and reflected waves as shown in Eq. (10.5.8).

The standing wave ratio is measured from the magnitudes. To obtain a complete description of the transmission line load impedance, it is necessary to have phase. The phase is determined in the following way. The probe measures a quantity proportional to V. We will call this quantity V_{probe} as in

$$|V_{\text{probe}}| = |t + r|. \qquad (10.5.10)$$

Impedances are transformed in passing along a transmission line as is explained by Moore.[6] We must, therefore, take a reference point for measurement. In microwave apparatus this reference point is arbitrary. We shall assume here that the reference is taken at some particular point, $Z = 0$, which we shall term the transmission line load, that is, all elements, including perhaps a portion of the transmission line, which are beyond this point are termed a portion of the load impedance. Referred from the load position back to the position of the probe, a distance $-z$ to the left of the load position, one has

$$|V_{\text{probe}}| = |te^{-i\beta z} + re^{+i\beta z}| = |t||1 + \varrho e^{+i2\beta z}|. \qquad (10.5.11)$$

The magnitude of the last term on the right depends upon the vector addition of unity with $\varrho e^{+i2\beta z}$. One can think of the vector addition then as a constant vector of unit length with another vector, always of length less than or equal to unity (since the reflected wave can never exceed the transmitted wave), at an angle ϕ, which is taken to be the phase angle of ϱ, plus $2\beta z$.

[6] R. K. Moore, "Traveling Wave Engineering." McGraw-Hill, New York, 1960.

As one moves the probe, the magnitude of the probe indication will vary as the small vector rotates around the end of the large vector. $| V_{\text{probe}} |$ is minimum when $\phi + 2\beta z = (2n + 1)\pi$ and is a maximum when $\phi + 2\beta z = 2n\pi$. Therefore, by determining the position of the maximum or the minimum of the probe indication, one can determine the phase angle ϕ, the angle of the reflection coefficient, since one need only subtract $2\beta z$ from the appropriate multiple of π. Ordinarily, one uses the minimum since this is more accurately found experimentally.

The procedure for measuring an arbitrary impedance terminating any transmission line is to find the VSWR, which will determine the magnitude of the reflection coefficient, and the position of the minimum or maximum, which will determine the phase of the reflection coefficient. One of the various circle diagram methods will then enable one to determine load impedance or admittance (i.e., the Smith chart[7]).

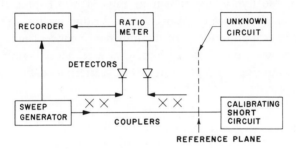

FIG. 1. Reflectometer block diagram.

Broadband or continuous frequency coverage standing wave measurements, better known as reflectometer measurements, makes possible direct and rapid measurement of VSWR and the magnitude of the reflection coefficient. This measurement technique, shown in Fig. 1, samples the incident and reflected waves from the generator and the unknown circuit load, respectively, through directional couplers and calculates their ratio via a direct reading ratio meter. The rf sweep generator is modulated at a 1 kHz rate since the ratio meter is normally designed for operation at this frequency. Before the unknown circuit to be tested is placed at the reference plane, the reflectometer is calibrated with a reference short circuit to obtain a known ratio level of 1.0 at the ratio meter (other known loads, each possessing different reflection coefficients, may also

[7] P. H. Smith, "Electronic Applications of the Smith Chart." McGraw-Hill, New York, 1969.

be utilized to calibrate the ratio meter more accurately). The short circuit is then replaced with the unknown circuit for measurement.

Insertion loss or transmission measurements of two-port microwave devices can be made easily with a reflectometer test setup. In this case the unknown circuit is placed between the two directional couplers in Fig. 1; the output coupler is reversed in direction such that both couplers sample only incident waves and finally a well-matched load is inserted in place of the calibrating short circuit. In this case the zero loss or reference level is determined before the circuit to be tested is inserted between the directional couplers.

In addition to the standing wavemeter and reflectometer measurements of impedance, it is possible to measure impedance or scattering coefficient by bridge methods very much as one measures impedance in low-frequency electrical technology by use of bridges. As pointed out in Section 10.2.3, the hybrid tee shown in Fig. 5a of Chapter 10.2 is fully equivalent in all necessary respects to a bridge circuit. Although the bridge circuit is used in terms of the impedance concept most commonly at low frequencies, the ideas are somewhat more clear here if one makes use of the scattering concept. In terms of scattering coefficients, one can write the matrix[8]

$$S = \frac{1}{\sqrt{2}} \begin{bmatrix} 0 & 1 & i & 0 \\ 1 & 0 & 0 & i \\ i & 0 & 0 & 1 \\ 0 & i & 1 & 0 \end{bmatrix} \tag{10.5.12}$$

where the element s_{ij} is defined by

$$r_i = \sum_j s_{ij} t_j. \tag{10.5.13}$$

One denotes by r_i the reflected, or more generally outgoing wave, on line i and by t_j the transmitted, or generally the ingoing, wave on line j.

The zeros in the matrix express the bridge properties of the hybrid tee. For instance, a wave transmitted in on line 1 produces no wave out on line 4 since $s_{14} = 0$. The device is also matched since the transmitted wave in on line 1 produces no outgoing wave on line 1; in other words, there is no reflected wave. Similar interpretations can be given the other elements in the matrix.

[8] C. G. Montgomery, R. H. Dicke, and E. M. Purcell, *in* "Principles of Microwave Circuits," p. 448. McGraw-Hill, New York, 1960.

The arrangement of the device to measure standing wave ratio is as follows: referring to Fig. 5a of Chapter 10.2 one arranges a power source on line 1; a perfect termination, that is, a device which will absorb all energy incident upon it without reflecting any on line 2; a matched detecting section, that is, a detector which will absorb all energy incident upon it without reflection, on line 4; and the unknown on line 3. Examination of the scattering matrix, Eq. (10.5.12), shows that the scattering coefficient s_{34} is equal to $1/\sqrt{2}$; thus a wave transmitted inward on line 3 produces a wave traveling outward on line 4 of exactly the same phase and half the power. If the wave transmitted inward on line 3 is the reflected wave from an unknown impedance terminating that line, then one has a measure of this reflected wave in line 4.

Note that if there is no reflected wave from line 3, there is no output on line 4. A calibration may be effected by putting a perfect reflector on line 3. That is, one places on line 3 a device, as for instance a flat plate closing off the waveguide, which reflects all of the incident wave. One measures then, on line 4, the complete reflection amplitude. This may be compared with reflection from any arbitrary impedance thereafter.

10.5.2. Power Measurements

One measures power in microwave apparatus on either an absolute or a relative basis. Of course the definition of absolute or relative depends upon the particular system of units in use. We shall refer here to calorimetric measurements as absolute, while measurements made relative to electrical power will be termed relative.

10.5.2.1. Absolute Power Measurements. A microwave calorimeter power measuring device can be made by arranging a transmission line or waveguide so that all of the transmitted wave is absorbed in a power dissipating element on the end of the line. Two possible arrangements are shown in Fig. 2.

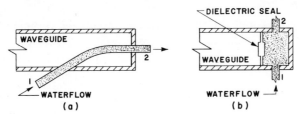

FIG. 2. Waveguide calorimeters.

Water or some other fluid is entered into the system at point 1 and leaves at point 2. Water flowing is heated by absorption of microwave energy so that the temperature at point 2 is greater than the temperature at point 1. By measuring rate of flow and temperature rise, and taking proper care that heat is not lost in the surroundings between the two thermometers, one can calculate the average energy input into the water and hence the average power flow into the transmission line on the assumption that all power in a transmitted wave is absorbed in the water. One arranges for no reflection by having the water tube cross the transmission line at a slant, for instance, starting in a region of low field and working towards the center or region of high field. By starting the mechanical discontinuity in a region of low electric field, there is only a small electrical discontinuity and hence a small reflected wave.

The arrangement in Fig. 2b is basically similar except that the water here is contained in a small cavity at the end of the transmission line. This cavity may be matched by perhaps an iris or a diaphragm at the entrance. A water seal, of course, is needed at this point.

Microwave calorimetric measurements have inherent in them all the troubles of calorimetry in general. Calorimetric measurements are most suitable for incident power in the range of several watts to several hundreds of watts. For powers less than a few watts it is very difficult to avoid heat loss in the surroundings and consequential errors in the determination. For powers of more than several hundred watts there is a serious heat dissipation problem. Calorimeter methods are used infrequently because of the difficulties of the measurement and because of the nuisance of flowing liquid.

10.5.2.2. Relative Power Measurements. Bolometers, thermocouples, and thermistors are the principal elements utilized in microwave apparatus for the measurement of power on a relative basis. The thermocouple operates in the same manner as is common throughout the electromagnetic spectrum by heating a small wire which produces a thermal emf between two dissimilar wires attached to the junction.

The bolometer and the thermistor operate on the principle that their electrical resistance is a function of temperature. If they are made very small compared to the wavelength and are arranged in a waveguide or transmission line in such a manner that all incident power is absorbed in the resistance of the element, the temperature rise will then be a function of the power absorbed. By placing the bolometer or thermistor in a Wheatstone bridge one can measure either the unbalance or the resistance and one can, by a proper calibration procedure, determine the amount of microwave power absorbed in the nonlinear resistance element.

A common method of performing the calibration is to start the measurement with no microwave energy absorbed in the bolometer or thermistor but with some fixed reference amount of direct current or 60-cycle electrical energy being absorbed. On applying the microwave energy, the bridge will become unbalanced. By then determining how much direct current or low-frequency power must be removed in order to rebalance the bridge, one has a direct measure of microwave power absorption. Accuracy of measurement can be to 0.05 dB.

A bolometer is ordinarily built as a very short piece of extremely fine wire with a resistance of perhaps 100 to 300 Ω at low frequencies. The length is made very short in comparison to a wavelength and, of course, this means that the wire itself must be exceedingly fine. It is, therefore, a rather sensitive element and is easily destroyed by an application of too much energy. A thermistor is a somewhat more rugged device which consists of a small glob of nonlinear resistance element between two wires. The resistivity of this material is high, it is larger and heavier, more rugged, and is harder to burn out.

Crystal detectors are sometimes used as power measuring devices. They are, however, rather inaccurate and the crystal characteristics tend to change somewhat erratically with time. They do not ordinarily measure power directly but must be calibrated; they are inaccurate, but very sensitive. One could also use a microwave receiver, perhaps of the superheterodyne type, as a power measuring device. It is necessary that the receiver circuits then be extremely stable and that some ready reference for power level be available. Very small powers in the microwatts and smaller range are ordinarily measured in this manner.

10.5.2.3. Attenuation Measurements. The attenuation of a microwave element is defined, as at low frequencies for transmission lines and networks, as the attenuation loss introduced by the element when operating between matched input and output. The loss introduced under any other conditions, that is, when input and output are not matched, is termed the insertion loss. For further elaboration on these ideas, see Ginzton.[9]

Any power measuring device can, of course, be used for the measurement of attenuation. One can measure the power level before the unknown device and then after it, and the difference in power levels is the

[9] E. L. Ginzton, "Introduction to Microwave Theory and Measurements," pp. 462–476. McGraw-Hill, New York, 1957.

attenuation in question. By the comparison method one substitutes a calibrated attenuator for the unknown device. By matching power levels at
the output by varying the attenuation introduced by the calibrated attenuator, one can read directly the insertion loss of the unknown element.

Note that insertion loss is specified in the preceding sentence. One
must state it this way because the unknown device may not be properly
matched at its ends; one assumes that the standard attenuator itself will
be well-matched. For example, suppose that an unknown device is inserted in an otherwise uniform matched transmission line and one attempts to measure its attenuation. If there is a large reflected wave at
its input and output ends, then in each instance this reflected wave is
lost to the wave transmitted beyond that point. If these reflected waves
are not too large in comparison with unity, one can simply assume that
this fraction of the wave is reflected at each point and there is no interaction. A complete treatment takes account of the multiple reflection
(it is here, of course, that the impedance concept comes into its own—the
impedance formulation automatically accounts for the reflected waves).

In the presence of reflected waves then, the true transmission attenuation is not as great as one would assume by the simple replacement of
the unknown element with a calibrated attenuator adjusted to the same
output reading; in this case one has neglected the reflection losses. The
standard attenuators might be one of those described in Section 10.2.4.
Accuracies with proper care range from 0.05 to 0.2 dB. The smallest of
these corresponds to approximately a 1% error in the power transmitted.

10.5.3. Frequency Measurements

The techniques for microwave frequency determination are basically
the same as at lower frequencies (see Chapter 9.2). The wavemeters, if
such are used, are apt to assume rather different forms since one would
use cavity resonators instead of coil–condenser combinations. The
accuracy of measurement using cavity resonators ranges from 0.005 to
1.0% depending upon the cavity Q, the temperature, the stabilization,
the tuning care, etc. For greater accuracy one would use heterodyne
methods. These are substantially identical to the heterodyne methods
described in Section 9.2.3.

One particular note of caution should be sounded, however. Since the
frequency in the microwave region is so very high, the standard, which
is apt to be a crystal oscillator, must be followed by several multipliers
to raise the frequency to the microwave region. One must exercise par-

ticular care in the multiplication chain since the multiplying factor is apt to be very large. A particularly satisfactory multiplying arrangement is the semiconductor crystal. Utilizing this as a nonlinear device, one can feed energy in at one frequency and pick off some multiple of this frequency in a different transmission line or waveguide. The device itself can be made resonably stable, and if high power at the output is not required, very high multiplying factors may be achieved. With a large multiplication in one element there are fewer total elements in the string to contribute to the instability of the measurement. One may also use atomic standards; see Vol. 3, Section 2.1.10.

For measuring the signal distribution in a selected portion of the rf or microwave frequency band, a spectrum analyzer may be used which presents a panoramic display of the frequency components that make up a given signal wave. This display takes the form of a plot of amplitude versus frequency, usually on the screen of a cathode ray oscilloscope. Hence much useful information can be obtained, such as the presence or absence of signals of interest, their frequencies, frequency differences, relative amplitude, and the nature of their modulation.

10.5.4. Phase Measurements

With the advent of sophisticated microwave communications systems the determination of the phase and time-delay characteristics of two-port microwave devices has become increasingly important because any circuit that exhibits a departure from a linear frequency-phase characteristic will produce signal distortion. Hence, there are several phase shift measurement techniques based on the use of a calibrated precision short circuit. Figure 3 shows one technique whereby phase measurements are made by keeping the slotted section probe fixed in position while introducing phase shift with the circuit under test (by changing the frequency of the signal generator) and maintaining the null at the probe of the slotted section through adjustment of the precision calibrated short circuit. Accurate calibration is a function of how well the source and circuit under test are matched to the transmission line.

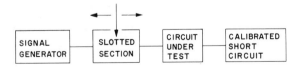

FIG. 3. Slotted section phase measurement circuit.

10.5.5. Field Strength Measurements

As mentioned in preceding paragraphs, one measures essentially field strength or something related to it in all microwave measurements. The problem which is different in the present section is that one interprets these measurements to mean field strength measurement in the absence of confining walls, that is, one is measuring field strengths for wave propagation in relatively unconfined space. Field strengths are ordinarily measured in volts per meter, and one assumes that this field strength is that of a plane wave propagating through the space at the position of measurement. A common measurement method is to use a receiver of adequate sensitivity coupled to an antenna. The antenna is positioned to pick up energy from the incident wave. To determine the magnitude, a calibration procedure must be followed. One radiates a known amount of power as measured by one of the preceding power measurement devices from an antenna with a known amount of gain as determined in the following section. By knowing the gain of the transmitting antenna, the power transmitted, and the gain of the receiving antenna, the field strength may be calculated on the basis of far-field spherical wave propagation. One assumes that the spherical waves are of sufficient low curvature to be approximated by plane waves.

10.5.5.1. Antenna Patterns and Gain. To measure an antenna pattern one invokes the reciprocity relation, that is, one assumes that the transmitter and receiver may be interchanged. Thus, no matter whether the given antenna is to be used for transmitting or receiving, it is set up, for instance, as a receiver, and rotated while receiving signals from a distant antenna fixed in one position. The power received versus angle to the unknown antenna then gives the radiation pattern of the antenna. Since these patterns are almost always assumed to be far-field patterns, one must make certain the actual measuring setup satisfies the appropriate criterion. It is easy to show that two apertures of diameters D_1 and D_2, one used as a transmitter and one for reception, must be spaced a distance apart greater than or equal to R as given in the following relation:

$$R \geq 2D_1D_2/\lambda. \tag{10.5.14}$$

If inequality (10.5.14) is satisfied, the phase error due to finite distance between the two antennas will be of the order of a sixteenth of a wavelength or less. This produces a few percent error in the measured received amplitude.

In addition to the spacing criterion above, one must be careful of ground reflection, that is, signals reflected off the ground or buildings or shrubbery in the vicinity which will arrive at the receiving antenna and interfere either constructively or destructively with the correct signal, hence producing an erroneous indication. Adequate sensitivity and dynamic range in the reception apparatus must also be provided—30 dB range is necessary for most antennas and 50 dB is very desirable.

For the actual received power measurement, one sometimes uses a calibrated receiver, that is, one whose output amplitude as a function of power input is known, or, a substitution method may be used in which a calibrated attenuator is varied to keep the detector indication constant. The attenuator reading is then the complement of the antenna pattern.

Antenna gain is defined by

$$G(\theta, \phi) = \frac{P(\theta, \phi)}{(1/(4\pi))P_t} \tag{10.5.15}$$

where θ and ϕ are spherical coordinates centered at the unknown antenna. $P(\theta, \phi)$ is power transmitted in the given direction θ and ϕ from the unknown antenna, while P_t is the total power transmitted integrated over a complete sphere. By convention, gain is often specified to be the gain in the direction of maximum power transmission.

One can measure gain by making use of two identical antennas. In this situation the power received is given by

$$P_r = \frac{P_t A_r G_t}{4\pi R^2}; \qquad A_r = \frac{\lambda^2 G_r}{4\pi}. \tag{10.5.16}$$

The term A_r is known as the effective area of the receiving antenna.[10] On the assumption that the two antennas are identical, then $G_t = G_r = G$. Solving for G yields the following:

$$G = \frac{4\pi R}{\lambda} \sqrt{\frac{P_r}{P_t}}. \tag{10.5.17}$$

Care must be exercised as noted above in relation to proper spacing of the antennas and elimination of ground reflection. One can also perform the measurement without two identical antennas if any three antennas are available. By making three separate measurements one obtains three

---

[10] S. Silver, "Microwave Antenna Theory and Design," Massachusetts Inst. Technol. Radiat. Lab. Ser., Vol. 12, pp. 37–60. McGraw-Hill, New York, 1949.

independent equations and by solving the three equations the gains of the three separate antennas may be determined.

Ground reflection or "cleanliness" of antenna range can be improved for some antenna aperture-wavelength relationships through the use of anechoic chamber measurements. Rf anechoic chambers are completely enclosed rooms, constructed such that, with rf absorbing material covering all of the inside wall space, any wall reflection appears negligible relative to the direct rf wave between transmitting and receiving antennas. The primary limitation on these chambers is their practical size which is approximately governed by the far-field relation of inequality (10.5.14).

11. MISCELLANEOUS ELECTRONIC DEVICES

11.1. Photoelectric Devices[1]*

11.1.1. Vacuum Phototubes

11.1.1.1. Principle of Operation. Vacuum phototube behavior is related to the laws of photoelectricity. (1) The number of electrons released per unit time at a photoelectric surface is directly proportional to the intensity of the incident light. (2) The maximum energy of the electrons released at a photoelectric surface is independent of the intensity of the incident light, but increases linearly with the frequency of the light as exemplified by Einstein's photoelectric equation: $mv^2/2 = h\nu - W$.

A vacuum phototube consists of a photocathode, an anode, a transparent envelope, and electrical terminals. The cathode collects light passing through the glass envelope and emits electrons which are collected by the anode.

11.1.1.2. Characteristics. The spectral response characteristic of a phototube is a display of the photoelectric current per unit incident radiant power as a function of wavelength. An alternate presentation may provide the spectral response information in terms of quantum efficiency as a function of wavelength. These two types of spectral response characteristics are related through the size of the quanta ($h\nu$), which varies inversely with the wavelength. Thus, for one photoelectron (e) per quanta—100% quantum efficiency—the response of the device must be $e/h\nu = e\lambda/hc = \lambda/1239.85$ A/W, where λ is expressed in nanometers.

[1] V. K. Zworykin and E. G. Ramberg, "Photoelectricity and its Application." Wiley, New York, 1949; A. H. Sommer, "Photoelectric Tubes." Methuen, London, 1951; D. Mark, "Basics of Phototubes and Photocells." Rider, New York, 1956; A. H. Sommer, "Photoemissive Materials." Wiley, New York, 1968; A. H. Sommer and W. E. Spicer, "Photoelectronic Materials and Devices" (S. Larach, ed.). Van Nostrand, Princeton, New Jersey, 1965; see also Vol. 6B of this series, Chapter 12.4 as well as Vol. 4A, Sections 2.1.3 and 2.3.1.

---

* Chapter 11.1 is by Ralph W. Engstrom.

TABLE I. Nominal Composition of Various Photocathodes: Their Designations, Envelope Materials, Typical Sensitivity Figures, and Typical Dark-Emission

Key to Fig. 1	Nominal composition	Response designation	Type of photocathode	Envelope material[a]	Conversion factor[b] (lm/W) at λ_{max}	Luminous sensitivity (μA/lm)	Wavelength of maximum response, λ_{max} (nm)	Sensitivity at λ_{max} (mA/W)	Quantum efficiency at λ_{max} (%)	Dark emission at 25°C (A/cm²)
1	Ag—O—Cs	S-1	Opaque	0080	92.7	25	800	2.3	0.36	$900. \times 10^{-15}$
2	Ag—O—Rb	S-3	Opaque	0080	285	6.5	420	1.8	0.55	—
3	Cs₃Sb	S-19	Opaque	SiO₂	1603	40	330	64	24	0.3×10^{-15}
	Cs₃Sb	S-4	Opaque	0080	1044	40	400	42	13	0.2×10^{-15}
	Cs₃Sb	S-5	Opaque	9741	1262	40	340	50	18	0.3×10^{-15}
4	Cs₃Bi	S-8	Opaque	0080	757	3	365	2.3	0.77	0.13×10^{-15}
5	Ag—Bi—O—Cs	S-10	Semitransparent	0080	509	40	450	20	5.6	70×10^{-15}
6	Cs₃Sb	S-13	Semitransparent	SiO₂	799	60	440	48	14	4×10^{-15}
	Cs₃Sb	S-9	Semitransparent	0080	683	30	480	20	5.3	—
	Cs₃Sb	S-11	Semitransparent	0080	808	60	440	48	14	3×10^{-15}
	Cs₃Sb	S-21	Semitransparent	9741	783	30	440	23	6.7	—
7	Cs₃Sb	S-17	Opaque[g]	0080	667	125	490	83	21	1.2×10^{-15}
8	Na₂KSb	S-24	Semitransparent	7056	1505	43	380	64	23	0.0003×10^{-15}
9	K—Cs—Sb	—	Semitransparent	7740	1117	80	400	89	28	0.02×10^{-15}
10	(Cs)Na₂KSb	—	Semitransparent	SiO₂	429	150	420	64	18.9	0.4×10^{-15}
	(Cs)Na₂KSb	S-20	Semitransparent	0080	428	150	420	64	19	0.3×10^{-15}

11	(Cs)Na$_2$KSb	S-25	Semitransparent	0080	276	160	420	44	13	—
12	(Cs)Na$_2$KSb	ERMA[c]	Semitransparent	7056	169	265	575	45	10	$1.\times10^{-15}$
13	Ga—As		Opaque[h]	9741	148	250	450	37	10	0.1×10^{-15}
14	Ga—As—P		Opaque[i]	Sapphire	310	200	450	61	17	0.01×10^{-15}
15	InGaAs—CsO[d]		Opaque[h]	0080	266	260	400	71	22	$1.\times10^{-15}$[d]
16	Cs$_2$Te		Semitransparent	LiF	—[e]	—[e]	120	12.6	13	—[f]
17	CsI		Semitransparent	LiF	—[e]	—[e]	150	24	20	—[f]
18	CuI		Semitransparent	LiF	—[e]	—[e]	150	13	10.7	—[f]

[a] Numbers refer to the following glasses: 0080 – Corning Lime Glass
    9741 – Corning Ultraviolet Transmitting Glass
    7056 – Corning Borosilicate Glass
    7740 – Corning Pyrex Glass
    SiO$_2$ – Fused Silica (Suprasil—Trademark of Engelhard Industries, Inc., Hillside, New Jersey)

[b] These conversion factors are the ratio of the radiant sensitivity at the peak of the spectral response characteristic in amperes per watt to the luminous sensitivity in amperes per lumen for a tungsten test lamp operated at a color temperature of 2854°K.

[c] An RCA designation for "Extended-Red Multialkali."

[d] An experimental photocathode, private communication from B. F. Williams, RCA Laboratories, Princeton, New Jersey. The dark emission indicated is a calculated value based on a bandgap of 1.1 eV for the particular composition studied. See also B. F. Williams, *Appl. Phys. Lett.* **14**, 273 (1969).

[e] Not relevant.

[f] Data unavailable; expected to be very low.

[g] Reflecting substrate.

[h] Single crystal.

[i] Polycrystalline.

In the United States, the electronic industry has adopted a designation system of S numbers to identify spectral response characteristics. Spectral response is determined not only by the photocathode material, but by the transmission characteristics of the envelope enclosing the phototube, particularly the ultraviolet cutoff characteristic. Table I provides an identification of photocathode materials, windows, S numbers and other related information. The number of spectral responses tabulated is rather more complete than those which are readily available in vacuum phototubes. It is intended that Table I provide an indication of possible developments, and that it be referred to in the consideration of photomultiplier tubes for which a wide variety of spectral response characteristics are available.

In Fig. 1 a selected group of spectral response characteristics are illustrated. When photocathodes are available in various envelopes which modify the ultraviolet cutoff, only one of these spectral response characteristics is illustrated—usually that one which is available with the most ultraviolet transmitting envelope. The result of other glass and photocathode combinations may be inferred from a comparison of the transmission of various glasses used in the fabrication of phototubes, as shown in Fig. 2.

A typical current-voltage characteristic for a vacuum phototube is illustrated in Fig. 3. At low voltages there is only partial collection of electrons due to the small size of the anode and the finite electron emission velocities. The relatively flat operating characteristic at higher voltage, where essentially all of the electrons are collected, permits the use of high-load impedances with vacuum phototubes.

Since the emission of electrons is directly proportional to the light flux, vacuum phototubes are characteristically linear devices. However, at high light levels, excess current and heat may permanently damage the photosurface. At low voltages and low currents, the bulb potential may vary erratically and modify the output currents. This is caused by photoelectrons not collected by the anode which strike the envelope. The envelope either collects electrons and charges negatively, or charges positively by the secondary emission of electrons from the inside surface of the envelope. A hysteresis-like loop may be observed on the current-voltage characteristic on some phototubes when the voltage is varied in the range 20–60 V. If accurate linearity between current and light flux is required, it may be advisable to place a conductor connected to cathode potential in contact with the outside of the glass surface with an aperture for the light; the anode voltage should usually exceed 90 V.

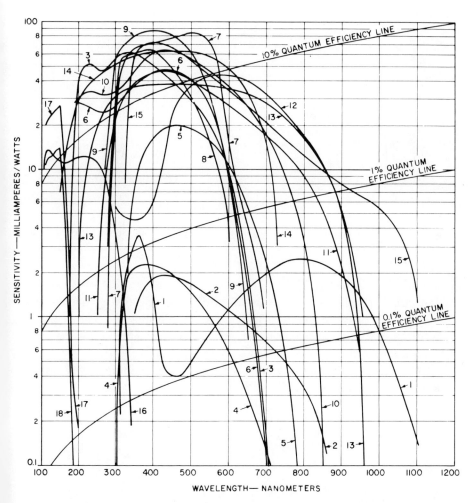

FIG. 1. Typical spectral response characteristics of photocathodes used in phototubes and photomultiplier tubes. (See Table I for related parameters.)

No delay has ever been measured in the photoemission process[2]; the dynamic response of a vacuum phototube is principally limited by the transit time and the spread in transit time of the electrons crossing from cathode to anode. Although these limiting times are of the order of a few nanoseconds only, it is difficult to realize a corresponding rapid response in practice because of coupling and amplifier limitations. The capacitance associated with the phototube elements may be about 2 pF,

[2] E. O. Lawrence and J. W. Beams, *Phys. Rev.* **29**, 903 (1927).

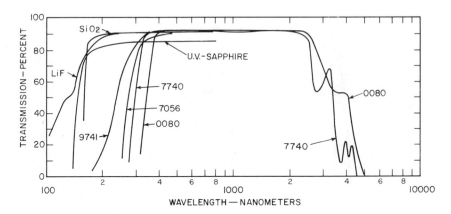

FIG. 2. Transmission characteristics of various glasses as envelopes for phototubes and photomultiplier tubes. All curves are for one-millimeter thickness. The numbers refer to Corning glass designations.

but circuit coupling elements usually bring the effective capacitance to 10 pF. For a load resistance of 1000 Ω, the associated RC time constant is 10^{-8} sec. But even 1000 Ω is not a practical load value unless the light level is quite high.

A vacuum phototube operated under normal conditions will suffer only moderate loss in sensitivity in many thousands of hours. For most

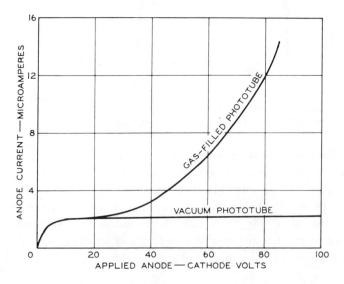

FIG. 3. Typical current-voltage characteristics for vacuum and gas-filled phototubes.

stable operation, the photocurrent should be small and the tube kept in the dark when not in use. Blue and U.V. radiation particularly should be avoided during storage.

Dark currents in phototubes limit detection of low light levels; they are frequently of nearly ohmic character, being due to leakage across the stem or the base. Some phototubes minimize leakage by bringing one lead out of the top of the bulb. In this case, the dark current may be primarily thermionic emission from the photocathode. Typical thermionic emission currents from various photocathodes are included in Table I.

11.1.1.3. Applications. Vacuum-type phototubes are perhaps most useful under the following conditions of service: where reliability, stability, and long life are important; where amplification of small currents is not a problem; and where large load impedance is not objectionable.

11.1.1.4. Types and Ratings. A large variety of vacuum phototubes are available from a number of manufacturers.[†] Specific tube selection can be made from their catalogs. In selecting a tube for a specific purpose one should consider the expected environment of the tube—space, temperature, humidity, vibration—as well as the amount and color of the light available. Tubes having S-1 spectral response are generally rated to 100°C, while those having S-4 are only rated to 75°C. On the other hand, dark current on S-4 types is usually lower than on S-1 types. For low dark current requirement it is generally advisable to select a tube with a double-ended contact arrrangement; it will also be observed that some tubes employ special nonhygroscopic base material to avoid leakage in humid conditions.

11.1.2. Gas-Filled Phototubes

11.1.2.1. Characteristics. Gas-filled phototubes were devised to overcome the problem of low output current of the vacuum phototube. An amplification factor of 5–10 is achieved by avalanche multiplication of

---

[†] Some of the manufacturers of vacuum and gas-filled phototubes are listed together with distribution offices in the U.S.A.: RCA (415 South Fifth Street, Harrison, New Jersey); Philips (Distributor: Amperex, 230 Duffy Avenue, Hicksville, Long Island, New York); Continental Electric Company (Cetron) (Geneva, Illinois); Hamamatzu TV Company, Ltd. (Distributor: Kinsho-Mataichi Corporation, 80 Pine Street, New York, New York 5); ITT (Fort Wayne, Indiana).

the photoelectrons (see Vol. 5A, Chapter 1.3 or Vol. 7B, Chapter 6.1). The filling is usually a noble gas at a pressure of about 0.1 Torr.

A typical current-voltage characteristic is shown in Fig. 3. At low voltages, the behavior is essentially the same as that of a vacuum phototube. Amplification occurs above the ionization potential (15.7 V for Argon). Operation above 90 V is usually avoided because of unstable behavior and the possibility of damaging the tube with the onset of a self-sustained discharge (see Part 4). However, for low light levels, higher amplification can be utilized by providing a stable voltage supply and a protective load resistance to limit the maximum current to a fraction of a microampere.

The current output from a gas-filled phototube is proportional to the light flux for currents less than a few microamperes. The sensitivity increases for high light levels because of a positive ion space charge near the cathode which provides a more efficient field distribution for multiple ionization than the undistorted field.

The frequency response of a gas-filled phototube is limited to the audio range because of the slow transit of the ions and secondary effects produced by even slower diffusing metastable atoms.[3]

The stability of gas-filled phototubes is not as great as that of vacuum phototubes, although with normal usage, a gas-filled phototube will operate for many thousands of hours. Deterioration is caused by ion bombardment of the photocathode. Longer life is obtained for low-voltage and low-current operation.

Since the gas amplification process is relatively noise-free, gas tubes have a practical advantage in improving the signal level relative to Johnson and amplifier noise.

11.1.2.2. Applications. High sensitivity is the principal advantage of gas-filled phototubes compared to vacuum phototubes. Loss of high-frequency response and linearity may be expected. The gas-filled tube should not be used as a reference standard nor in circuits requiring a large voltage drop across the load resistance.

11.1.2.3. Types and Ratings. Manufacturers of gas-filled phototubes are listed under Section 11.1.1.4. Their catalogs should be consulted for specific choice of tube type. Most gas-filled tubes are designed for a maximum of 90 V. Types with S-1 spectral response are rated generally to 100°C compared with the 75° rating of S-4 types. Stability during

---

[3] R. W. Engstrom and W. S. Huxford, *Phys. Rev.* **58**, 66 (1940).

life is usually better for types having S-1 spectral response than for those with S-4 spectral response.

11.1.3. Photomultiplier Tubes[4]

11.1.3.1. Principle of Operation. Although photoelectric emission is a relatively efficient process on a per-quantum basis, the actual photocurrent for low light levels is so small that special amplification techniques are required. In the photomultiplier tube, the amplification problem is solved by means of secondary-electron emission. Photoelectrons are electrostatically directed to a secondary emitting surface. At normal applied voltage, 3–6 secondary electrons are emitted per primary electron. These secondaries are focused to a second dynode (or secondary emitting surface) where the process is repeated. In commercial photomultiplier tubes there may be as many as 14 or more dynodes, although 9 or 10 is the usual number. Typical electron gain is of the order of 10^6. Following the last dynode stage is an anode. In addition, a photomultiplier tube may contain one or more electrodes used to improve electrical focus, to reduce space charge, or to reduce transit time effects.

The channel photomultiplier is another type of photomultiplier which uses a hollow cylinder or "channel" as a continuous-strip secondary emitter. The cylinder is provided with an internal coating giving a finite resistance. Voltage applied to the terminals of the cylinder provides a uniform gradient along the length of the channel. The inside surface of the channel is sensitized to yield a reasonable number of secondary electrons per primary. When electrons are introduced at one end of the channel and impact the inside wall, an avalanche of secondaries is created as electrons drift across the channel and are accelerated down its length. Gain of the channel is a function of the applied voltage and of the length-to-diameter ratio of the channel which may be of the order of 50:1. A photocathode may be mounted in proximity to the entrance of the channel or it may be incorporated as part of the channel. In some applications bundles of channels serve to collect electrons from a larger area photocathode. Channels are usually curved to avoid feedback effects.

---

[4] S. Rodda, "Photoelectric Multipliers." MacDonald, London, 1953; N. O. Chechik, S. M. Fainshtein, and T. M. Lefshets, "Electron Multipliers." Distributed by Four Continent Book Corp., New York, 1957. In Russian; F. Boeschoten, J. M. W. Milatz, and C. Smith, *Physica* **20**, 139 (1954); D. E. Persyk, New photomultiplier detectors for laser applications. *Laser J.* **1**, No. 1, 21 (1969); A. T. Young, Photometric error analysis. IX: Optimum use of photomultipliers. *Appl. Opt.* **8** (2431) 1969.

Channel photomultipliers[5] are relatively new and are used in rather special applications. While the photocathode is usually small, this may be no drawback in the case of laser applications. The device has fewer connections than a conventional photomultiplier tube and can be made very rugged. Pulse risetime for channel diameters of the order of 40 μm can be in the subnanosecond region, but statistics of amplification are not as good as in the latest conventional photomultipliers. In a saturated output operating mode, single photoelectrons can be counted, but this does not provide discrimination of multiple photoelectron inputs, which is possible in conventional photomultipliers with high-gain first dynodes.

Channel electron multipliers can also be used as detectors for various particles by omitting the photocathode and window to the channel entrance. By operating the channel in a saturated pulse mode, detection of electrons over an energy range 250 eV to 10 keV is estimated to be greater than 50%.[6] Windowless channel electron multipliers have been used in sounding rocket and satellite research. The device may also be used to detect ions with energy as low as a few hundred electron volts and to detect short wavelength U.V. and x-ray radiation.

11.1.3.2. Characteristics. A wide variety of photocathodes are employed in photomultipliers. The typical photocathode is one applied to the inner surface of the window to the tube so that radiation is applied to one side and electrons are ejected from the other. This transmission type of cathode is ideal in applications demanding efficient and wide-angle light collection. Other photocathodes are of the reflecting type; that is, light is incident and electrons are ejected from the same side. Spectral response characteristics and other related parameters are provided in the previous discussion on vacuum phototubes, Table I, and Fig. 1.

Secondary emission increases with voltage for practical materials up to several hundred volts. At higher voltages, the secondary emission per primary electron decreases, probably because the penetration of the primaries begins to exceed the escape depth of the secondaries. This description applies particularly to such commonly used secondary emitters as cesium–antimony, silver–magnesium, and copper–beryllium. However,

[5] G. Goodrich and W. Wiley, Continuous electron multiplier. *Rev. Sci. Instrum.* **33**, 761 (1962); G. W. Goodrich and J. Love, A 10,000 G photomultiplier. *IEEE Trans. Nucl. Sci.* **NS-15**, No. 3, 193 (1968); K. C. Schmidt and C. F. Hendee, Continuous channel electron multiplier operated in the pulse saturated mode. *IEEE Trans. Nucl. Sci.* **NS-13**, No. 3, 100 (1966); W. G. Wolber, The channel photomultiplier—a photon counting light detector. *Res. Developm.* **19**, No. 12, 18 (1968).

[6] D. S. Evans, *Rev. Sci. Instrum.* **36**, 375 (1965).

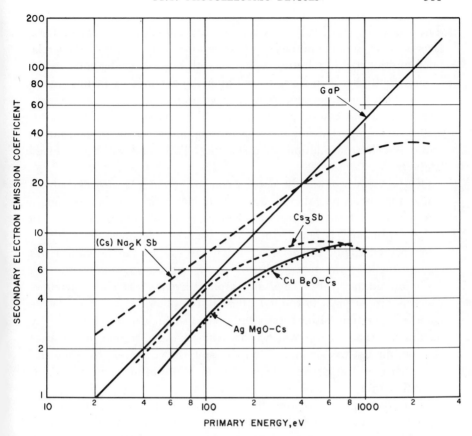

FIG. 4. Typical secondary emission characteristics for various materials used in photomultiplier tubes.

a new material GaP(Cs) has recently been introduced[7] for use in photo-multipliers. This secondary emitter is prepared with *P*-type GaP and treated with Cs vapor. Band-bending occurs near the surface in such a way that the material has, in effect, a negative electron affinity. In GaP(Cs) therefore, the escape depth is greatly increased and consequently also the secondary emission yield, especially at high primary energies.

Typical secondary emission characteristics are illustrated in Fig. 4 for various materials, including GaP(Cs), which are used in photomultipliers.

[7] R. E. Simon and B. F. Williams, *IEEE Trans. Nucl. Sci.* **NS-15**, No. 3, 167 (1968); G. A. Morton, H. M. Smith, Jr., and H. R. Krall, *IEEE Trans. Nucl. Sci.* **NS-16**, No. 1, 92 (1969); R. E. Simon, A. H. Sommer, J. J. Tietjen, and B. F. Williams, *Appl. Phys. Letters*, **13**, No. 10, 15 (1968).

The curve for $(Cs)Na_2KSb$ is a secondary emission yield curve for a layer similar to the multialkali photocathode.[8] At present, commercial photomultiplier tubes are not available with multialkali dynodes.

Photomultiplier gain characteristics may be obtained by taking the nth power of the per-stage gain, where n is the number of stages of secondary emission. In the case of $GaP(Cs)$ only the first stage may be of this material; its importance is not in achieving an over-all high gain but in improving electron multiplication statistics. This point will be discussed later.

Of the more commonly used secondary emitter materials, silver–magnesium or copper–beryllium (which have very similar secondary emission characteristics) are more stable at high-current densities than cesium–antimony, although the latter has the advantage of a higher secondary emission. Either silver–magnesium or copper–beryllium will tolerate higher temperatures during processing than cesium–antimony.

Photomultiplier tubes have an output current—less dark current—which is proportional to the light flux on the photocathode over a wide range of operation.[9] Linearity is terminated at high output currents by the onset of space charge which usually blocks the operation in the space between the last two dynodes. By increasing the voltage on the tube and particularly on the last few stages, the range of output current without saturation may be increased.[10]

Contrary to the above remarks, certain individual tubes and certain types have shown some lack of linearity in scintillation counting applications.[†] For example, when the number of counts is increased from 1000 per minute to 10,000 per minute by increasing the strength of the nuclear radiation, the photomultiplier output pulse height may change slightly. A typical change is an increase of less than 1%. The cause of this slight nonlinearity is the changing electric fields between dynodes resulting from insulator charging. Tubes designed to avoid the effect have special shields or conductive coatings applied to critical insulators.

Anode dark current determines the lower level of light detection. There are several sources of dark current in multiplier phototubes: ohmic

[8] O. B. Vorob'yeva, A. A. Mostovskiy, and G. B. Stuchinski, *Radio Eng. Electron. Phys.* **10**, No. 3, 414 (1965).

[9] R. W. Engstrom, *J. Opt. Soc. Amer.* **37**, 420 (1947).

[10] W. Widmaier, R. W. Engstrom, and R. G. Stoudenheimer, *IRE Trans. Nucl. Sci.* **NS-3**, No. 4, 137 (1956).

[†] See also Vol. 5A, Chapter 1.4 and Sections 2.2.1.2.2 and 2.2.3.3.

leakage, thermionic emission, and regenerative effects. Ohmic leakage predominates at low operating voltages and is due to the imperfect insulating properties of the glass, the base, and supporting members. It can be minimized by keeping the terminal structure of the tube clean and dry. Coating the base and socket with a nonhygroscopic material may be worthwhile; ceresin wax or a noncorrosive silicone such as G. E.'s RTV-11 with primer are recommended.

As the voltage is increased, the character of the dark current changes and parallels the approximately exponential increase of gain with voltage. The source of this dark current is primarily thermionic emission of electrons from the photocathode. (For typical values see Table I.) Since thermionic emission is random in time, and secondary emission is a statistical process, the output dark current of the photomultiplier tube consists of randomly spaced, unidirectional pulses of variable height. If the tube is cooled, the initiating thermionic emission decreases at first very rapidly with the temperature,[9] but at lower values of temperature the dark emission fails to be reduced as one might expect from Richardson-type reciprocal temperature plots.[11] A possible explanation may be patches of variable activation on the photocathode surface. Another source of dark emission may be residual radioactive elements in the photomultiplier parts which cause scintillations in the glass envelope; ^{40}K, for example, is a common contaminant of most glasses.

At higher voltage, the dark current increases rapidly and becomes very erratic. Contributing mechanisms may be sparking, ionization of impurity gases or alkali vapors remaining from the sensitization process, and feedback by light generated by electron impact on insulators in the tube.[12] Figure 5 shows a typical curve for the equivalent-anode-dark-current-input versus sensitivity. This quantity is the luminous flux (from a tungsten lamp source operating at 2854°K color temperature) required to produce an output current equal to the anode dark current and is simply a convenient figure of merit for comparing multiplier phototubes even of very different gain figures. Ohmic limitation may be observed at the left, thermionic in the center and regenerative breakdown at the right of the characteristic.

It is not the dc value of the dark current but rather its random variation which is the limitation to the detection of a light signal. If the

[11] J. A. Baicker, *Proc. Scintillation Counter Symp., 7th IRE Trans. Nucl. Sci.* **NS-7**, No. 2-3, 74 (1960).

[12] H. R. Krall, *IEEE Trans. Nucl. Sci.* **NS-14**, No. 1, 455 (1967).

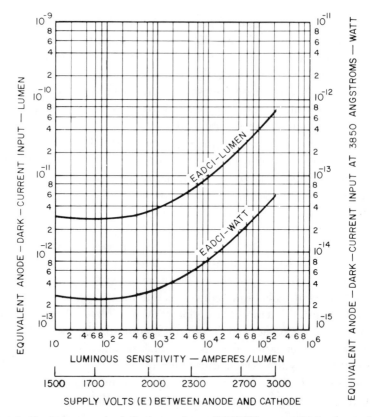

FIG. 5. Equivalent anode dark-current input (EADCI) vs. sensitivity for a photo-multiplier with K_2CsSb photocathode (twelve-stage RCA Type 8850) having a gallium-phosphide first dynode.

photomultiplier tube is operated at very low light levels and with a sensitivity equivalent to the minimum of the curve, Fig. 5, the principal component of noise would be the shot noise associated with the thermionic or dark emission (i_t). The rms value of this noise current would be

$$\sqrt{\overline{I^2_{\Delta f}}} = \mu \sqrt{2ei_t \, \Delta f} \qquad (11.1.1)$$

where e is the electron charge; μ, the gain of the tube; and Δf, the bandwidth of the observation. This equation does not take account of the noise resulting from the variable nature of secondary emission. For a Poisson distribution for secondary emission, the rms noise current may be written

$$\sqrt{\overline{I^2_{\Delta f}}} = \mu \left[2ei_t \left(1 + \frac{1}{\delta_1} + \frac{1}{\delta_1 \delta_2} + \frac{1}{\delta_1 \delta_2 \cdots \delta_n} \right) \Delta f \right]^{1/2} \qquad (11.1.2)$$

where δ_1, δ_2, ..., δ_n are the secondary emission for the 1st, 2nd, ..., nth stages of the photomultiplier. If the δ's are all the same, one may closely approximate Eq. (11.1.2) by

$$\sqrt{\overline{I_{\Delta f}^2}} = \mu\left[\frac{2ei_t\,\Delta f}{(1-1/\delta)}\right]^{1/2}. \tag{11.1.3}$$

Actual noise measurements indicate somewhat greater noise[13] than would be expected from (11.1.2) or (11.1.3) due to deviations from Poisson statistics and from less than 100% collection of photoelectrons at the first dynode. Note, however, the importance of maintaining a high secondary emission for the first dynode. With the new GaP dynodes which are used in the first stage of some photomultipliers, the secondary emission may be 30 or more and contribute a negligible amount to the total noise.

A figure of merit which describes the limitation of detection by noise is the equivalent-noise-input. ENI represents the light flux required to develop an anode signal just equal to the dark noise of the tube measured in a bandpass of 1 Hz. For optimum detection of low light levels, a moderate voltage operation is best. As the voltage is increased, the relative noise increases as does the relative dark current (Fig. 5). The non-regenerative range can sometimes be extended by wrapping or painting the exposed outside glass surface of the tube with a conductor connected to the negative end of the voltage supply. This prevents bulb charging by secondary emission of electrons from the glass and reduces feedback.

In the detection of low light levels, it is often advantageous to modulate the light by means of a "chopper" and to couple the multiplier photo-tube to an amplifier having a narrow bandpass at the frequency of "chop." In this way the dc component of the dark current is eliminated and the inherent signal-to-noise ratio of the multiplier phototube is more readily realized. Another method of observing very small light signals is to count the output current pulses from the multiplier photo-tube which correspond to individual photoelectrons. The limit of detection is set when the number of such pulses observed during a given period with the light on is just significantly larger than the number observed in a comparable period in the dark. All pulses are not of the same size because of the statistical nature of secondary emission. There are also a

[13] G. A. Morton, H. M. Smith, and H. R. Krall, Pulse-height resolution of high-gain first-dynode photomultipliers. *Appl. Phys. Lett.* **13**, 356 (1968); R. Foord, R. Jones, C. J. Oliver, and E. R. Pike, The use of photomultiplier tubes for photon counting. *Appl. Opt.* **8**, 1975 (1969).

significant number of small pulses in the dark emission which represent electrons originating from electrodes other than the cathode. The limit of detectability may be reduced by discriminating against these smaller pulses.

When the photocurrent is well in excess of the thermionic emission, measurement precision is limited by the randomness of photoemission and secondary emission. Examples of this type of limitation are the detection of a star against the background of the sky where the modulated signal is produced by scanning back and forth across the star; the detection of small marks on scanned paper. The expression for the rms noise current output is identical to Eq. (11.1.1) or Eq. (11.1.2) except that the average cathode photocurrent i_k is substituted for i_t. In a situation where i_k and i_t are comparable, they must be added together in the appropriate expression.

In an application of the sort requiring the ultimate detection capability in the presence of a light background, the most important parameter in order to maximize signal-to-noise is photocathode sensitivity. Gain is generally unimportant except for convenience and at the first stage to minimize secondary emission statistics as indicated in Eq. (11.1.2).

Scintillation counting[14] is another application where the statistics of the photomultiplier operation are important. In the application to nuclear spectrometry, the pulse height is proportional to energy. Resolution of pulse heights depends upon the statistics of the scintillation, the optical coupling efficiency, and the related statistics of the photomultiplier.

A common method of comparing photomultipliers for their efficiency in scintillation counting applications is to measure the pulse-height-resolution (PHR) using ^{137}Cs (661 keV gamma ray) and a NaI(Tl) crystal coupled to the photocathode of the photomultiplier. Figure 6 shows a typical distribution of pulse heights obtained with ^{137}Cs and NaI(Tl) using a multichannel analyzer coupled to the photomultiplier output. PHR is defined as the ratio of the full width at half-maximum to the pulse height at maximum count rate expressed in percent. Thus,

$$PHR = \frac{FWHM}{PH} \cdot 100. \tag{11.1.4}$$

[14] J. B. Birks, "Scintillation Counters." McGraw-Hill, New York, 1953; G. A. Morton, *RCA Rev.* **10**, 525 (1949); C. E. Croutham, "Applied Gamma-Ray Spectrometry." Pergamon, Oxford, 1960; J. H. Neiler and P. R. Bell, *in* "Alpha, Beta, and Gamma-Ray Spectroscopy" (K. Siegbahn, ed.). North-Holland Publ., Amsterdam, 1965; N. S. Wall and D. E. Alburger, *in* "Nuclear Spectroscopy" (F. Ajzenberg-Selove, ed.). Academic Press, New York, 1960; see also the references to Vol. 5A given earlier in this section.

Typical PHR in the case of ^{137}Cs and NaI(Tl) for a good scintillation counter is between 7 and 8%. The important attributes of a photomultiplier necessary to obtain good PHR are: (1) a photocathode with high quantum efficiency in the part of the blue spectrum which matches the radiation from the scintillator, (2) efficient electron-optical construction which directs the emitted photoelectrons to the surface of the first dynode, (3) a photocathode of uniform sensitivity across the area which is coupled to the scintillator, and (4) a high secondary emission at the first dynode to minimize statistical variations.

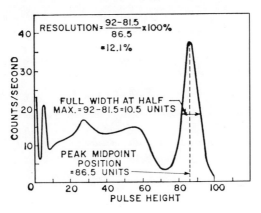

FIG. 6. Pulse-height distribution for a scintillation counter using ^{137}Cs and a NaI(Tl) crystal.

In the case of ^{137}Cs and other radioactive sources which produce relatively high-energy gamma rays, the number of photoelectrons resulting from a single scintillation may be quite large—of the order of 6000 for ^{137}Cs, with the bialkali (K$_2$CsSb) photocathode. In this case, the interference from the background count of the photomultiplier tube is negligible. However, in the case of liquid scintillation counting used for low-energy beta emitters such as tritium (^{3}H) and ^{14}C, the number of photoelectrons created per scintillation may be small. It then becomes important to differentiate them from the dark-current spectrum of the tube itself.

A typical background pulse spectrum for a photomultiplier having a GaP first dynode is shown in Fig. 7. The peak shown at the left of the figure corresponds to thermionic electrons emitted from the photocathode. The resolution of this peak is only possible because of the high gain of the first dynode. At the right of the figure and down several orders of

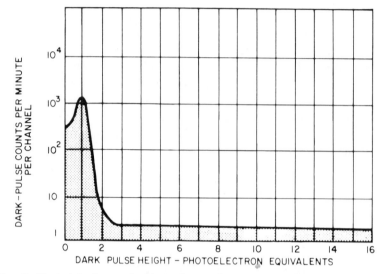

FIG. 7. Typical background pulse spectrum for a photomultiplier with a K_2CsSb photocathode and GaP·first dynode (RCA Type 8850). In this plot, one photoelectron equivalent pulse height is equal to 8 counting channels. The integration time is 10 μsec. The total count summing all channels from $\frac{1}{8}$ photoelectron equivalent through 16 photoelectron equivalents is 9000 counts/min. The data were taken after 24 hours operation in darkness.

magnitude from the numbers of thermionic electrons are a class of large pulses. These may originate from several sources: after-pulses[15] initiated by light feedback or ions; scintillations in the glass envelope of the tube itself resulting from radioactive contaminants such as ^{40}K or from external radiations such as cosmic rays.

One way to minimize the interference of photomultiplier background pulses with those originating in the liquid scintillator is to couple two photomultipliers to the liquid scintillator and count only those pulses which are coincident in the two photomultipliers. This is quite an effective technique but is somewhat inefficient at very low pulse amplitudes because of the statistics of small numbers of photoelectrons.

The introduction of the GaP dynode mentioned earlier has a significant bearing on the problems related to the identifying of very small pulses.[16] The improved statistics of the amplification process result in much better

[15] G. A. Morton, H. M. Smith, and R. Wasserman, *IEEE Trans. Nucl. Sci.* **NS-14**, No. 1, 443 (1967).

[16] G. A. Morton, H. M. Smith, Jr., and H. R. Krall.[7]

identification of the number of electrons in each pulse. In Fig. 8 is illustrated a typical photoelectron pulse-height spectrum showing distributions corresponding to single, double, triple, and quadruple electron emissions occurring during the resolving time of the pulse-height analyzer. The source of light in the experiment was selected so that, on the average, only a small number of photoelectrons would be emitted in the resolving time of the analyzer. With a conventional first dynode all of these peaks would have been smeared together.

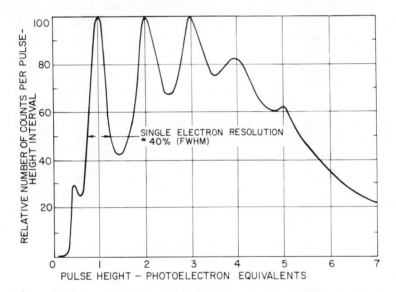

FIG. 8. Photoelectron pulse-height spectrum showing resolution of single and multiple electron events for a photomultiplier with GaP first dynode (RCA Type 8850). The light level was adjusted in this experiment to provide the particular ratios of 1, 2, 3, 4, and 5 photoelectrons observed per integration period. The explanation for the small peak near one-half photoelectron equivalent pulse height is unknown.

Photomultiplier tubes are exceedingly rapid in their response to light. No time delay has ever been measured for the emission of electrons by photons or by secondary electrons; for the latter, the delay has been shown to be less than 3×10^{-11} sec.[17]

Electron transit time effects, however, do limit time resolution capability in photomultiplier tubes. The actual delay in the arrival of electrons is usually not of prime importance; rather, it is the statistical fluctuation in

[17] M. H. Greenblatt and P. H. Miller, *Phys. Rev.* **72**, 160A (1947).

electron-arrival time which limits the measurement of short time intervals.[18] The error in determining the arrival time of a pulse is proportional to the reciprocal square root of the number of photoelectrons in the event. The photocathode-to-first-dynode transit is usually where the major contribution to time jitter occurs, both because the spacing is usually larger than that from dynode to dynode and because, at this point, the number of electrons in the pulse is the least and therefore of maximum importance in the statistics of arrival time spread.

The output design of the photomultiplier is also important in time measurements. The anode must be coupled to the external leads of the tube so that impedance mismatches are avoided. Photomultipliers designed primarily for high-speed measurements usually have a transmission or coaxial line output of 50 Ω characteristic impedance. Pre- and after-pulses are also possible if the anode is subject to multiple transits of electrons as, for example, through an anode of the shape of a grid. The distribution in arrival time for electrons in the photomultiplier tube can be minimized by increasing the operating voltage, and by proper electron-optical design of the dynode structure. Withdrawal fields for secondary electrons should be large and focusing should be such that transit times are essentially the same for different points on the electrodes. Thus, focusing-type dynodes are to be chosen rather than "venetian blind"-type or bucket-type.

In utilizing the anode output pulse for time discrimination,[19] one common technique is to choose a point on the rising characteristic of the pulse. Good results are obtained by discriminating at a point of approximately 20% of the pulse height; however, if the pulses to be observed are quite variable in height, choosing a fixed amplitude may give rather poor results. For variable-height pulses, it may be useful to differentiate the pulse and observe the time for the zero crossing which corresponds to the peak of the original pulse. Another technique is to form a negative delayed pulse by reflection in a transmission line stub which, when combined with the positive pulse, produces a zero crossing independent of the amplitude of the pulse.

By careful techniques, photoevents can be discriminated to the order of 10^{-10} sec depending upon the magnitude of the pulse.

[18] C. R. Kerns, *IEEE Trans. Nucl. Sci.* **NS-14**, No. 1, 449 (1967).

[19] L. G. Hyman and R. M. Schwartz, Study of high-speed photomultiplier systems. *Rev. Sci. Instrum.* **35**, 393 (1964); E. Gatti and U. Svelto, Revised theory of time resolution in scintillation counters. *Nucl. Instrum. Methods* **30**, 213 (1964); F. J. Lynch, Improved timing with NaI(Tl). *IEEE Trans. Nucl. Sci.* **NS-13**, No. 3, 140 (1966).

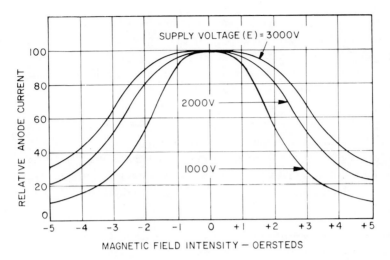

FIG. 9. Sensitivity dependence of a photomultiplier resulting from an axial magnetic field variation (RCA Type 8850).

To some degree, all photomultiplier tubes are sensitive to the presence of magnetic fields. Typical variation of sensitivity in the presence of a magnetic field is shown in Fig. 9. The loss of gain results from the deflection of electrons from their normal path between stages. Different characteristics result from different orientations of the magnetic field. Tubes with long electron-path lengths, as from the photocathode to the first dynode in tubes for scintillation counting, are the most susceptible to magnetic fields.

If photomultiplier tubes are to be used in the presence of magnetic fields, it may be essential to provide magnetic shielding around the tube. High-mu-material shields are generally available commercially. In some experiments, the earth's magnetic field may be critical, especially if the tube is moved about.

Magnetic fields may cause a slight magnetic polarization of some of the internal structure of the tube. In this case the tube may be degaussed by placing it in an alternating magnetic field and gradually withdrawing it. A coil producing 100 G on a 60 Hz alternating current is usually sufficient. Magnetic fields may be used to modulate the output current of the photomultiplier or to limit the photocathode area from which electrons are focused onto the first dynode.

Circuits for dividing the voltage between stages of multiplier photo-tubes are usually designed to minimize the power drain and to provide the output current range required. Generally, a voltage divider current of ten times the maximum anode current expected is adequate except for very strict linearity requirements.[20]

If the average output current requirement is modest but high-pulse currents are required, it may be sufficient to provide for peak currents by means of capacitors shunting the divider resistors.

Since the gain of multiplier phototubes varies so rapidly with voltage, special consideration should be given to the design or purchase of the power supply. Since the current drawn from the multiplier phototube is usually small, important characteristics of a power supply are minimum output voltage variation with line voltage or temperature.

11.1.3.3. Application. A photomultiplier tube is particularly useful for applications involving low light levels (10^{-5} down to 10^{-11} lm or less with refrigeration) or wide bandpass. Typical applications are in stellar photometry, spectrometry, scintillation counting, nuclear event timing, industrial controls, and laser ranging and communications.

Photomultiplier tubes are generally not used as a reference standard because of their susceptibility to aging effects; they are better employed as comparators between light sources. Fairly good stability may be expected if the anode current is kept small. For currents of the order of $10 \mu A$, changes of the order of 20% in several hours of operation are typical.

11.1.3.4. Types and Ratings. There are so many types and ratings of photomultipliers that a detailed catalog of them is impractical in this volume, and would soon become obsolete by the rapid pace of developments. Table II provides a listing of some of the more prominent manufacturers of photomultiplier tubes together with an indication of the type of photomultiplier manufactured by each one. For more particular data and a listing of types, it is suggested that the individual manufacturer's catalogs be consulted.

The following comments provide some guidance as to the merits of particular classes of tubes. Tubes with box and grid dynodes are excellent for electron collection efficiency, but rather poor for transit time spread. The circular cage is good for transit time spread but because of its small entrance aperture is at a disadvantage in any but small cathode types. Venetian-blind dynodes are excellent for collection efficiency and low dark current. Staggered linear arrays of semicylindrical dynodes are good

---

[20] R. W. Engstrom and E. Fischer, *Rev. Sci. Instrum.* **28**, 525 (1957).

for transit time spread and electron collection efficiency except for very large photocathodes.

Of the dynode materials, GaP has the highest gain and is most useful as a first dynode to improve statistics of multiplication; Cs_3Sb dynodes have higher gains than Ag-Mg or Cu-Be but are not as stable as the latter materials, especially at high currents.

11.1.4. Solid-State Photocells[21][†]

11.1.4.1. Principle of Operation. *Photocell* is a general term referring to any of a variety of photosensitive semiconductor devices including photoconductive cells, photodiodes, phototransistors, photoelectromagnetic detectors, and photovoltaic cells, in which transport of charge takes place through a solid.

It has only been within the last two decades, with the emphasis on semiconductor research, that a real understanding of the basic mechanism of photocells has been achieved, although Becquerel discovered a photovoltaic effect in electrolyte in 1839, and Willoughby Smith discovered the photoconductivity of selenium in 1873. Most insulators and semiconductors show photoconductive effects, but only a few of these are useful as photoelements and then only after careful processing.

The basic mechanism in a photoconductive cell is the absorption of a photon and the excitation of an electron to a higher energy level within the solid.[‡] Conduction may take place either by electron or hole

---

[21] Solid-State Electronic Issue, *Proc. IRE* **43** (12), (1955); R. G. Breckenbridge, B. R. Russel, and E. E. Hahn, "Photoconductivity Conference." Wiley, New York, 1956; R. H. Bube, "Photoconductivity of Solids." Wiley, New York, 1960; A. Rose, "Concepts in Photoconductivity and Allied Problems." Wiley (Interscience), New York, 1963; C. P. Hadley *et al.*, Phototubes and photocells. "Encyclopedia of Chemical Technology" (Kirk-Othmer, eds.), Vol. 15, 2nd ed. Wiley, New York, 1968; J. J. Brophy, "Semiconductor Devices." McGraw-Hill, New York, 1964; H. Levinstein (ed.), "Photoconductivity." Pergamon, Oxford, 1962; Luminescence and Photoconductivity Issue, *RCA Rev.* **20** (4), (1959); D. L. Greenway and G. Harbeke, "Optical Properties and Band Structure of Semiconductors." Pergamon, Oxford, 1968; M. Ross, "Laser Receivers." Wiley, New York, 1966; Special ISSCC Issue on optoelectronic circuits and solid-state microwave circuits, *IEEE J. Solid-State Circuits*, **SC-4**, No. 6 (1969); Special issue on solid-state imaging, *IEEE Trans. Electron Devices* **ED-15**, No. 4 (1968); *Rep. Int. Conf. Photoconductivity, 3rd, Stanford Univ., August 1969.* Pergamon, Oxford, 1970.

[†] See also Vol. 6B, Chapter 12.1.

[‡] For what follows, see also the relevant chapters of Vol. 6.

TABLE II. List of Photomultiplier Manufacturers

Manufacturer identification	Address of outlet in U.S.A.	Photomultiplier product
Bendix	The Bendix Corporation Electro-Optics Division 1975 Green Road Ann Arbor, Mich. 48107	Channel
CBS	CBS Laboratories High Ridge Road Stamford, Conn. 06905	Customized; image section for limiting the effective photocathode area
Centronic	The Bailey Co. 5919 Massachusetts Ave. Washington, D. C. 20016	Venetian-blind
DuMont	DuMont Laboratories 750 Bloomfield Ave. Clifton, N. J. 07015	General with box and grid, and venetian-blind dynodes
E.E.V.	Calvert Electronics, Inc. 220 East 23rd St. New York, N. Y. 10010	Side-on
EMR	Electro-Mechanical Research, Inc. Box 44 Princeton, N. J. 08540	Customized; venetian-blind; rugged
Hamamatsu	Hamamatsu TV Co., LTD c/o Kinsho-Mataichi Corp. 80 Pine St. New York, N. Y. 10005	General

ITT	Electron Tube Division, ITT Fort Wayne, Ind. 46803	General; Image section for limiting the effective photocathode area
Philips	Amperex Electronic Comp. Electro-Optical Devices Div. Statersville, R. I. 02876	General; high-speed focused
RCA	RCA Electronic Components Harrison, N. J. 07029	General; costumized; high-speed focused; venetian-blind; GaP dynodes; III-V cathodes
Space Research Corp.	Space Research Corp. 1525 Kings Highway Fairfield, Conn. 06430	Venetian-blind
Sylvania	Sylvania Electronic Systems–Western Div. P. O. Box 188, Mountain View, Calif. 94040	High-speed cross-field
Toshiba	Toshiba America, Inc. 530 Fifth Ave. New York, N. Y. 10036	General
Twentieth Century	Bailey Company 5919 Massachusetts Ave. Washington, D. C. 20016	General
Varian/EMI	Gencom Division Varian/EMI 80 Express St. Plainview, N. Y. 11803	General; high-speed cross-field

movement. This increase in conductivity remains until the extra carriers disappear. An electron may also be lost by combining with a hole, either directly or indirectly by first being captured into a bound state. In addition to such bound states, there exist trapping states which may capture an electron or hole and hold it until thermal agitation again frees the carrier. These states are associated with impurities, vacant lattice sites, interstitial atoms, and crystal defects.

Spectral response, especially the long wavelength cutoff, is primarily determined by the energy separation of the conduction band and the ground state for an intrinsic photoconductor. For an extrinsic photoconductor, the threshold wavelength is determined by the energy separation of the acceptor level from the filled band in a P-type material or by the energy separation of the donor level from the conduction band in an N-type material. Sensitivity is determined by the absorbed fraction of the radiant energy in the photoconductor and by the accumulated time an electron or hole remains mobile before recombination occurs. Speed of response is determined by the time an electron spends in the conduction band and in trapping states.[22]

Numerous other characteristics of photocells—infrared quenching, dark current, nonlinear response with light, temperature characteristics, dependence on previous history—are determined by the type and abundance of the bound and trapping states.

In a photoconductive cell, it is customary to increase sensitivity by using interdigitated electrodes which, in effect, provide many cells in parallel with an increase in the electrical field on each. For a given illumination, applied voltage, and cell size, the output current increases approximately as the square of the number of interdigitated electrodes.

Photodiodes or junction photoconductive cells are photocells with rectifying junctions such as P-N junctions in germanium. When voltage is applied in the reverse direction, only a small current flows in the dark. Excitation by light near the barrier junction produces holes which move through the junction to the negative electrode and electrons which move to the positive electrode. Because of the nature of the rectifying junction, current does not continue to flow after the holes and electrons reach their respective collectors. For this reason the current-voltage characteristic shows a saturated characteristic similar to that of a vacuum phototube. An efficiency of one electron and hole pair per incident quantum is approached in some cells.

[22] A. Rose, *RCA Rev.* **12**, 362 (1951); R. H. Bube,[21]; A. Rose.[21]

A phototransistor combines transistor action with light-injected electron-hole pairs. For example, a silicon phototransistor may have two junctions to form an *NPN* device. Contrary to the case of a transistor, there are only two contacts—no electrical signal being applied to the *P*-section. The *P*-section acts in the manner of a light-activated grid, passing an amplified current when light-activated charges are created.[23] A phototransistor has higher sensitivity but larger dark current than a junction photocell. In addition to the phototransistor, similar solid-state techniques have been used to produce a variety of similar devices such as a light-gated rectifier employing *PNPN* silicon or a silicon optical hybrid device using a light-sensitive diode and a transistor.

A photo-electro-magnetic detector employs a semiconductor in which a magnetic field is applied at right angles to the incoming radiation and a potential difference is developed at right angles both to the magnetic field and the direction of the radiation.

Photovoltaic cells[24] develop a voltage usually of the order of 0.1 to 0.2 V when irradiated. Current flow depends upon the external circuit load and the level of illumination; it can be of the order of several milliamperes. Functioning of a photovoltaic cell depends upon some sort of electrical barrier or rectifying junction in the cell. This barrier may be a *P-N* junction as in the case of a germanium- or silicon-type cell. It may be a heterojunction-type contact between a metal and a semiconductor as is the case of the cadmium sulfide cell, the cuprous oxide cell, or the selenium photovoltaic cell. The maximum open-circuit voltage which can be developed (neglecting surface-state effects) is the difference in work functions of the two components of the cell.

The silicon solar cell, so common in space applications, usually consists of a single crystal of *P*-type silicon with a layer of *N*-type material diffused into it. The cell may have a series of narrow conductors on the surface to reduce series resistance. It may also have a nonreflective coating to increase the output.

An interesting new type of photodiode is the avalanche photodiode.[25] These cells are typically made of germanium or silicon. Multiplication

[23] W. Shockley, M. Sparks, and G. K. Teal, *Phys. Rev.* **83**, 151 (1951).

[24] P. G. Witherell and M. E. Faulhaber, The silicon solar cell as a photometric detector, *Appl. Opt.* **9**, 73 (1970); S. W. Angrist, "Direct Energy Conversion." Allyn and Bacon, Boston, Massachusetts, 1965; IEEE Photovoltaic Specialists Conf. Record, November 19-21, 1968, 68 C 63 ED.

[25] J. R. Baird and W. N. Schaumfield, Jr., *IEEE Trans. Electron Devices*, **ED-14**, No. 5, 233 (1967); H. Ruegg, *IEEE Trans. Electron Devices* **ED-14**, No. 5, 239 (1967).

of the fundamental photogenerated current is obtained by extra carrier generation in an avalanche mode in which a relatively high field is applied to a reverse-biased junction. As the bias voltage approaches breakdown voltage, a fairly high useful gain of the order of 200 is obtained. Although the useful photosensitive area is quite small, of the order of 0.25 mm in diameter, these devices are of particular interest because of their very high speed. Risetimes for a pulsed signal current has been reported to be less than 10^{-10} sec. The cells have potential usefulness in a variety of laser applications.

11.1.4.2. Characteristics. A great range in characteristics is obtained with photocells. Not only are there many semiconductor materials which may be used, but there are wide ranges of treatments, impurities, etc., which may be used to obtain desired characteristics. Although photocells have been known and used for many years, a better understanding of their physics has resulted in a rapid proliferation of types, materials, and uses.

Photocell spectral responses extend over a wide range of wavelengths, from the ultraviolet to the infrared. A most useful spectral response, particularly for illumination measurement, is that of the selenium photovoltaic cell used in most photographic exposure meters. Germanium or silicon junction photodiode cells have the interesting property that over much of their spectral sensitivity range they have almost unity quantum yield. Military interest in infrared sensors has stimulated a number of remarkable developments.[26] Among the earlier developments were PbS, PbSe, and PbTe. More recent developments include InSb, InAs, HgCdTe and various formulations of Ge and Si. The latter two materials have been rather thoroughly studied as a function of added impurity atoms which control the long wavelength cutoff corresponding to impurity ionization energies.

Figure 10 displays spectral responses of the more common commercial photocells as well as of some of the new promising materials.

Although the fundamental process of liberating an electron to the conduction band by the absorption of a quantum would imply a linear

---

[26] See Vol. 6 of this series, where numerous chapters constitute excellent background reading. Quantum yields of photoconductors, photocells, and other devices are the subject of critical reviews by R. C. Jones, *Advan. Electron. Electron Phys.* **5**, 1 (1953); **11**, 88 (1959); P. W. Kruse, L. D. McGlauchlin, and R. B. McQuistan, "Elements of Infrared Technology." Wiley, New York, 1962; F. Kneubuhl, Diffraction grating spectroscopy. *Appl. Opt.* **8**, 505 (1969); Special issue on infrared physics and technology, *Proc. IRE* **47**, No. 9 (1959); "Semiconductors and Semimetals" (R. K. Willardson and A. C. Beer, eds.), Vol. 5, Infrared Detectors. Academic Press, New York, 1970.

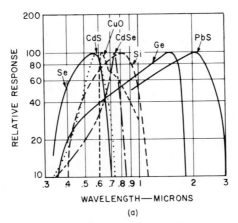

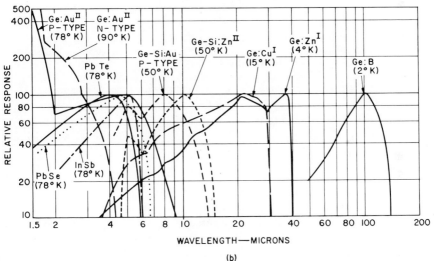

FIG. 10. Relative spectral characteristics of a number of photocells (commercial and developmental). All data refer to room temperature (25°C) operation except as indicated.

relationship between response and light, linearity is not always observed. For a selenium photovoltaic cell, the output current approaches linearity as the external circuit resistance approaches zero. Broad-area CdS and CdSe photoconductive cells exhibit ranges of linearity, sublinearity, and superlinearity.[27] Germanium junction-diode cells operated as photoconductive devices show good linearity.

[27] R. H. Bube, *Proc. IRE* **18**, 1936 (1955); C. P. Hadley and E. Fischer, *RCA Rev.* **20**, 635 (1959).

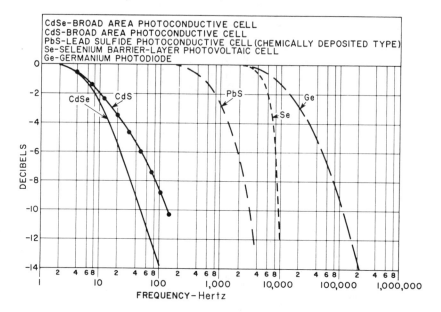

FIG. 11. Frequency response characteristics for several types of photocells.

Current-voltage characteristics are similarly varied. Junction-type photodiodes show a saturated current characteristic when the voltage is sufficient to collect the holes and electrons generated. Some CdS photoconductive cells have a photocurrent which increases linearly with voltage. Other CdS cells and most CdSe cells show a variation of current which is superlinear.

In general, photocells are quite limited in speed of response. Transistor-derived photocells are generally useful to somewhat beyond 100 kHz; photovoltaic cells, to several thousand hertz. Photoconductive cells, on the other hand, are usually quite slow, being useful to only tens or hundreds of hertz. Some typical frequency response characteristics for photocells are shown in Fig. 11.

Photocell properties are markedly affected by temperature as might be expected for semiconductor devices. Dark current in photojunction cells and in photoconductive cells increases exponentially with temperature. The internal resistance of a photovoltaic cell is similarly affected. Sensitivity is usually more independent of temperature.

Noise in photocells[28] originates from a number of causes (see Part 13)

[28] Al van der Ziel, "Noise," p. 204 ff. Prentice-Hall, Englewood Cliffs, New Jersey, 1954.

and varies widely in different types of cells. At best, the relative noise in some cells may approach theoretical limits; at worst, various excess noise sources may be present and the cell may be quite useless as a low-level detector.

Particularly in the case of infrared detectors, it has become useful to compare the relative merits of detectors by means of figures of merit such as noise equivalent power (NEP), detectivity (D), or a special detectivity figure, D* (pronounced "dee-star"). NEP is the rms value of the sinusoidally-modulated incident radiant power in watts which gives unity signal-to-noise ratio under specified conditions which include the modulation frequency, the frequency bandwidth of the measuring system (often taken as 1 Hz), the detector operating temperature, the detector angular field of view, the background temperature, and the spectral content of the test radiation. Specification is sometimes referred to radiation at the peak of the spectral response. Detectivity, D, is the reciprocal of the NEP value. D* has become a common designation because it takes account of the sensitive area of the particular detector permitting the comparison of detectors of different areas. In terms of NEP, D* is defined as

$$D* = \frac{A^{1/2}(\Delta f)^{1/2}}{NEP}$$

where A is the working area of the cell, in cm^2, Δf is the bandwidth of the measurement in Hz, and NEP is the noise equivalent power in watts (most frequently at 500°K black body radiation).

An ideal infrared detector is limited by the statistical fluctuations of the background radiation from which the signal radiation must be distinguished. Thermal background radiation from surfaces within the angle of view of the detector causes excitation of carriers in the detector as does the signal radiation. The D* values corresponding to the background radiation limit for an ideal detector have been calculated[29] for a detector having a sharp cutoff wavelength, λ_0, and unity quantum efficiency at all shorter wavelengths, for various scene background temperatures and a 180° angular field of view (2π sr). These are shown in Fig. 12; the value of D* in this figure corresponds to monochromatic radiation at the cutoff wavelength. Some detectors have D* values within a factor of 2 or 3 of the background-limited values shown in Fig. 12. Detectors usually are provided with cooled apertures which limit the

[29] P. W. Kruse, L. D. McGlauchlin, and R. B. McQuistan.[26]

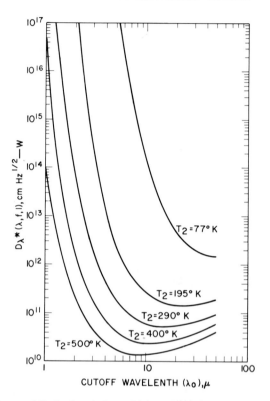

FIG. 12. Background-limited peak D* values for an ideal photodetector as a function of cutoff wavelength for several background temperatures (T_2) and a 180° field of view.

field of view to that actually required. For an angular aperture of θ, assumed to be conical with a half-angle of $\theta/2$, the value of D* is given by

$$D^*(\theta) = D^*(180°)/\sin(\theta/2).$$

Well-made photocells are quite stable over long periods of time. This is particularly so for selenium cells which are frequently used as reference standards. Care should be taken in not overexposing selenium and other type of cell to excess light as a temporary fatigue will set in. Some photoconductive cells are quite sensitive to previous light history. Temporary sensitivity values are related to previous exposures through distributions of electrons and holes in various energy levels of the semiconductor.[30] These distributions affect the probability of recombination, the carrier

[30] R. H. Bube, *J. Phys. Chem. Solids* **1**, 234 (1957).

lifetime and, hence, the sensitivity. Under the new light conditions, a redistribution takes place and the sensitivity gradually assumes a characteristic value. These effects may be quite reproducible so that under controlled conditions the cell may be quite stable—this is true for certain CdS and CdSe photoconductive cells. Some cells change characteristics upon exposure to the atmosphere, particularly water vapor. Better stability is, therefore, obtained in glass-sealed envelopes than in plastic envelopes, which are somewhat pervious to moisture.

11.1.4.3. Applications. Development of new and improved photocells has been and is proceeding at a rapid pace. It is very likely that there will be a great increase in the applications of these semiconductor devices. Most of these devices are remarkable for their high sensitivity and the consequent simple circuit requirements.

Broad-area CdS photoconductive cells are nicely suited to relay operation in applications such as street light turn-on, automatic door opener, industrial counting, and various smoke and fire alarms. Together with a Triac and a solid-state light source, an all-solid state relay system can be devised using a CdS cell to control fairly large amounts of power. CdS cells are also used as automatic TV brightness and contrast controllers, as iris control in cameras, and as a means of automatically maintaining focus in slide projectors.

Because CdSe is faster than CdS, CdSe cells are used in counters requiring higher speed. They are also used in an interesting application of photochopper for converting dc to ac to simplify amplification problems. This is accomplished by using the cell as a load resistance whose value is modified by an ac-operated light source. Important considerations are sensitivity, linearity, lack of polarization effects, and time response sufficient to follow the light modulation.

With silicon technology becoming more and more common, one finds silicon junction cells as the sensor in motion picture sound replacing the old gas-filled phototube. Variations of silicon cells are used in card reading or punched tape reading for computers. Silicon-cell arrays are now being used in a character recognition application.

The germanium cell is not used as much as it was a few years ago, being replaced by silicon; but for laser detection, it is still useful because of the longer wavelength sensitivity.

Photovoltaic cells are particularly useful as portable light meters because of the elimination of battery requirements. The silicon cell has been widely used as a power source for space vehicles. It has been found that N-on-P-type silicon cells have more resistance to nuclear radiation than P-on-N-type which makes the N-on-P-type more suitable for critical satellite applications.[31]

11.1.4.4. Types and Ratings. There are so many suppliers and so many varieties of photocell and the field has been developing so rapidly that no listing is provided here. It is suggested that the literature or well-known manufacturers be consulted for specific information regarding types and ratings.

[31] J. A. Baicker and B. W. Faughnan, *J. Appl. Phys.* **33**, 3271 (1962).

11.2. Cathode-Ray Devices*

11.2.1. Electron-Ray Indicator Tubes ("Magic Eye" Tubes)[1-4]

The electron-ray indicator tube is a vacuum tube which gives a visible indication of the magnitude of a controlling voltage. The visible indication is in the form either of a shadowed sector of a bright circular pattern, the shadow angle being proportional to the input voltage, or a linear band whose length is varied by the input. This type of tube is often used as a tuning indicator in radio receivers and as a volume-level indicator in recording equipment. It is also used as an indicator in many types of scientific instruments, particularly as a null indicator in bridge circuits. This type of tube provides a relatively inexpensive, yet sensitive, voltage indicator having a high input impedance (little loading of the circuit measured).

A typical electron-ray indicator tube is the 6E5. It provides, within a conventional receiving-tube envelope, a triode amplifier section whose anode is directly coupled to the control electrode of an indicator section at the top of the tube envelope. The indicator is viewed head-on through the top of the tube. The indicator has a separate cathode which emits electrons to a phosphor-coated metal anode disc. The ray-control electrode is a blade inserted at one point between the cathode and the anode. When the ray-control electrode is at a lower potential than the anode it deflects the electrons and casts a dark shadow on the anode; the width of the shadow depends on the potential of the ray-control electrode.

The various types of indicator tubes differ in the characteristics of the

---

[1] RCA Receiving Tube Manual RC-27, p. 79. RCA Electron. Components, Harrison, New Jersey, 1970.

[2] F. E. Terman (ed.), "Radio Engineers' Handbook," p. 320. McGraw-Hill, New York, 1943.

[3] L. C. Waller, *RCA Rev.* **1**, 111 (1937).

[4] F. Langford-Smith, Radiotron Designer's Handbook, 4th ed., p. 1134. RCA Electron. Components, Harrison, New Jersey, 1953.

---

* Sections 11.2.1, 11.2.2, and 11.2.4 are by **Robert P. Stone**; Section 11.2.3 is by **Paul D. Huston** (deceased).

built-in amplifier section, or the absence of a built-in amplifier, and in the type and number of independent visible displays in a single tube. The 6E5 incorporates a sharp cutoff triode with a single, angular-type display. This design provides high sensitivity and a nearly linear relationship between triode control-grid voltage and shadow angle. A 7-V signal reduces the shadow angle to zero; a 0.1-V change produces a readily discernable indication.

The 6U5 is similar in over-all structure, but incorporates a remote cutoff triode which produces a nonlinear (nonsaturating) characteristic. The 6AF6-G has two identical and independent indicators and no amplifier section. The EM84/6FG6 incorporates a remote cutoff triode and is mounted in a miniature tube envelope. The indicator is a bright band on the side of the tube envelope, parallel to the tube axis. The sensitivity is about 1 mm of shadow length elongation per volt of signal. The 6AL7-GT contains three independent indicators in a single bulb and no amplifier sections. The display consists of two parallel bright bands, viewed through the top of the tube. A single deflecting electrode controls one end of each band simultaneously. The other end of each band is controlled independently by a separate deflecting electrode.

11.2.2. Cathode-Ray Tubes[5-8]

11.2.2.1. General Principles. A cathode-ray tube is an electron device with electrical input and simultaneously displayed visible output. It is characterized by an evacuated envelope containing at least one electron gun and a phosphor screen. The envelope is typically funnel-shaped and is usually formed from glass. The electron gun is located within the small-diameter neck of the funnel with its leads sealed through the stem at the base of the funnel. The stem also contains the exhaust tubulation, which is sealed off after the tube is evacuated. The large-diameter end of the envelope funnel is closed with a glass plate on the inner surface of which is deposited a thin layer of a cathodoluminescent phosphor.

In operation, the electron gun produces a thin, pencil-like stream of electrons. This electron stream is controlled in intensity (modulated)

[5] K. R. Spangenberg, "Vacuum Tubes." McGraw-Hill, New York, 1948.

[6] V. K. Zworykin and G. A. Morton, "Television," 2nd ed., Wiley, New York, 1954.

[7] RCA Electron Tube Handbook HB-3. RCA Electron. Components, Harrison, New Jersey.

[8] H. Moss, "Narrow Angle Electron Guns and Cathode Ray Tubes," Suppl. 3, Advances in Electronics and Electron Physics. Academic Press, New York, 1968.

by the gun, focused for minimum spot size on the phosphor screen, and deflected to the desired region of the screen. These functions are described in turn below, and the manner in which they are combined for specific application is then discussed.

The electron gun produces and controls the electron beam of the cathode-ray tube. A typical gun is composed of cylindrically symmetrical elements located along the axis of the tube, and produces a beam of circular cross section. The cathode is usually a small disc of thermionically emitting material on the closed end of a small-diameter cylinder. This disc is maintained at an operating temperature of about 1050°K by an electric heater within the cylinder. Closely spaced in front of the cathode is the control grid, a disc with an aperture somewhat smaller in diameter than the active area of the emitting material. The control grid is often supported mechanically by a cylinder of larger diameter than the cathode cylinder and concentric with it. Spaced a short distance further from the cathode is an accelerating grid, which consists of an apertured disc and cylindrical supporting section. This grid completes the basic structure of the triode gun, but a tetrode gun has additional accelerating electrodes. The final accelerating electrode is usually operated at the anode or "ultor" voltage, which is the highest positive voltage in the tube. When a voltage positive with respect to the cathode is applied to the first accelerating electrode, the magnitude of the flow of electron current can be modulated, or controlled, by the magnitude of the negative voltage applied to the control grid. The beam current decreases as the control-grid bias voltage increases, and a sufficiently large negative voltage completely cuts off the electron beam.

The electrostatic fields in the gun bend the paths of the individual electrons emitted from the cathode toward the axis, and accelerate them in the axial direction. The electrons cross the axis in the neighborhood of the control-grid aperture, and then diverge as they progress toward the phosphor screen. This point of minimum beam diameter is called the beam crossover. It is this crossover which is usually imaged as the spot on the phosphor screen.

In order to produce the small spot on the phosphor screen it is necessary to confine and redirect the electron beam when it diverges beyond its crossover. This process is directly analogous to the problem of optical focus and is accomplished by the use of electron lenses. There are two distinct forms of electron lenses, those which use appropriately shaped electrostatic fields, and those utilizing magnetic fields. Both types are used in cathode-ray tube practice.

The *Einzel* lens is a commonly used form of electrostatic focus lens. It consists of three elements, either apertured discs or short cylindrical sections. In the cathode-ray tube these elements become a part of the gun structure. The first lens element may be incorporated into the accelerating grid and operated at its potential; the adjustable focus voltage is applied to the central element; and the third element is operated at the same potential as the first element. The focus voltage is adjusted so that the beam crossover is imaged on the phosphor screen to produce the minimum spot size. For a given structure there are two voltages which will produce focus, one below and one above the potential of the end electrodes. The lower focus potential is usually used.

Whereas the electrostatic lens is usually a part of the electron-gun structure, the magnetic lens is generally formed by an electromagnet placed outside the tube envelope. The electron-gun structure then terminates with the last accelerating electrode. The magnetic focus coil, when placed over the neck of the tube with the coil axis coincident with the tube axis, produces a magnetic field which converges the electron beam in a spiral path. The current through the focus coil is adjusted to a value which images the beam crossover on the phosphor screen.

A focused spot size as small as 12 μm is possible in special high resolution tubes, and diameters as large as 1.25 mm are obtained in some situations. The spot attained in a given situation is determined by display requirements such as screen area, brightness required, and data input rate.

As there are two alternate means of focus of the electron beam, so there are two means of deflection: electrostatic and electromagnetic. Electrostatic deflection is accomplished by deflecting electrodes which are a part of the electron-gun structure. Two essentially parallel electrodes are located one on either side of the tube axis, between the focus elements and the phosphor screen. One electrode is made positive and the other negative with respect to the beam potential. When the electron beam enters the region between the deflecting electrodes on the tube axis, it is attracted toward the positive electrode, and repelled by the negative electrode, so that it bends away from the axis in a parabolic path. On leaving the deflection region, it continues to the screen in a straight-line path tangent to the parabola. One-dimensional deflection is thus obtained by varying the voltages on the deflecting electrodes. A second set of deflecting electrodes is then placed with its axis rotated by 90° about the tube axis to obtain the second dimension of deflection. In practice, the deflecting electrodes are not parallel but are flared to match the contour of the deflected beam in order to achieve maximum deflection sensitivity.

With magnetic deflection, the components are usually mounted outside the tube envelope. The simplest deflection unit consists of a coil with its axis perpendicular to the tube axis. When a current passes through the coil a magnetic field is produced parallel to the coil axis. The electron beam is deflected in a circular path in a plane normal to the coil axis. On leaving the deflection region, it proceeds to the screen on a path tangential to this circle. A second coil with its axis rotated 90° about the tube axis provides deflection in a direction orthogonal to that of the first coil.

The phosphor screen converts the electron beam input to the visible luminous output. A large variety of inorganic crystalline solids exhibit the property of cathodoluminescence, i.e., give off light under electron bombardment. Most commercial phosphors are prepared by synthesis and include controlled amounts of impurities called activators. The phosphors differ in color and in persistence of luminescence after excitation. The typical luminous efficiency of a P31 phosphor is 50 lm/W. The light output varies directly with the current and as about the second power of the accelerating voltage, for very thick screens.

Phosphors are available to produce almost any color in the spectrum. When the electron beam exciting the phosphor is removed, the luminous output decays in an exponential or power-law fashion. The time constant of this decay is a measure of the persistence. For different phosphors the persistence varies from microseconds to several seconds. Table I lists the properties of the most common commercial phosphors; a complete listing is given by JEDEC Electron Tube Council.[9]

Because the phosphor is usually an electrical insulator, some care must be taken to establish its surface potential. The phosphor may be simply deposited on the glass and the inside of the adjacent bulb portions made conducting with a graphite coating. The maximum, or anode, potential is applied to this bulb coating. If the phosphor has a secondary-emission ratio greater than unity at the anode voltage, it maintains itself at the anode potential by secondary emission. If, however, the anode potential is raised to a value above the so-called second crossover of the phosphor (beyond which the secondary-emission ratio is less than unity), the phosphor is not maintained at anode potential but "sticks" at its second

[9] "Optical Characteristics of Cathode Ray Tube Screens." Publ. No. 16B, JEDEC Electron Tube Council, New York, 1970.

TABLE I. Principal Screen Materials Used in Cathode-Ray Tubes

Designation	Composition	Color	Spectral maximum (Å)	Persistence (sec)	Luminous efficiency (lm/W)
P1	Zn_2SiO_4:Mn	Yellow-green	5250	$25 \cdot 10^{-3}$	30
P2	ZnCdS:Cu	Yellow-green	5350	$50 \cdot 10^{-6}$	32
P4	ZnS:Ag + ZnCdS:Ag	White	—	$60 \cdot 10^{-6}$	—
P7	ZnS:Ag + ZnCdS:Cu	Yellow-green	5550	0.4	38
P11	ZnS:Ag	Blue	4600	$30 \cdot 10^{-6}$	27
P14	ZnS:Ag + ZnCdS:Cu	Yellow-orange	6000	$5 \cdot 10^{-3}$	38
P16	$2CaO \cdot MgO \cdot 2SiO_2$:Ce:Li	Near U.V. (Bluish-purple)	3830	$0.1 \cdot 10^{-6}$	0.1
P20	ZnCdS:Ag	Yellow-green	5600	$60 \cdot 10^{-6}$	62
P22	ZnS:Ag	Purplish-blue	4500	$22 \cdot 10^{-6}$	8
	ZnCdS:Ag	Green	5150	$60 \cdot 10^{-6}$	55
	YVO_4:Eu	Red	6190	$1 \cdot 10^{-3}$	10
P24	ZnO:Zn	Green	5100	$1 \cdot 10^{-6}$	10
P31	ZnS:Cu	Green	5220	$40 \cdot 10^{-6}$	50

crossover potential. The light output then corresponds to the second crossover potential and not to the anode potential.

Because the second crossover may occur at 5 to 10 kV for many phosphors, the light output obtainable with high-voltage operation may be limited unless steps are taken to avoid this effect. This difficulty can be avoided if a thin layer of aluminum is evaporated on top of the phosphor. The layer must be thin enough to permit easy penetration by high-velocity electrons, yet establish continuous electrical conductivity over the surface and be connected to the anode potential. At higher voltages the loss in efficiency due to absorption of electrons in the aluminum film is more than offset by a gain in luminous output because the smooth mirror-like film reflects light which would otherwise be lost in the interior of the tube.

11.2.2.2. Applications. The oscilloscope is a laboratory or test instrument producing, on the face of a cathode-ray tube, a two-dimensional display of illuminated lines corresponding to the electrical inputs. Separate inputs are usually available for the electrical input to the horizontal and vertical deflecting electrodes. Alternate provision is usually made to provide internally a horizontal deflection linear with time. In addition to the cathode-ray tube, the oscilloscope usually contains the power supplies, deflection amplifiers, sweep generators, and synchronizing means.

The 5UP1 is typical of the standard oscilloscope cathode-ray tube. It uses electrostatic focus and deflection and has a total deflection angle of 30°. The bulb has a maximum diameter of 13 cm and is about 36 cm long. The tube is operated with a maximum anode voltage of 2500 V and has a deflection sensitivity of about 30 V/cm of deflection at this anode voltage.

Electrostatic focus and deflection are usually used for this type of service so that higher frequency response can be obtained. Response to 15 MHz is usual with standard techniques, and response to about 250 MHz is possible. At higher frequencies, the transit time of the electron through the deflecting electrodes becomes appreciable, and more sophisticated techniques must be used. These techniques involve deflection systems of the transmission-line variety, where the wave velocity is made equal to the electron velocity.[10] Sampling techniques may also be used with repetitive waveforms.[11] Frequencies greater than 10,000 MHz, or pulses shorter than 10^{-10} sec, may be displayed by these techniques.

[10] J. E. Day, *Advan. Electron. Electron Phys.* **10**, 252 (1958).
[11] R. Sugarman, *Rev. Sci. Instrum.* **28**, 933 (1957).

In wide bandwidth applications, in order to achieve high brightness with reasonable deflection sensitivity, it is desirable to accelerate the electron beam in the region between the deflecting electrodes and the phosphor screen. The simplest embodiment of postdeflection acceleration, involving graphite wall coatings at successively higher voltages, introduces a series of electron lenses which distort the reproduced pattern. This may be corrected by a field forming mesh at the end of the deflection region. Another solution utilizes a continuous fine resistive spiral along the neck of the bulb which minimizes the lens action. The 5BGP1 is a tube of this type.

The phosphor chosen for the oscilloscope cathode-ray tube depends upon the type of observation to be made of the trace. For visual observation of repetitive phenomena, a phosphor whose output falls in a sensitive region of the human eye, such as the P31 phosphor, is used. A long-persistence phosphor such as the P7 is used to display transient data. For photographic recording, a phosphor with high actinic brightness, such as P11, is used.

The flying-spot cathode-ray tube may be used to generate a video signal from a slide or transparency. The 5ZP16 is typical of this type of tube. This tube produces a small, intense, and fast-decaying spot of light which is scanned over the transparency. The transmitted light is modulated by the density of the film and its intensity is measured by a phototube which collects light from any point of the film. The output of the phototube is the desired video signal.

The family of cathode-ray tubes intended for the display of television pictures is known as picture tubes. Because of the wide deflection angle required for large screen size, magnetic deflection is almost universally used. Magnetic focus was used in early tubes, but electrostatic focus is now used because it minimizes the external components required. The P4 phosphor is universally used for black-and-white picture reproduction because of the white color produced. The phosphor screen is usually aluminized. This process not only increases the brightness but also eliminates the ion-trap magnet used with early tubes. Some negative ions are generated in the region of the gun and are accelerated toward the phosphor screen. Because of their larger mass they are relatively unaffected by magnetic deflection and are concentrated near the center of the screen. Such ions soon discolor an unprotected phosphor and produce an "ion spot" at the center of the screen. In early tubes the gun was skewed and an additional magnet used to direct only electrons to the phosphor screen. The use of aluminized coatings makes this arrangement unnecessary because the heavier ions do not penetrate the aluminum film.

The 22VABP4 is a picture tube typical of current design. It utilizes electrostatic focus and magnetic deflection with 110° total deflection angle. This tube has a rectangular glass bulb with a very short neck. The large end of the bulb is about 40 cm × 50 cm (which, with the rounded corners, gives a screen diagonal slightly greater than 56 cm (22 in.) hence the new standard 22 V designation) and the total depth of the bulb is less than 38 cm. The maximum operating voltage is 23.5 kV, and an aluminized phosphor screen is used. In normal use the tube is operated with an average highlight brightness of over 100 ft-L (340 cd/m²).

The usual picture tube is intended for direct viewing of the phosphor screen. Tubes are available, however, which produce an extremely bright image which may be optically projected onto a large viewing screen. The 5AZP4, operating at 40 kV, projects a picture of several foot-lamberts brightness onto a 1.8 × 2.4 m screen. The 7WP4, operating at 80 kV, produces a similar brightness on a 4.5 × 6.0 m screen.

For reproduction of television pictures in full color, the color picture tube has been developed. The shadow mask principle is almost universally used, and a typical version of this type is the 25VABP22, which has a useful screen area of 0.20 m². The 25 V designation is the new standard method of specifying the tube by its useful screen diagonal. This color picture tube is essentially three tubes in one, with a separate electron gun and system of phosphor dots for each primary color. A "shadow mask" permits each electron gun to illuminate only the appropriate family of phosphor dots. Recent developments include a black matrix between the phosphor dots. This lowers the reflectivity of the screen to ambient light, so that higher transmission faceplates can be used. This permits a design for higher light output without sacrifice of contrast.

In radar applications, there is often a period of several seconds between successive excitations of a given spot on the face of the cathode-ray tube. Use is then made of a long-persistence phosphor such as the P7 to retain a visible, though dim, display between scans. With a short-persistence phosphor the eye would see only a moving spot of light, because the repetition rate is well below the flicker frequency, and would retain little sense of the over-all display. Magnetic deflection is usually used, and angular deflection is often introduced into the display by physically rotating the deflection yoke on the tube neck.

A typical tube for radar use is the 5FP7A. This tube uses magnetic focus and magnetic deflection. It is approximately 13 cm in diameter and 28 cm long and has a total deflection angle of 53°. It operates at a maximum of 8 kV.

The above families of cathode-ray tubes are available in a wide variety of sizes and bulb configurations. Oscilloscope and radar display tubes may have round or rectangular faceplates with diameters from less than 2 cm to as large as 56 cm. For special readouts or recording, a variety of fiber optic and wire lead faceplates are available to transfer light or charge without lateral spreading, as well as thin strip membranes which are electron permeable. Picture tubes are now generally produced with rectangular faceplates having a diagonal dimension of from 7.5 to 70 cm.

Cathode-ray tubes with multiple independent electron guns may be used for the simultaneous display of concurrent signals. Tubes with two electron guns are commonly available, and tubes have been produced with as many as ten guns.

In addition to the physical variations, there are several tubes which are functionally quite different. Among these is the flat cathode-ray tube. One form[12] of the flat tube consists of a thin rectangular envelope, with a phosphor on one rectangular face and a series of parallel deflecting strips on the opposite face. An electron gun introduces a beam at one corner of the bottom of the tube and parallel to the bottom edge of the phosphor screen. The beam is first deflected upward by a series of electrodes at the bottom of the tube, then turned toward the phosphor screen by the deflecting strips on the back face of the tube. These deflecting strips, as well as the phosphor screen, may be made transparent so that the viewer looks through the tube, seeing the cathode-ray display superimposed upon the background seen by direct vision.

Another type of special cathode-ray tube generates an electron beam whose cross section is some readily selectable character. In one form[13] of this tube, a stencil pattern with cutouts corresponding to the desired characters is inserted in the electron gun. The usual round cross-section electron beam is deflected to illuminate the desired character on the stencil and the beam transmitted through the stencil assumes the desired cross section. The beam is then focused and deflected to the appropriate point on the phosphor screen. This type of character display has application in fast computer readouts.

For computer readout and radar display application, increasing use is being made of cathode-ray tubes whose phosphor screens have a luminous

[12] W. R. Aiken, *Proc. IRE* **45**, 1599 (1957).
[13] C. R. Corpew, *SID IDEA Symp. Digest* 98 (1970).

output whose color is voltage dependent. Multiple color displays can be obtained either by switching the voltage applied to a single electron gun or by utilizing multiple guns operating at different voltages.

11.2.3. Camera Tubes

11.2.3.1. General Principles. A camera tube is an electron device with an optical image input and providing an electrical output. The optical image is transformed into an electrical image by a photoelectric process. The surface area containing the electrical image is scanned in a predetermined and regular format so that at a given moment the electrical output is derived from a given incremental area (picture element) in the image. Thus the output, the video signal, is a time-varying current which describes the brightness of each picture element in turn.[14] The video signal when applied to the Z-(intensity) input of a cathode-ray tube results in a reproduction of the original image, displayed on the phosphor screen. The electron beam of the cathode-ray tube must be deflected in synchronism with the scanning action in the camera tube. When this sequence is accomplished very rapidly the eye perceives the complete image apparently simultaneously, and when the sequence is continuously repeated the result, as in cinematography, is a rapid sequence of images seen as a continuous moving picture. Thus we have television.

Camera tube classification. Two camera tube characteristics are particularly significant and can be used as a basis for classification. One characteristic is whether or not a storage mechanism exists. The other classifies according to the type of photoelectric effect employed.

11.2.3.2. The Storage Mechanism. Even in its simplest form the camera tube combines at least two distinct functions, i.e., photodetection and area dissection or scanning.

The possibility of storage[15] is one of the most significant considerations resulting from the scanning function. If a storage mechanism is not present, the camera tube output will, at any moment, be a function of the illumination on the picture element (incremental area) then being addressed by the scanning process. Illumination on that area prior or subsequent to its interrogation is wasted energy, yet, in most camera situations all parts of the image area are continuously illuminated. The use of a storage mechanism in conjunction with the photodetector permits

---

[14] IEEE Std. 160.

[15] M. Knoll and B. Kazan, "Storage Tubes and their Basic Principles." Wiley, New York, 1952.

the effect of continuing illumination to be integrated and thereby provides a greater output current. It can be shown that the output improvement factor, i.e., sensitivity increase, can be roughly the ratio of total image area to picture element area. Customarily this ratio exceeds 100,000.

11.2.3.2.1. NONSTORING CAMERA TUBES. Only one form of nonstoring camera tube is generally obtainable. It is the image dissector which is discussed in a subsequent section. All other camera tubes described here employ storage.

11.2.3.2.2. STORAGE CAMERA TUBES.[16] Several forms of camera tube exist in which storage is employed. Representative forms are described in the following sections. All (1) employ an electron beam and therefore contain an electron gun; (2) supply or require a means of deflecting the electron beam to accomplish scanning; and (3) contain a capacitance-type storage surface—customarily called the target.

11.2.3.2.3. TARGETS. Target constructions and materials are many and varied. All have in common the ability to behave as though they were a mosaic-like structure—in effect an array of isolated picture-element-sized individual capacitors. Such a target can retain an electrical pattern or image because the lateral motion of charge carriers is inhibited.

Silicon targets. One of the more recent target constructions consists of a wafer of single-crystal silicon semiconductor into one surface of which is fabricated an array of diodes. When reverse-biased each diode is an isolated storage capacitor. Typical arrays have diode densities of 500,000 to 1,000,000 per square centimeter.

Dielectric film targets. Most varieties of camera tube storage targets contain very thin films of high-resistivity material (10^{12} Ω-cm is typical). The absence of conductive coatings on at least one side of such a target film results in the desired mosaic-like behavior because of the very high lateral leakage resistance. These films are usually much less than 10 μm in thickness.

11.2.3.2.4. TARGET ACTION. The target is positioned in the camera tube so that one surface faces the electron gun but at a distance sufficient to allow for an intervening deflection section. See Figs. 1, 2, and 3. A conductive electrode adjacent to the opposite surface of the target is connected to the outside of the tubes as "the target connection." The

[16] B. Kazan and M. Knoll, "Electronic Image Storage," Sect. V, pp. 321–411. Academic Press, New York, 1968.

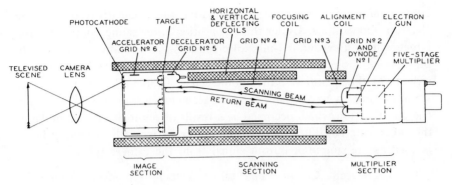

FIG. 1. Cross section of image orthicon camera tube.

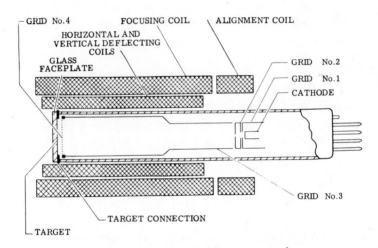

FIG. 2. Cross section of vidicon camera tube.

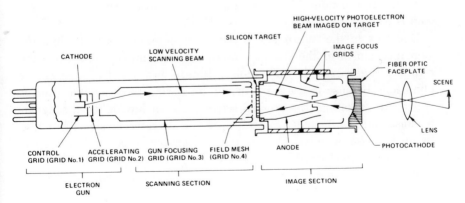

FIG. 3. Cross section of SIT camera tube.

target connection is operated at a potential positive with respect to the electron-gun cathode. Under these conditions the scanned portion of the target will be a capacitor (mosaic) charged so as to show approximately the voltage applied to the target connection. This charge is selectively reduced by the operation of the photoelectric effect employed. A repetitive scanning process will replace, in each elemental area, ideally all of the charge reduction which occurred as a result of photoelectric action since the previous scan. The charging current taken from the scanning beam is used as the source of output signal from the camera tube.

11.2.3.3. Output Circuit. The target voltage employed is generally limited by other considerations to a value sufficiently low so that the secondary emission ratio of the scanned target is much less than unity. Under this condition, the target in being scanned accepts electronic charge until the surface potential is approximately equal to that of the electron-gun cathode.

In full operation, the optical image is presented to the photoelectric surface. The resulting electrical image is stored by having it discharge the storage target. Repetitive scanning of the target restores the charge. The beam current accepted by the target constitutes a video signal and appears in the target connection lead.

11.2.3.3.1. RETURN BEAM SIGNAL. Any beam charge not accepted by the target is repelled back toward the electron gun. This repelled residue, called the return beam, is a differential video signal in that it is the difference between the outbound beam current and the current taken by the target to restore equilibrium therein. This differential video signal, still in the form of a beam in vacuum, can be intercepted and directed into an electron-multiplier structure wherein substantial and noise-free gain is to be had.

One flaw in the return beam usage just described is that while the gain mechanism of the electron multiplier is essentially noise-free, the return beam carrier is quite noisy. Return beam noise is especially obnoxious because it reaches a maximum during the scanning of areas which are darkest in the original scene. Thus at low lights, the signal-to-noise ratio is degraded not only by the decrease of signal but also by an increase of noise! Despite the undesirable aspect of beam noise, the return beam-derived video output signal has been used extensively—in particular in the image orthicon (Fig. 1). The image orthicon,[17] devel-

---

[17] A. Rose, P. K. Weimer, and H. B. Law, *Proc. IRE* **34**, 424 (1946).

oped during World War II, has been manufactured virtually unchanged for over thirty years and during most of that period reigned unchallenged as the epitome of high-performance camera tubes.

11.2.3.3.2. ISOCON-MODE RETURN BEAM.[18,19] Recent advances in both materials technology and in computing capacity have enabled the economic development of a long-recognized subtlety in the return beam mechanism, the isocon mode. In the isocon system a nondifferential target-derived return beam is used for output and, by virtue of its separation from the scanning beam, bears little of the latter's noise.

By suitable target processing, the return beam can contain an additional video component made up of target-scattered current. By suitable orientation of the electron gun, the scanning beam—including its reflected component—can be given an alignment different from that of the target-scattered return beam. There are now two return beam components with different alignments. The two can be simultaneously deflected so that only one enters the electron-multiplier output circuit. By so choosing the scattered beam, the isocon mode is achieved.

11.2.3.4. Photoelectric Effects.

11.2.3.4.1. PHOTOCONDUCTIVE TARGETS. VIDICON. In many cases the storage targets also possess a useful photoelectric property, usually photo-conductivity.[20] Camera tubes employing photoconductive storage targets are called vidicons[21] and are widely employed. See Fig. 2. Vidicon target materials in general use include antimony trisulfide,[22] lead oxide,[23] cadmium selenide[24] (all are polycrystalline or amorphous films), and the silicon wafer/diode array.[25] Most vidicons are used under conditions (substantial light exposure) where the target charging current is sufficiently large to allow its direct use as the output signal. Such is the vidicon of Fig. 2. So-called return beam vidicons are available when the added gain of the built-in beam electron multiplier is needed.

[18] P. K. Weimer, *RCA Rev.* **10**, 366 (1949).
[19] R. L. Van Asselt, *Proc. Nat. Electron. Conf.* pp. 333–338 (1968).
[20] A. Rose, *RCA Rev.* **12**, 303, 362 (1951).
[21] P. K. Weimer, S. V. Forgue, and R. R. Goodrich, *RCA Rev.* **12**, 306 (1951).
[22] S. V. Forgue, R. R. Goodrich, and A. D. Cope, *RCA Rev.* **12**, 335 (1951).
[23] E. F. de Haan, A. van der Drift, and P. P. M. Schampers, *Philips Tech. Rev.* **25**, 133 (1963/64).
[24] K. Shimizu, O. Yoshida, S. Aihara, and Y. Kiuchi, *IEEE Trans. Electron Devices* **ED18-11**, 1058 (1971).
[25] M. H. Crowell, T. M. Buck, E. F. Labuda, J. V. Dalton, and E. J. Walsh, *Bell Syst. Tech. J.* **46** (2), 491 (1967).

11.2.3.4.2. PHOTOEMITTERS. The first widely used camera tube, the image orthicon (Fig. 1), uses a photocathode to convert the optical input to an electrical pattern—as does the image dissector. When, as in the former, a storage target is to be used, it is separated from the photo-cathode and operates as the anode for the photocathode in a substructure termed the "image section." By control of electric and (sometimes) magnetic fields, the electron pattern emitted by the photocathode is transferred to the target where the continually arriving photoelectrons are integrated and sequentially scanned by an electron beam as described earlier.

Choice of photocathode. Several different photocathodes have been adapted to camera tubes.[26] The choice is dictated by such factors as spectral response and cost considerations. Among the more widely used photocathodes are the silver–bismuth–cesium–oxygen surface (usually identified as S-10), the cesium–antimony surface (S-11), the sodium–potassium–cesium–antimony (S-20 or multialkali) and the potassium–cesium–oxygen–antimony surface (dubbed the "bialkali"). For characteristics, see Section 11.1.1.2.

Storage targets.[27] Targets for use with separate photocathodes must possess the mosaic or quasimosaic property described above but need (and should) not be photosensitive. They must, however, somehow manage to accept incoming photoelectrons and cause them to reduce a charge which also had been established by incoming electrons (the scanning beam). This challenge is met by accelerating the photoelectrons across a high-voltage field so that they arrive at the storage target with sufficient energy to generate charge-pairs (i.e., induce secondary emission). The secondary electrons are conducted away and the positive component (ions or holes) operate to discharge the storage target.

Targets used in image orthicons (Fig. 1) and related tubes contain (1) a transparent (to high-energy photoelectrons) collector (of secondary electrons) usually in the form of a metal mesh, closely spaced to (2) a high-resistivity membrane which constitutes the dielectric of the storage capacitor mosaic. Materials employed for the membrane include glass (from bubbles blown nearly to bursting), aluminum oxide, and magnesium oxide—listed in ascending order of both secondary emission and fragility.

More recently high-gain targets have been made of potassium chloride[28]

[26] A. H. Sommer, "Photoemissive Materials," pp. 234–235. Wiley, New York, 1968.
[27] Kazan and Knoll,[16] Sect. I and II.
[28] R. S. Filby, S. B. Mende, and N. D. Twidly, *Advan. Electron. Electron Phys.* **22A,** 273 (1966).

and also of the same silicon wafer diode array used in the vidicon. The silicon target, now protected from direct illumination, is the basis of the SIT[29] (silicon intensifier-target) tube pictured in Fig. 3. Using silicon targets, high-voltage photoelectron acceleration has shown gains (hole-electron pairs per incident photoelectron) in excess of 2000 in practical devices.

11.2.3.5. Image Dissector. The image dissector operates on the principle of an electron-collecting aperture which is scanned by an electron image from a photoemitting surface upon which an optical image is focused. The aperture collects the electrons emitted by the photosensitive material of the photocathode. The electron streams forming the electron image from the photocathode are caused to move past the aperture by externally applied magnetic deflection fields. The electrons collected by the aperture are introduced into the first stage of an electron multiplier and subsequently amplified by secondary multiplication to a sufficiently high level to develop a signal appreciably higher than the noise generated in the following (conventional) video-amplifier stage.

The streams of electrons emitted by the photocathode of the image dissector are brought to a focus on a plane passing through the multiplier aperture and perpendicular to the axis of the tube. This is accomplished by the action of an axial magnetic focusing field and the electrical field produced by the accelerator rings. Therefore, as the entire raster is deflected across the multiplier aperture, the aperture intercepts the sharply focused stream of electrons produced by each illuminated area and translates the light image into a stream of electrons forming the video-signal information.

The useful area of the commercial image-dissector photocathode is a circle of $2\frac{3}{4}$-in. diameter, permitting the use of a scanned area of 2.2 in. $\times$ 1.65 in. The light from the scene to be televised is focused directly on the photocathode, which is deposited directly on the inside of the faceplate of the image dissector. The lens should be designed to cover this area, since the resolution varies directly with the size of the image used on the photosurface.

The photocathode of the image dissector (which is usually custom-made) is of the cesium–silver oxide or the cesium–antimony type. These two types of surfaces are different in spectral response and efficiency. The cesium–silver oxide surface has its highest response in the red region,

[29] R. W. Engstrom and R. L. Rodgers III, *Opt. Spectra* **5**, No. 2, 26 (1971).

while the cesium–antimony surface has its highest response in the blue region and is nearly equivalent to the S-11 response of a phototube. These photosurfaces are relatively insensitive to radiation outside the visible spectrum, except that the silver–cesium oxide surface has some infrared response. This is a desirable feature in some industrial applications.[30]

11.2.3.5.1. GENERAL USAGES. The image-dissector tube is a camera tube that is useful for industrial television systems where high light levels are available. A particular feature of this tube is its usefulness for scanning in one direction only and letting the object being scanned produce the scanning motion in the other direction by mechanical motion. Such applications are inspecting sheet material or other material as it passes through a machine. It can also be used as a counting device as it scans a moving conveyer or film track, or to determine the position or count excursions of an ink trace on a recording tape or chart, etc. This tube is particularly suited for random access type of scanning where a punched tape or card or another similar information-storage device having a two-dimensional array of information is to be interrogated.

11.2.3.6. Applications of Camera Tubes. Television broadcasting as a means of entertaining and informing the populace represents an obvious and significant use of camera tubes. The devices and techniques developed in these applications have been suitably combined, modified, and otherwise adapted to create viewing systems in a myriad of situations. A partial listing of specific uses includes:

Hospital room (patient) observation

In x-ray systems—medical and industrial—to permit remote viewing and to increase sensitivity (reduce dosage)

An instructional aid to allow many to view a small-area demonstration or other procedure

Centralized traffic monitoring on roads, bridges, in tunnels, etc.

Picture telephones and intercommunication sets

More comfortable and/or multiple viewing of microscope and electron-microscope images

"Electronic periscopes" and through-the-hull viewing on deep submersible marine vehicles

Weather mapping from earth-orbiting satellites

[30] D. G. Fink (ed.), "Television Engineering Handbook," Sect. 5. McGraw-Hill, New York, 1957.

Pictures from Earth's moon, Mars, etc.

Pictures from balloon- and satellite-borne telescopes

Remote sighting of weapons

Inspection of the interiors of installed pipelines, sewers, etc.

Interior views of chemical and nuclear reactors

Multipoint viewing of transportation status boards, e.g., in airports

Ship and aircraft training simulators—in which a television camera sails or flies in a scale model of the sea or landscape. Its pictures are presented lifesized to a pilot-trainee in a vehicular mockup whose controls manouever the camera

Pattern analysis—in which a camera tube views a stained tissue section, a fingerprint, handwriting, or the output of a prism or diffraction grating. The resulting patterns are presented as electrical signals to computer-type analyzing circuitry for counting, classifying, and/or comparing with previous inputs.

11.2.3.6.1. FACTORS IN THE CHOICE OF TYPE. Many laboratory requirements can be met by a choice from the profusion of types readily available. To arrive at a suitable selection, one should consider each of the following:

Illuminance (or irradiance) levels to be encountered. In practice, available camera tubes fall into sensitivity classes. The more sensitive, i.e., "low light level" types, are generally capable of useful performance in television service where the illumination is below twilight. They employ separate photocathodes and storage targets and include the image orthicon, image isocon, SEC, and SIT types. The less sensitive group contains the vidicon in its several forms and also the image dissector.

Spectral response requirements. Most device development has emphasized performance in the visible portion of the spectrum (400–700 nm). Operation in the ultraviolet region requires tubes especially made using U.V. transmitting image entrance windows instead of the optical glass usually employed. Operation in the near infrared is possible only with a limited number of photosensitive materials, including at present the S-1 photocathode, the silicon diode array vidicon, some selenium compound vidicon targets, and potentially some III–V compound crystalline photocathodes (notably GaAs) currently being developed for imaging purposes.

Cycling or frame times. As noted in Section 11.2.3.2, the storage type of camera tube permits charge integration and hence greater useful output current. Most storage types camera tubes exhibit an optimum storage time—usually in the tens or hundreds of milliseconds—an interval suited to the need of broadcast televison. Attempts to use substantially different frame times should be reviewed in advance with the tube maker. Among the options perhaps available are the use of storage media having different resistivities—and also the operation of tubes at depressed or (less likely) elevated temperatures in order to alter the effective resistivity of the target.

Scanning or addressing techniques. Storage-type camera tubes depend upon periodic scanning to sensitize the storage target. This class of tube should be operated only in a scanning mode which uniformly addresses the entire active target area.

Associated components. The electron optics which influence the cathode ray-electron beam are defined not only within the camera tube but also by the character of magnetic fields in associated structures. The effect of the external magnetic structure on primary performance characteristics —resolution and geometric uniformity—should not be ignored.

11.2.3.6.2. COLOR CAMERAS. The inclusion of color as an information item in television involves, at the present, essentially a triplex system. This consists of channels for signals representing each of three primary colors. Each channel contains a camera tube which views the scene through an optical filter which transmits only a band of colors centered on the primary color of that channel. At the output end of the system the signals separately excite phosphors which fluoresce in the appropriate primary color. It is left to the observer's eye to combine the three colored images and synthesize the original tint—or an approximation thereof! The output is most often a single electron tube device; nonetheless, the three channels are kept separate to the point that three separate phosphor screens are employed—structured so that the phosphor components for each remain isolated.

At the camera end of the system, the channel separation is, if anything, even more complete. Photosensitive surfaces with sufficient inherent color selectivity have not been developed—much less combined in a single structure. Current practice is to separate the incident light with optical filters directing each of the spectral segments to a separate camera tube. The most common configuration uses three essentially identical camera tubes, but some camera designs have added a fourth tube to view the scene sans filter and to produce a signal containing the fine detail. This "separate luminence channel" system reduces the need for precise registration of the images in the color ("chrominence") channels.

More recently camera systems have been shown that use but two and sometimes only one camera tube. In all cases these systems continue to employ optical color separation techniques in conjunction with camera tubes essentially unchanged from those used for black-and-white pickup.

11.2.4. Storage Tubes[31-33]

A storage tube is an electron device whose output is obtained at some arbitrary time after input, or which continuously presents its output following momentary excitation. The input is usually in the form of electrical signals; the output may be either electrical signals or a visible display.

The mechanism of storage is often a pattern of electric charge on the surface of a thin insulating film which acts as a capacitor. Different amounts of charge are stored on different portions of the insulator surface, corresponding to the input information to be stored. Because the film is assumed to have uniform capacitance per unit area, the charge pattern manifests itself as a variation of voltage from point to point over the surface. This voltage variation in turn controls, or modulates, the output.

The process of establishing the charge pattern is termed "writing." It is carried out by a focused and deflected electron beam from a gun similar to the cathode-ray tube gun. The mechanisms for establishing the charge pattern include secondary emission from the surface of the insulator layer and electron-bombardment-induced conductivity through the thickness of the layer.

For electrical output, or "reading," a focused and deflected electron beam is also used. The same gun may be used for both writing and reading, or separate electron guns may be provided. If the same gun is used for both functions, writing and reading must be done sequentially. If separate guns are used, writing and reading may be simultaneous.

With visible output, the tube is a display storage tube. A viewing gun continuously illuminates the storage layer with an unfocused flooding beam. The transmission of this flooding beam to the phosphor screen is modulated point by point over the screen by the charge pattern set up during writing.

Typical storage tubes cover a wide variety of types. The Radechon,[34]

[31] B. Kazan and M. Knoll, "Electronic Image Storage." Academic Press, New York, 1968.

[32] RCA Electron Tube Handbook HB-3. RCA Electron. Components, Harrison, New Jersey.

[33] M. D. Harsh, *Electron. Ind.* **25**, 54 (1966).

[34] A. S. Jensen, *RCA Rev.* **16**, 197 (1955).

or barrier-grid storage tube, type 6499, uses a single gun which establishes a charge pattern by secondary emission on a thin sheet of metal-backed mica. A fine-mesh barrier grid placed directly in front of the mica surface prevents the lateral redistribution of secondary electrons, and a consequent loss of resolution, as well as establishing the equilibrium potential to which the surface charges. The Alphecon and the silicon target storage tube are more recent additions to this general family. They are useful for obtaining a delayed output signal, integrating coherent signals in the presence of noise, obtaining a difference signal between an input signal and one previously stored, and changing the time scale of a recorded signal.

The Graphechon[35] tube, type 7539, has separate electron guns for writing and reading which face opposite sides of a thin-film target. By electron-bombardment-induced conductivity, the writing beam alters the potential distribution on the reading side of a thin film. The potential pattern thus set up modulates the reading beam, as the reading beam reestablishes an equilibrium potential. The Graphechon is intended for scan-converter service, such as converting radar PPI signals (plan-position indication, a polar coordinate display) to standard TV scanning (a rectangular coordinate display).

The signal-converter storage tube,[36] or recording storage tube, may take a physical form similar to that of the Graphechon, with separate writing and reading guns facing opposite sides of a target. In this tube, however, the storage layer is in the form of a thin dielectric coating on a fine metal mesh. The writing beam establishes a charge pattern on this surface by secondary emission. This charge pattern in turn controls the penetration of the reading beam through the mesh to an output electrode. This type of tube may be operated so that the process of reading does not disturb the stored charge pattern, and many output copies may be obtained.

One form of visible-output tube is the display storage tube.[37] The storage layer is a thin dielectric coating on a fine metal mesh, as in the recording storage tube described above. The focused and deflected writing beam lays down a charge pattern on the storage layer. This pattern controls, point by point over the surface, the transmission of a flooding-type viewing beam to an adjacent phosphor screen. Type 7268B is a

[35] L. Pensak, *RCA Rev.* **10**, 59 (1949).

[36] R. C. Hergenrother, A. S. Luftman, and C. E. Sawyer, *Electron. Ind. Tele-Tech.* **15**, 82 (1956).

[37] M. Knoll and B. Kazan, *Advan. Electron.* **8**, 447 (1956).

high writing speed tube with electrostatic focus and deflection. The viewing screen is about 10 cm in diameter. Type 7315 is similar, but is designed for slower writing. Type 7183A produces a similar display, but is designed for magnetic deflection. Many of these tubes are capable of a continuous output brightness of 2500 ft-L. They are thus useful where displays of transient data, such as radar, must be viewed under conditions of high ambient illumination. Other versions of this type of tube have been available in diameters from 7.5 to 53 cm.

The display storage tube reproduces a halftone display; that is, the output luminance may be continuously controlled in amplitude from zero to maximum with all intermediate levels reproduced.

If only the presence or absence of a signal is to be noted, with no intermediate levels, a different form of operation can be used. The Memotron[38] is designed for this bistable type of operation, and will store traces indefinitely. This is accomplished by the use of a storage layer whose surface, point by point, is held at one of two stable equilibrium potentials while controlling the flow of the reading beam to the phosphor screen. This tube is useful for the oscilloscope display of transient phenomena. The Typotron[39] operates on the same principle as the Memotron, producing a bistable display. The writing beam, however, is shaped in cross section to a desired character form by passing through a mask, as in the Charactron[40] tube. The Typotron then produces a stored display of selectable characters, and is useful in computer readouts.

Recently, simplified storage cathode-ray tubes have been described in which the phosphor screen itself, suitably embedded in a structure which serves as the collector grid, performs the bistable storage function.[41] This type of tube is incorporated in the commercially available storage oscilloscopes. A typical instrument of this type has an 8 cm × 10 cm display area which can be used in the storage mode or as a standard nonstorage cathode-ray display, or as a display of mixed stored and nonstored traces. The stored display can be viewed for as long as one hour.

The dark-trace tube,[42] such as the 4AP10, produces a stored pattern by quite a different mechanism. This tube is similar to a standard cathode-

[38] S. T. Smith and H. E. Brown, *Proc. IRE* **41**, 1167 (1953).

[39] H. M. Smith, *IRE Nat. Convention Record* Part 4, p. 129 (1955).

[40] C. R. Corpew, *SID IDEA Symp. Digest* 98 (1970).

[41] F. B. Mootry and R. A. Frankland, *SID IDEA Symp. Digest*, 58 (1970).

[42] S. Nozick, N. H. Burton, and S. Newman, *IRE Nat. Convention Record* Part 3, 121 (1954).

ray tube, but the usual phosphor screen is replaced by a potassium chloride screen. This material has the property of visibly darkening under electron bombardment and slowly recovering when heated.

Other inorganic cathodochromic materials are presently under investigation. When used as the screen of a cathode-ray tube, these cathodochromic materials may be written (selectively darkened) in a few TV scanning frames.

This type of tube produces a halftone display which is retained even if the power is turned off. The display is viewed by reflected light, so it may be easily seen in high ambient conditions, and projection displays are possible. Controllable erasure can be achieved by internal heating of the screen, or by flooding it with light.[43,44] Photochromic materials may also be incorporated into special cathode-ray tubes to perform similar functions. This type of tube has been used for radar and oscillograph displays, and for computer output.

[43] P. M. Heyman, I. Gorog, and B. W. Faughnan, *IEEE Int. Electron Devices Meeting* paper 21.2 (October 1970).
[44] I. Gorog, *Appl. Opt.* **9**, 2243 (1970).

11.3. Magnetic Amplifiers and Other Magnetic Devices*

The devices to be discussed here make use of the nonlinear B-H relations of ferro- and ferrimagnetic materials to obtain such functions as amplification, detection, or memory in many different circuits. Magnetic amplifiers enjoyed a great deal of popularity before the introduction of a large variety of semiconductor devices. Control inputs of less than 10^{-12} W and outputs in the megawatt range are possible with magnetic amplifiers, with power gains up to 10^{12}. Magnetic amplifiers provide isolation between multiple inputs and outputs and hence enjoy certain advantages over semiconductor devices in some applications. A magnetic core device is also extremely rugged and much less sensitive to radiation than semiconductors. In this brief treatment only a few typical circuits will be described, and the reader is referred to the literature for other circuit arrangements and more detailed methods of analysis.[1-3] Digital magnetic devices and tape recording will be discussed elsewhere (see Sections 9.6.6 and 9.10.5).

11.3.1. Magnetic-Core Materials

In recent years a great variety of core materials have been developed and improvements have been made in the silicon irons which are most commonly used in transformers and larger magnetic amplifiers and magnetic power devices. To obtain more abrupt saturation, oriented silicon iron laminations are usually employed, and air gaps are kept at a minimum by suitably lapping the laminations. In the smaller sizes "gapless" core constructions become practical. Most commonly the material is reduced to the required thickness by cold rolling, split to the desired width,

[1] W. A. Geyger, "Nonlinear-Magnetic Control Devices." McGraw-Hill, New York, 1964.

[2] Magnetic Amplifier Bibliography 1951–1956, AIEE Committee Rep. *Trans. AIEE* **77**, Part I, 613 (1958).

[3] 1957 Magnetic Amplifier Bibliography, AIEE Committee Rep. *Trans. AIEE* **77**, Part I, 1051 (1958).

---

* Chapter 11.3 is by F. J. **Friedlaender.**

and then wound on a cylindrical form. After annealing and processing, the core is sealed in a plastic or aluminum case on which insulated copper wire windings are then placed by means of a toroidal winding machine. Grain-oriented 50% Ni-Fe alloys find the widest application in magnetic amplifiers of medium and smaller sizes and are available under a number of trade names (Deltamax, Orthonol, Hypernik 5, Permenorm 5000Z, Sendelta, etc.). In low-level magnetic amplifiers and other devices, the low coercive force permalloys are used to advantage. For high-frequency applications, above the audio range, ultrathin Ni-Fe tapes or ferrites have to be employed to avoid excessive eddy current losses in the magnetic material.

11.3.2. Principles of Operation and Terminology

Magnetic devices make use of the nonlinear properties of the magnetic cores in various ways. In its simplest form, the operation of the magnetic amplifier (or really, saturable reactor, Fig. 7a) depends on the change of inductance of windings placed on one or more cores connected in series with a load. Since the "inductance" can be controlled by means of dc currents, this provides an easy means of controlling the current that flows through the load circuit.

An idealized B-H relation of a magnetic amplifier or memory device core is shown in Fig. 1. The "squareness" of the hysteresis loop will depend on the core material used, the processing of the material, and the core geometry. Partial traversal of the hysteresis loop (minor loop) plays an important part in the operation of many devices. Adequate predictions of such minor loops are often not possible from simple tests or a knowledge of magnetic material properties. Hence device perfor-

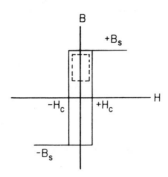

FIG. 1. B-H relationship of idealized "square-loop" core, with a minor loop shown in dotted lines.

mance has to be predicted from empirical data or greatly simplified models of core behavior. From the core characteristics, a flux-ampere turn relationship is readily derived, since the flux $\phi = BA$ where A is the cross-sectional area of the core, and $lH = \sum_j N_j i_j$ where l is the length of the magnetic path, and ampere turns $N_j i_j$ are summed over all the windings linking the core.

In most common magnetic-amplifier circuits, a given core in the circuit will be reset to a certain flux level by the action of the control winding together with other windings on the core, in a given half-cycle of the applied power source voltage (the "control half-cycle"), and will gate (i.e., allow load currents to flow) during the succeeding half-cycle (the "gating half-cycle"). When a core is not saturated—its operating point lies somewhere between $+B_s$ and $-B_s$ in Fig. 1—flux changes can occur with a very small number of ampere-turns applied to the core, and hence "large" voltages can be sustained across windings with very small currents. On the other hand, when a core is in saturation ($+B_s$ or $-B_s$) very little flux change is possible, and hence currents can take on any value required by the external circuit, as long as they are in a direction so as to keep the core in saturation.

Load current will start to flow at a certain instant in the gating half-cycle, when the controlling core reaches saturation. The instant at which saturation occurs depends on the flux level to which the core was reset during the previous half-cycle. The gating current (which may also be the load current, depending on the particular circuit arrangement used) has the waveshape of a truncated sine wave, as found in controlled-rectifier and thyratron circuits (see Chapter 5.2). Changes in the amount of flux reset will cause a change in the firing angle and, hence, also in the length of the interval during which the gate currents flows.

In some magnetic-amplifier circuits which will not be described here in detail, two (or more) cores gate during the same half-cycle in such a manner that output current is obtained only during the interval between the gating of the two cores. In such circuits pulse outputs are obtained with the input signal controlling the pulse width.

The voltage required to cause a flux change from negative to positive saturation (or vice versa) in a core in one half-cycle of the applied voltage is known as the *normal voltage* for that winding of the core and its half-cycle average value E_a can readily be found, using Faraday's law,

$$E_a = \frac{4N\phi_s}{T} = \frac{4NAB_s}{T}$$

where T is the length of a cycle, with $T = 1/f$, where

  f is the frequency,
  N is the number of turns,
  ϕ_s is the saturation flux,
  B_s is the saturation flux density,
  A is the cross-sectional area of the core.

For a sine wave $1.11E_a = E_{rms} = 4.44fN\phi_s$.

11.3.3. Single-Core Circuits

The circuit shown in Fig. 2a is known as a voltage reset circuit and was first suggested by Ramey.[4] The gating circuit contains the load, here assumed to be resistive and represented by R_L. During the gating half-cycle (the half-cycle when the power supply voltage e_g is positive for the sign convention shown in Fig. 2a) the flux in the core increases from ϕ_r toward saturation. The value ϕ_r had been reached at the end of the previous half-cycle, the control half-cycle, as will be explained below. During the time that the flux is increasing toward saturation the gate current is $i_g = lH_c/N_g$ if the B-H relationship of Fig. 1 is assumed to hold for the core, with H_c the coercive force of the core material and l the mean magnetic path length. The voltage drop $i_g(r_g + R_L) = (lH_c/N_g)$ $\times (r_g + R_L)$ is usually so small that it can be neglected compared with e_g. The auxiliary transformer provides a voltage $e_g N_2/N_1 = e_g N_c/N_g$ which, with the control circuit input voltage, acts in such a direction as to block the rectifier γ_2 during the gating half-cycle.

When the flux in the core reaches saturation, it will no longer sustain an induced voltage across its windings, and hence the voltage across N_g will now become $i_g r_g$, the voltage drop due to the winding resistance r_g, with $i_g = e_g/(R_L + r_g)$ for the remainder of the gating half-cycle. If saturation is reached at $\omega t = \alpha$, then, from Faraday's law, $e_g = N_g\, d\phi/dt$ for $0 < \omega t < \alpha$ and

$$\int_0^{\omega t=\alpha} e_g\, dt = \frac{1}{\omega}\int_0^\alpha E_{gm}\sin\omega t\cdot d(\omega t) = N_g(\phi_s - \phi_r)$$

where ϕ_s is the flux when the core is "saturated" (it is assumed that in the idealized core characteristic of Fig. 1, $B_s = \mu_0(H + M_s) \simeq \mu_0 M_s$ so that $\phi_s = AB_s \simeq \mu_0 AM_s$ for a core with cross-sectional area A). Thus α

[4] R. A. Ramey, On the mechanics of magnetic amplifier operation. *Trans. AIEE* **70**, Part II, 1214 (1951).

is given by

$$(E_{gm}/\omega)(1 - \cos \alpha) = N_g(\phi_s - \phi_r)$$

where E_{gm} is the amplitude of the applied voltage.

During the next half-cycle, the control or reset half-cycle ($e_g < 0$), rectifier γ_1 is blocked so that $i_g = 0$ but rectifier γ_2 unblocks when

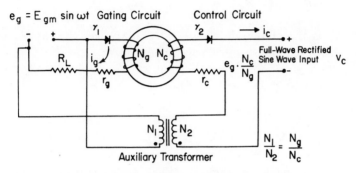

FIG. 2a. Voltage reset circuit.

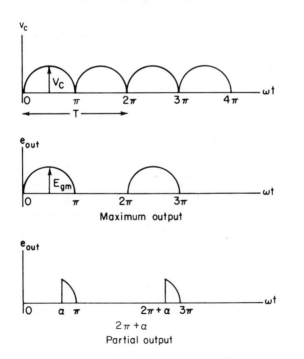

FIG. 2b. Waveshapes of control voltage v_c and gross output voltage $e_{out} = i_g(R_L + r_g)$. The magnetizing current lH_c/N_g has been neglected.

$e_g N_c/N_g$ exceeds v_c. The control voltage v_c is a full-wave rectified signal derived from the controlling element through a suitable full-wave bridge rectifier. v_c has an amplitude equal to or less than that of $e_g N_c/N_g$ and is "in-phase" with it. Neglecting $r_c i_c = r_c l H_c/N_c$ we have

$$N_c \frac{d\phi}{dt} = e_g \left(\frac{N_c}{N_g} \right) - v_c = E_{gm}\left(\frac{N_c}{N_g} \right) \sin \omega t - V_c \sin \omega t$$

where V_c is the amplitude of v_c. Integrating over the interval of the control half-cycle $\pi < \omega t < 2\pi$ we obtain

$$N_c(\phi_r - \phi_s) = -2 \cdot (E_{gm}/\omega)(N_c/N_g) + 2(V_c/\omega).$$

Therefore

$$N_g(\phi_s - \phi_r) = (2/\omega)(E_{gm} - V_c N_g/N_c).$$

The flux change during the control half-cycle is obviously equal to the flux change during the gating half-cycle as indicated in Fig. 3. It is also obvious from this figure that the integral of the voltage e_g during the output part of the gating half-cycle is equal to the area under the $v_c N_g/N_c$ curve during the control half-cycle.

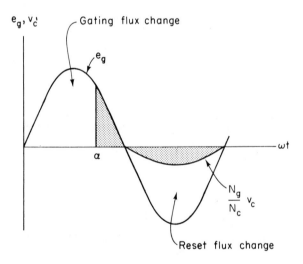

FIG. 3. Gating and control voltages in voltage reset circuit of Fig. 2a. Core saturates at $\omega t = \alpha$ when output commences. The white areas which represent gating and reset flux changes are equal, as are the shaded areas. Magnetizing current voltage drops have been neglected.

Using $\int_0^{\pi/\omega} e_g\,dt = 2E_{gm}/\omega$ we can readily find E_{out}, the half-cyclic average of the "gross output voltage" $e_{out} = (R_L + r_g)i_g$ (the net output is $R_L/(R_L + r_g)e_{out})$.

$$E_{out} = \frac{1}{T/2}\cdot\{(2E_{gm}/\omega) - (E_{gm}/\omega)(1 - \cos\alpha)\}$$

$$= 2f\cdot\{(2E_{gm}/\omega) - N_g(\phi_s - \phi_r)\}$$

$$= 2f\cdot\{(2/\omega)V_c(N_g/N_c)\}$$

$$= (2/\pi)V_cN_g/N_c = \langle v_c\rangle N_g/N_c.$$

Hence, it can be shown in general that the half-cyclic average gross output voltage is equal to $\langle v_c\rangle N_g/N_c$, the half-cyclic average of v_cN_g/N_c, for any waveshape which provides the proper blocking and unblocking of the rectifiers—this places restrictions on the relationship between v_c and e_g which must, of course, both be periodic but need not be sinusoidal.

Normal voltage is usually chosen as gate voltage. For half-wave output as shown in Fig. 2b, the actual average gross output is, of course, only half of the average half-cyclic output.

It follows that voltage gains in this type of circuit are of the order of N_g/N_c. Current and power gains are very high, but a precise definition requires a more careful analysis. For the circuit shown, the control circuit would always receive power. If a full-wave rectifier bridge is used to obtain the waveshape of v_c shown in Fig. 2b, a resistor will have to be placed across the bridge output to allow i_c to flow at all. Then, finite power and current gains can be calculated. A detailed discussion of this problem is beyond the scope of this review, but the current gain is of the order of the ratio of load current to magnetizing current.

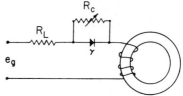

FIG. 4. Simpler form of circuit in Fig. 2a.

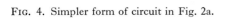

In Fig. 4, an even simpler form of this circuit is shown. R_c is varied to allow different amounts of flux reset during the reset half-cycle. An analysis of the circuit becomes more difficult, since a detailed knowledge of the *B-H* relation is required. An idealization of the type shown in Fig. 1 is no longer useful.

For instance, if R_c is a photoresistor, very little reset will be obtained with R_c in the dark when it has a very high value. As R_c is exposed to increasing amounts of light, more reset current will flow since the resistance of R_c will decrease. This circuit will provide a full output with no light input with the core remaining in saturation. With large light input (R_c small) only the (negligible) magnetizing current will flow, and the core flux will just oscillate between $+\phi_s$ and $-\phi_s$, reaching these values at the end of each half-cycle if e_g is adjusted to normal voltage.

In the circuits described above, a change in the control voltage will cause a corresponding change in the output in the following half-cycle, without further transient delay.

11.3.4. Multicore Circuits

In Fig. 5 three circuits are shown which provide dc or full-wave ac output. These circuits are essentially current controlled as will be shown later (Fig. 6), as opposed to the voltage controlled circuit of the previous section, and the control input can be a dc voltage or current. But ac or pulse inputs can be used in modified multicore circuits. The input-output relations for these circuits are shown in Fig. 6. Half-cyclic average values are usually used in describing the behavior of magnetic amplifiers, since, where analytical expressions can be derived, they are expressed in terms of half-cyclic average quantities.

Only the operation of the doubler circuit will be briefly described here. Voltage equations are readily written for both the gate and control circuits. Through integration over the control half-cycle for a core it is shown that the "gross half-cyclic average output voltage" (load voltage plus iR drops in windings and rectifier forward voltage drops) V can be related to the saturation (ϕ_s) and reset (ϕ_r) flux levels and the peak value of the applied sinusoidal gate voltage E_{gm} by Eq. (11.3.2):[5]

$$(2/\pi)E_{gm} - V = 2fN_g(\phi_s - \phi_r)$$

with normal voltage used as gate voltage. Hence, if ϕ_r could be related to the control current, the output voltage would be known in terms of

[5] L. A. Finzi and J. J. Suozzi, On feedback in magnetic amplifiers—Part I. *Trans. AIEE* **77**, Part I, 1019 (1958) (January 1959 section).

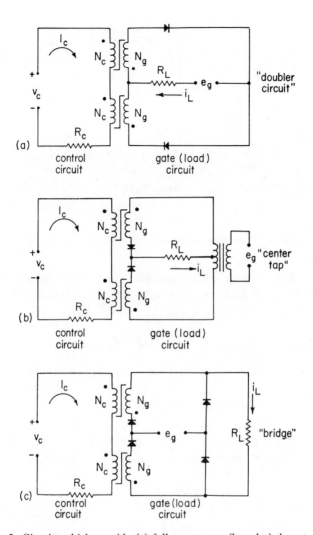

FIG. 5. Circuits which provide (a) full-wave ac or (b and c) dc outputs.

the control current. An examination of the equations obtained by writing Ampere's circuital law for each of the cores shows that the control current is determined by the magnetomotive force (mmf) requirements of the resetting core (if the gate rectifier of that core does not conduct during that half-cycle, a condition which is satisfied with resistive loads in this circuit). Again, an idealization of the core characteristic as shown in Fig. 1 will not be adequate. In fact, a direct experimental determination of the control characteristic of Fig. 6, either in the circuit shown in

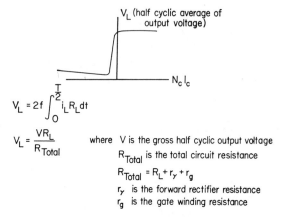

$$V_L = 2f \int_0^{\frac{T}{2}} i_L R_L dt$$

$$V_L = \frac{V R_L}{R_{Total}}$$ where V is the gross half cyclic output voltage

R_{Total} is the total circuit resistance

$R_{Total} = R_L + r_y + r_g$

r_y is the forward rectifier resistance

r_g is the gate winding resistance

FIG. 6. Typical input-output relations for circuits of Fig. 5.

Fig. 5a or some circuit with equivalent behavior, is most convenient. A discussion of this problem, together with a consideration of this circuit with inductive loads, can be found in Finzi and Jackson.[6]

11.3.5. Transient Behavior and Feedback Circuits[7]

To analyze both the transient behavior and the effect of feedback for the basic amplifier discussed above, it is assumed that a single-valued functional relationship exists between the reset flux, ϕ_r, and the half-cyclic average of the control mmf, $\mathscr{F}_r$, causing the reset ($\mathscr{F}_r = N_c I_c$ for the case of a single-control winding and ideal rectifiers which do not conduct during the reset half-cycle)

$$\phi_r = f(\mathscr{F}_r)$$

and

$$\frac{d\phi_r}{d\mathscr{F}_r} = K_\phi.$$

Then the gross output-voltage gain of the amplifier can be defined as

$$K_E = \frac{dV}{dV_c} = \frac{N_c}{R_c} 2f N_g K_\phi.$$

[6] L. A. Finzi and R. R. Jackson, The operation of magnetic amplifiers with various types of loads. *Trans. AIEE* **73**, Part I, 270, 279 (1954).

[7] L. A. Finzi and J. J. Suozzi, On feedback in magnetic amplifiers—Part II. *Trans. AIEE* **78**, Part I, 136 (1959).

The time constant τ of the amplifier can be shown to be approximately

$$\tau = \frac{N_{\mathrm{c}}^{2}}{R_{\mathrm{c}}} K_{\phi}$$

if K_{ϕ} is assumed to be a constant (a condition which is usually met over most of the control characteristic).

If there are additional control or bias windings, the time constant becomes

$$\tau = \sum_{i} \frac{N_{i}^{2}}{R_{i}} K_{\phi}.$$

The time constant can also be rewritten in terms of the voltage gain K_{E} as

$$\tau = \frac{1}{2f} \cdot \frac{N_{\mathrm{c}}}{N_{\mathrm{g}}} K_{E}$$

showing that increases in gain also lead to increases in time constant.

Various forms of feedback can be introduced to modify the gain, time constant, and input and output impedances of the magnetic amplifier.

Of particular interest in certain instrumentation circuits are magnetic amplifiers with both electric and magnetic feedback. An analysis of multiple-feedback circuits can be found in Finzi and Suozzi.[7]

11.3.6. Saturable Reactors and Other Devices

In Fig. 7a a series saturable reactor circuit is shown. In this circuit, as in those discussed in previous sections, cores are reset and gated in alternate half-cycles with a dc control input. In the resetting half-cycle, the reset flux level ϕ_{r} is reached at the instant when the other core reaches saturation (for R_{c} small). An analysis of the input-output relation is carried out simply by writing Ampere's circuital law for both cores and applying it appropriately. It follows immediately that

$$N_{\mathrm{c}} i_{\mathrm{c}} = \pm N_{\mathrm{g}} \mid i_{\mathrm{g}} \mid$$

at all times, if magnetizing currents can be neglected. Relating half-cyclic averages of control and gate current, the simple relationship shown in Fig. 7b results between control and gate mmfs. However I_{g} is clearly limited to the maximum value

$$I_{\mathrm{g}} \leqq \frac{(2/\pi) E_{\mathrm{gm}}}{R_{\mathrm{g}}}$$

where $R_{\mathrm{g}} = R_{\mathrm{L}} + 2 r_{\mathrm{g}}$ and r_{g} is the winding resistance of each gate winding.

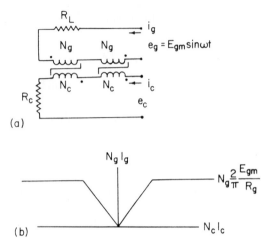

FIG. 7. (a) A series-saturable reactor circuit. (b) Simple relationship between half-cycle averages of control and gate currents. The latter is always positive, by convention.

The current i_c contains a large even harmonic component in this case, where the impedance in the control circuit is low.[8,9]

If the control circuit is supplied from a very high-impedance source (i.e., $R_c \rightarrow \infty$),

$$N_c i_c = \pm N_g \mid i_g \mid$$

still holds, but now i_c is dictated by the current source and the waveshape and amplitude of i_g are now dictated by i_c, except that i_g reverses its sign each half-cycle. If i_g is allowed to flow through a full-wave rectifier, the output current of the rectifier will be identical to $i_c(N_c/N_g)$ and, hence, a circuit of this type can be used as a "dc current transformer." The limitations of this circuit are discussed in Barker[9] and Storm.[10]

11.3.7. Even Harmonic Modulators, Frequency Multipliers

In the circuit shown in Fig. 7a, additional windings N_h can be added to both cores, connected in the same sense as the control windings. Since the two N_h windings are connected in such a manner that the

[8] H. F. Storm, Series-connected saturable reactor with control source of comparatively low impedance. *Trans. AIEE* **69**, Part II, 756 (1950).

[9] R. C. Barker, The series magnetic amplifiers, Parts I and II. *Trans. AIEE* **75**, Part I, 819 (1956).

[10] H. F. Storm, Series-connected saturable reactor with control source of comparatively high impedance. *Trans. AIEE* **69**, Part II, 1299 (1950).

voltages produced by the gate windings tend to cancel, no voltage will appear at the terminals as long as zero control current flows. However, as soon as $N_c i_c$ exceeds the coercive mmf of the cores, an "even harmonic" voltage appears at the terminals at twice the frequency of the gate voltage e_g. If the control current i_c is supplied from a source that has a high ac impedance (e.g., by placing a large inductor in the circuit) an output proportional to i_c can be obtained by using a suitable demodulator connected to the terminals of N_h. Very sensitive amplifiers can be constructed using this and similar schemes.

The device described above can also be used to obtain frequency multiplication. There are many schemes to obtain frequency doubling, tripling, or even higher order multiplication, by using the high harmonic content of the voltage waveform appearing across a winding on a saturating magnetic core. Which harmonics will appear depends on the number of cores used and the interconnections between them. A number of circuits are described and further references can be found in Geyger[1] and Johnson and Rauch.[11]

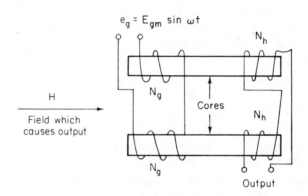

FIG. 8. Flux-gate magnetometer.

Returning to Fig. 7a it will be noted that the control current acts so as to aid the gate supply in one core and to oppose it in the other. External magnetic fields can have the same effect on suitably designed magnetic structures. In Fig. 8 winding N_g supplies cyclic magnetization to the two magnetically very soft open cores. No voltage appears across the output winding (N_h) terminals (in the ideal device) as long as the two cores are

[11] L. J. Johnson and S. E. Rauch, Magnetic amplifier multipliers. *Trans. AIEE* **73**, Part I, 448 (1954).

balanced, since this winding is connected differentially. As soon as an unbalance occurs, for example by placing the cores in a magnetic field, an even harmonic voltage will appear at the output terminals, with the output proportional to the field over some range. Such a device is called a flux-gate magnetometer and can be used to detect extremely small flux variations, when employed in more sophisticated circuits. Such flux-gate magnetometers have found applications in the detection of magnetic objects, in geomagnetism, and in crime prevention through the detection of guns (usually made in part of magnetic materials). Gordon and Brown[12] give an extensive review and bibliography.

11.3.8. Magnetic Analog Memories

Digital memories in computers have employed magnetic materials for many years. It is also possible to store analog information magnetically. In one form of magnetic analog memory,[13] a certain flux level is established in a core by means of a suitable positive pulse input which provides a low magnetizing field, just larger than the coercive field of the core. A partial flux reversal takes place by means of domain walls, with the amount of penetration of the domain walls, and hence the amount of flux reversal, determined by the duration of the low-field pulse. The flux state established is quite stable and will remain indefinitely in the absence of a field. To read out the flux state nondestructively (i.e., to read it without changing the flux level in the core permanently), a short, negative, high-field pulse (say ten times the coercive field or more) is applied and the peak voltage induced in a separate winding is recorded. This peak voltage is proportional to the flux state of the core (i.e., a maximum if the core is in positive saturation remanence decreasing with lower remanent flux levels, to a minimum at negative saturation remanence). The change in flux level due to the interrogating high-field pulse is reversible, and, depending on the core material used and the specific field values, the stored flux level will be restored after interrogation. A subcoercive bias field may be needed to eliminate the effect of the interrogating (read) field. If the bias field is carefully chosen, it will not affect the ability of the core to retain its remanent flux level indefinitely.

There are other schemes to read the flux level in a magnetic element

[12] D. I. Gordon and R. E. Brown, Recent advances in fluxgate magnetometry. *IEEE Trans. Magn.* **MAG-8**, 76–82 (1972).

[13] F. J. Friedlaender and J. D. McMillen, A magnetic core analog memory. *IEEE Trans. Magn.* **MAG-3**, 463 (1967).

nondestructively. Many of these depend on the use of a small constant amplitude ac signal which will produce different amplitude induced ac signals, depending on the flux state. In one such device,[14] the field due to the ac signal is applied in a direction orthogonal to the direction of the "stored flux." Second harmonic induced voltages are observed and are found to increase linearly (but with opposite polarities) as the remanent flux level increases in opposite directions (i.e., towards positive or negative saturation remanence) from zero. A permalloy plated wire is used as flux storage element.

Additional sources are listed in Milnes,[15] Meyerhoff,[16] Lynn et al.,[17] Geyger,[18] Schilling,[19] Storm,[20] and Katz.[21]

[14] S. Konishi, S. Sugatani, and Y. Sakurai, A high-speed NDRO analog memory using electrodeposited Permalloy wires. IEEE Trans. Magn. MAG-5, 14–18 (1969).

[15] G. A. Milnes, "Transducters and Magnetic Amplifiers." Macmillan, New York, 1957.

[16] A. J. Meyerhoff (ed.), "Digital Application of Magnetic Devices." Wiley, New York, 1960.

[17] G. E. Lynn et al., "Self-Saturating Magnetic Amplifiers." McGraw-Hill, New York, 1960.

[18] W. A. Geyger, "Magnetic Amplifier Circuits; Basic Principles; Characteristics, and Applications," 2nd ed. McGraw-Hill, New York, 1957.

[19] W. Schilling, "Transduktortechnik; Theorie und Anwendung steuerbarer Drosseln." Oldenbourg, Munich, 1960.

[20] H. F. Storm, "Magnetic Amplifiers." Wiley, New York, 1955.

[21] H. W. Katz (ed.), "Solid State Magnetic and Dielectric Devices." Wiley, New York, 1959.

12. FEEDBACK CONTROL SYSTEMS*

12.1. General Techniques[1-6]

12.1.1. Introduction

12.1.1.1. Feedback Control System Description. A *control system* is a system which regulates an output variable with the objective of producing a given relationship between it and an input variable. In a *feedback* control system, at least part of the information used to change the output variable is derived from measurements performed on the output variable itself.

Elements common to many feedback control systems are illustrated in Fig. 1. The value of the output variable is determined (and possibly operated upon) with a measuring element. The signal from the measuring element is fed back and compared with the input variable to provide an error signal. The error is operated upon by an amplifier and the amplifier output is used to drive the controlled element. The system output may also be influenced by disturbances applied to the controlled element. This type of *closed-loop* control is often used in preference to *open-loop* control (where the system does not use output-variable information to alter the output) since feedback can reduce the sensitivity of the system to externally applied disturbances and to changes in system parameters.

---

[1] G. S. Brown and D. P. Campbell, "Principles of Servomechanisms." Wiley, New York, 1948.

[2] J. G. Truxal, "Automatic Feedback Control System Synthesis." McGraw-Hill, New York, 1955.

[3] H. Chestnut and R. W. Mayer, "Servomechanisms and Regulating System Design," Vol. 1, 2nd ed. Wiley, New York, 1959.

[4] R. N. Clark, "Introduction to Automatic Control Systems." Wiley, New York, 1962.

[5] J. J. D'Azzo and C. H. Houpis, "Feedback Control System Analysis and Synthesis," 2nd ed. McGraw-Hill, New York, 1966.

[6] B. C. Kuo, "Automatic Control Systems," 2nd ed. Prentice-Hall, Englewood Cliffs, New Jersey, 1967.

---

* Part 12 is by James K. Roberge.

Familiar examples of feedback control systems include residential heating systems, most high-fidelity audio amplifiers, and the iris-retina system of the human eye.

A *servomechanism* is a feedback control system with at least one of the system variables a mechanical motion. The term *regulator* is used to describe a feedback control system with the primary objective of maintaining its output fixed with a known relationship to a constant input (or *reference*) independent of applied disturbances.

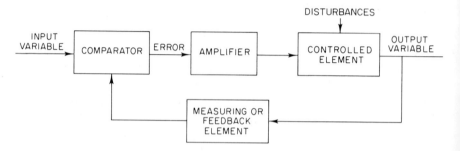

FIG. 1. Representation of a feedback control system.

12.1.1.2. Symbology. The relationships among variables in a feedback control system are frequently illustrated by means of a *block diagram*. A block diagram includes three types of elements.

1. A *line* represents a variable, with an arrow on the line indicating the direction of information flow. A line may split, indicating a single variable is supplied to two or more portions of the system.
2. A *block* operates on an input variable to produce an output variable.
3. Variables are added algebraically at a *summation point* drawn as follows:

One possible representation for the system of Fig. 1 is shown in block diagram form in Fig. 2. The notation used throughout this section is:

r = reference or input variable,
e = error or activating variable,
u = upset or disturbance,
c = controlled output variable.

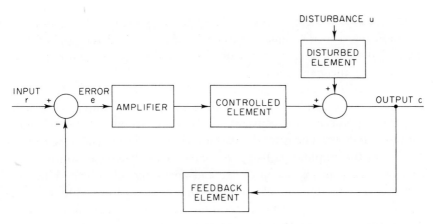

FIG. 2. Block diagram for system of Fig. 1.

12.1.1.3. Advantages of Feedback. The system shown in block diagram form in Fig. 3 is used to illustrate a major advantage of feedback systems. This diagram can be used to represent an electronic feedback amplifier. It is assumed that the feedback element is simply an attenuator which provides an output signal which is a fraction β of its input signal. The difference between this fraction of the system output and the input r is amplified by an amount A to produce the system output. The block diagram represents a set of equations:

$$c = Ae, \qquad (12.1.1a)$$

$$e = r - \beta c. \qquad (12.1.1b)$$

Combining these equations and solving for the *closed-loop gain* c/r yields:

$$c/r = A/(1 + A\beta). \qquad (12.1.2)$$

The quantity $A\beta$ in Eq. (12.1.2) is called the *open-loop gain*. The open-loop gain of many feedback systems can be determined as illustrated in Fig. 4. All system inputs and disturbances are set to zero and any line

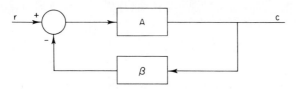

FIG. 3. Feedback amplifier.

inside the closed loop is broken. The open-loop gain is the *negative* of the ratio between the signal returned by the loop and the applied test input. Note that the open-loop gain is *not* necessarily the gain c/r of the system with the loop opened.

The closed-loop gain expression (Eq. (12.1.2)) shows that as the open-loop gain of the amplifier becomes large compared to unity, the closed-loop gain approaches the constant $1/\beta$. The significance of this relationship is as follows. The gain A may be relatively unpredictable since this portion of the amplifier is likely to contain active elements whose characteristics vary as a function of age and operating conditions. This uncertainty may be unavoidable in that elements are not available with the stability desired for a given application, or it may be introduced as a compromise necessary to achieve economic or other advantages.

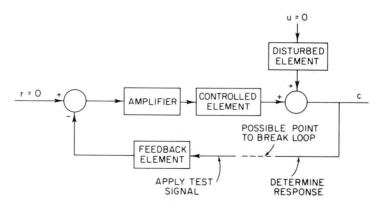

FIG. 4. Determination of open-loop gain.

However, the required feedback attenuation can presumably be achieved with resistors, and resistors with precisely controlled characteristics are readily available. If the open-loop gain is sufficiently high, the closed-loop gain can be made dependent on the precise attenuation characteristics of the resistive network.

This feature of feedback systems can be emphasized by calculating the fractional change in closed-loop gain $d(c/r)/(c/r)$ caused by a given fractional change in amplifier gain dA/A:

$$\frac{d(c/r)}{c/r} = \frac{dA}{A}\frac{1}{1+A\beta}. \tag{12.1.3}$$

Changes in A can be attenuated to insignificant levels if $A\beta$ is sufficiently large.

This result contrasts with the one-to-one dependence of input-output gain on amplifier gain if no feedback is used. It is also important to underline the fact that changes in the gain of the feedback element have direct influence on the gain of the system. It is concluded that it is necessary to observe or measure the output variable of a feedback system accurately in order to realize the advantages of feedback.

Analogous results are obtained when the influence of external disturbances applied to the output of a feedback system are calculated. It can be shown that the sensitivity to disturbances of a system such as that shown in Fig. 1 is proportional to the reciprocal of the open-loop gain.

The preceding discussion indicates that high open-loop gain is necessary to achieve the advantages associated with feedback. Unfortunately, the open-loop gain of a feedback system cannot be increased without bound, since sufficiently high gain invariably causes the system to become unstable. A *stable* system is defined as one for which a bounded output is produced in response to a bounded input. Conversely, an *unstable* system exhibits runaway or oscillatory behavior in response to a small input. Instability occurs in high-gain systems since small errors give rise to large corrective action. The propagation of signals around the loop is delayed by the dynamics of the elements in the loop, and as a consequence high-gain systems tend to overcorrect. When this overcorrection produces an error larger than the initiating error, the system is unstable.

Much of the design effort required for a feedback control system is devoted to achieving the combination of open-loop gain and dynamics which yield acceptable insensitivity to element variations and external disturbances without compromising stability.

12.1.2. Mathematical Preliminaries

12.1.2.1. Linear, Time-Invariant Differential Equations.[7] A system is called *linear* if the principle of *superposition* applies. If a system produces an output c_1 in response to an input r_1, and an output c_2 in response to an input r_2, linearity requires that the output is $c_1 + c_2$ in response to an input $r_1 + r_2$. While no physical system adheres to this ideal relationship for all inputs, linearity is a useful concept since many systems can be adequately represented as linear for a large class of inputs, and since the assumption of linearity greatly facilitates analysis. This discussion is initially limited to the consideration of linear systems, with two techniques for the analysis of nonlinear systems reserved until Section 12.1.6.

---

[7] W. Kaplan, "Advanced Calculus." Addison-Wesley, Reading, Massachusetts, 1952.

A system is called *time invariant* if the response to a particular input is independent of the time at which the input is applied. This discussion is also limited to time-invariant systems.

A linear, time-invariant element provides an output $x(t)$ in response to an input $y(t)$ related as

$$\sum_{i=0}^{n} a_i \frac{d^i x(t)}{dt^i} = \sum_{j=0}^{m} b_j \frac{d^j y(t)}{dt^j} \qquad (12.1.4)$$

where all a's and b's are constants. This expression is an example of a linear, time-invariant differential equation of order n. It can be solved if $y(t)$ is known for all t and additionally the values of $x(t)$ and all but one of its first n derivatives at some time t_1 are known.

The solution of a linear, time-invariant differential equation consists of two parts, the *homogeneous* and the *particular* solutions. The homogeneous solution is found by setting the right-hand side of Eq. (12.1.4) to zero. This solution consists of a linear combination of complex exponentials of the form

$$x_h(t) = \sum_{k=1}^{n} A_k e^{s_k t}. \qquad (12.1.5)$$

The s_k's are determined by evaluating the algebraic roots of the *characteristic* equation

$$\sum_{i=0}^{n} a_i s_k{}^i = 0. \qquad (12.1.6)$$

The particular solution $x_p(t)$ consists of terms proportional to the excitation $y(t)$ and its derivatives. By adding the particular solution to a homogeneous solution with appropriately selected A_k's, it is possible to satisfy the original differential equation subject to the given constraints on $x(t)$ and its derivatives at time t_1.

One particular solution of a differential equation which is of special interest in the study of linear systems is the impulse response, obtained by making the system excitation the Dirac delta function or unit impulse $\delta_0(t)$. The unit impulse, which can be considered a function with unit area and nonzero amplitude at only one instant, is defined by the relationships

$$\delta_0(t - a) = 0, \qquad t \neq a, \qquad (12.1.7)$$

$$\int_{-\infty}^{\infty} \delta_0(t - a)\, dt = 1, \qquad (12.1.8)$$

$$\int_{-\infty}^{\infty} f(t)\, \delta_0(t - a)\, dt = f(a). \qquad (12.1.9)$$

Linearity permits us to find the output of a system by summing the responses to individual components of the input. If the input to a system is approximated as a series of small area impulses, the output can be determined by appropriately delaying, weighting, and summing the impulse response of the system. The limiting integral approached by the sum is called the *superposition* or *convolution* integral and relates a system output $x(t)$ to its input $y(t)$ and impulse response $h(t)$ as

$$x(t) = \int_{-\infty}^{\infty} h(\tau)\, y(t - \tau)\, d\tau. \qquad (12.1.10)$$

If consideration is limited to *physically realizable* or *causal* systems which respond only after an excitation is applied ($h(t) = 0$ for $t < 0$) and to excitations which are zero for negative time, Eq. (12.1.10) reduces to

$$x(t) = \int_{0}^{t} h(\tau)\, y(t - \tau)\, d\tau = \int_{0}^{t} y(\tau)\, h(t - \tau)\, d\tau. \qquad (12.1.11)$$

The convolution operation is commutative and is denoted by an asterisk. Thus Eq. (12.1.11) can be written

$$x(t) = h(t) * y(t) = y(t) * h(t). \qquad (12.1.12)$$

12.1.2.2. Laplace Transforms.[8] Figure 5 illustrates a simple linear system with unity feedback. The forward gain element is characterized by its impulse response $g(t)$. The system inputs and outputs are related as

$$c(t) = e(t) * g(t) = [r(t) - c(t)] * g(t). \qquad (12.1.13)$$

This integral equation is often difficult to solve. The Laplace transformation offers an alternative method of solution which is usually computationally advantageous. The *Laplace transform* of a time function

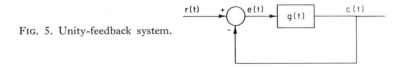

FIG. 5. Unity-feedback system.

[8] M. F. Gardner and J. L. Barnes, "Transients in Linear Systems." Wiley, New York, 1942.

and the *inverse Laplace transform* are defined as

$$\mathscr{L}[f(t)] \triangleq F(s) \triangleq \int_0^\infty f(t)e^{-st}\,dt,\qquad\qquad (12.1.14)$$

$$\mathscr{L}^{-1}[F(s)] = f(t) = \frac{1}{2\pi j}\int_{\sigma_1-j\infty}^{\sigma_1+j\infty} F(s)e^{st}\,ds\qquad\qquad (12.1.15)$$

where s is the complex variable $\sigma + j\omega$. The direct-inverse transform pair is unique[†] so that $\mathscr{L}^{-1}\mathscr{L}[f(t)] = f(t)$ if $f(t) = 0$ for $t < 0$, and if $\int_0^\infty |f(t)|\,e^{-\sigma_1 t}\,dt$ is finite for some constant σ_1.

A number of theorems useful in the analysis of dynamic systems can be developed. The more important of these theorems include:

1. *Linearity*:

$$\mathscr{L}[af(t) + bg(t)] = [aF(s) + bG(s)]$$

where a and b are constants.

2. *Differentiation*:

$$\mathscr{L}\left[\frac{df(t)}{dt}\right] = sF(s) - \lim_{t\to 0^+} f(t)$$

(the limit is taken by approaching $t = 0$ from positive t).

3. *Integration*:

$$\mathscr{L}\left[\int_0^t f(\tau)\,d\tau\right] = \frac{F(s)}{s}.$$

4. *Convolution*:

$$\mathscr{L}[f(t) * g(t)] = F(s)G(s).$$

5. *Time shift*:

$$\mathscr{L}[f(t - \tau)] = F(s)e^{-s\tau}$$

if $f(t - \tau) = 0$ for $t - \tau < 0$, where τ is a constant.

6. *Time scale*:

$$\mathscr{L}[f(at)] = (1/a)F(s/a)$$

where a is a constant.

[†] There are three additional constraints called the Dirichlet conditions which are satisfied for all signals of physical origin. The interested reader is referred to Gardner and Barnes.[8]

7. *Initial value:*

$$\lim_{t \to 0^+} f(t) = \lim_{s \to \infty} sF(s).$$

8. *Final value:*

$$\lim_{t \to \infty} f(t) = \lim_{s \to 0} sF(s).$$

Theorem 4 has important implications for feedback control system calculations. In particular, it states that the transform of the output of a linear element (assuming that the element has zero initial conditions prior to the application of an input) is the product of the transforms of the input and the impulse response of the element. (The transform of the impulse response of a system is frequently called its *transfer function*.) Application of this theorem and Theorem 1 permits simplification of Eq. (12.1.13) as follows:

$$C(s) = E(s)G(s) = [R(s) - C(s)]G(s). \qquad (12.1.16)$$

This equation is easily solved for the transform of the closed-loop transfer function

$$\frac{C(s)}{R(s)} = \frac{G(s)}{1 + G(s)}. \qquad (12.1.17)$$

The transforms of many time functions of interest can be obtained by direct integration using the defining relationship, Eq. (12.1.14). Table I lists some common transform pairs.

The time functions corresponding to ratios of polynomials in s not listed in the table can be evaluated by means of a *partial fraction expansion*. The function of interest is written in the form

$$F(s) = \frac{p(s)}{q(s)} = \frac{p(s)}{(s + s_1)(s + s_2)\ldots(s + s_n)}. \qquad (12.1.18)$$

It is assumed that the order of the numerator polynomial is equal to or less than that of the denominator. If all of the roots of the denominator polynomial are *first order* (i.e., $s_i \neq s_j$, $i \neq j$),

$$F(s) = \sum_{k=1}^{n} \frac{A_k}{s + s_k} \qquad (12.1.19)$$

where

$$A_k = \lim_{s \to -s_k} (s + s_k)F(s). \qquad (12.1.20)$$

TABLE I. Laplace Transform Pairs

$F(s)$	$f(t),\ t \geq 0$ $(f(t) = 0,\ t < 0)$
1	Unit impulse $\delta_0(t)$
$\dfrac{1}{s}$	Unit step $\delta_{-1}(t)$ $(f(t) = 1,\ t > 0)$
$\dfrac{1}{s^2}$	Unit ramp $\delta_{-2}(t)$ $(f(t) = t,\ t > 0)$
$\dfrac{1}{s^{n+1}}$	$\dfrac{t^n}{n!}$
$\dfrac{1}{s+a}$	e^{-at}
$\dfrac{1}{(s+a)^n}$	$\dfrac{t^{n-1}}{(n-1)!}\,e^{-at}$
$\dfrac{\omega}{(s+a)^2 + \omega^2}$	$e^{-at}\sin \omega t$
$\dfrac{s+a}{(s+a)^2 + \omega^2}$	$e^{-at}\cos \omega t$
$\dfrac{1}{s^2 + 2\zeta\omega_n s + \omega_n{}^2}$	$\dfrac{1}{\omega_n\sqrt{1-\zeta^2}}\,e^{-\zeta\omega_n t}\sin(\omega_n\sqrt{1-\zeta^2}\,t),\ 0 < \zeta < 1$

If one or more roots of the denominator polynomial are *multiple roots,* they contribute terms of the form

$$\sum_{k=1}^{m} \frac{B_k}{(s+s_i)^k} \qquad (12.1.21)$$

where m is the order of the multiple root located at $s = -s_i$. The B's are determined from the relationship

$$B_k = \lim_{s \to -s_i} \frac{1}{(m-k)!}\frac{d^{m-k}}{ds^{m-k}}[(s+s_i)^m F(s)]. \qquad (12.1.22)$$

The required time function is found by summing the contributions of all its components.

Laplace transforms offer a convenient method for the solution of linear, time-invariant differential equations, since they replace the integration and differentiation required to solve these equations in the time domain by algebraic manipulation.

Example 12.1. Consider the differential equation

$$\frac{d^2x}{dt^2} + 3\frac{dx}{dt} + 2x = e^{-t} \tag{12.1.23}$$

subject to the initial conditions

$$x(0^+) = 2 \qquad \frac{dx}{dt}(0^+) = 0.$$

The transform of Eq. (12.1.23) is taken using Theorem 2 (applied twice in the case of the second derivative) and Table I to determine the Laplace transform of e^{-t}:

$$s^2X(s) - sx(0^+) - \frac{dx}{dt}(0^+) + 3sX(s) - 3x(0^+) + 2X(s) = \frac{1}{s+1}. \tag{12.1.24}$$

Collecting terms and solving for $X(s)$ yields

$$X(s) = \frac{2s^2 + 8s + 7}{(s+1)^2(s+2)}. \tag{12.1.25}$$

Equations (12.1.19) and (12.1.21) show that since there is one first-order root and one second-order root,

$$X(s) = \frac{A_1}{s+2} + \frac{B_1}{s+1} + \frac{B_2}{(s+1)^2}. \tag{12.1.26}$$

The coefficients are evaluated with the aid of Eqs. (12.1.20) and (12.1.22):

$$X(s) = \frac{-1}{s+2} + \frac{3}{s+1} + \frac{1}{(s+1)^2}. \tag{12.1.27}$$

The inverse transform of $X(s)$, evaluated with the aid of Table I, yields

$$x(t) = -e^{-2t} + 3e^{-t} + te^{-t}. \tag{12.1.28}$$

12.1.3. Feedback System Response

12.1.3.1. Transient Response. The *transient response* of an element or system is its output expressed as a function of time following application of a specified signal. The test signals used to excite the transient response of the system may either be the actual input signals which are anticipated in normal operation, or they may be mathematical abstractions chosen because of the insight they lend to system behavior. Commonly used test inputs include the impulse defined by Eqs. (12.1.7), (12.1.8), and (12.1.9) and the time integrals of this function. The transient response of a control system is most easily evaluated by means of Laplace transforms.

The transient response of a complex control system can often be satisfactorily approximated by that of a much simpler system. Cases of particular interest are those where the closed-loop transfer function $C(s)/R(s)$ is dominated by one or two *poles*. (A pole is a complex frequency at which the denominator or *characteristic equation* of the transfer function is zero.) This behavior occurs if one or two poles of the closed-loop transfer function are located at significantly lower frequencies than all other poles and *zeros* (frequencies where the numerator is zero).

Example 12.2. Consider a unity-feedback system as shown in Fig. 5 with $G(s)$ (the Laplace transform of $g(t)$) equal to K/s. The closed-loop transfer function for this system is

$$\frac{C(s)}{R(s)} = \frac{G(s)}{1 + G(s)} = \frac{K/s}{1 + K/s} = \frac{1}{s/K + 1}. \qquad (12.1.29)$$

The system is excited with a unit step input. The unit step, $\delta_{-1}(t)$, is the time integral of the unit impulse, defined by

$$\delta_{-1}(t) = \int_{-\infty}^{t} \delta_0(t)\, dt = \begin{cases} 0, & t \leq 0 \\ 1, & t > 0. \end{cases} \qquad (12.1.30)$$

Table I shows that the Laplace transform of the unit step is $1/s$. If this function is substituted into Eq. (12.1.29), and the inverse transform of the resultant expression is taken to determine $c(t)$,

$$c(t) - 1 - e^{-Kt}. \qquad (12.1.31)$$

This transient response is plotted in Fig. 6. The general applicability of this simple result is appreciated when we recognize that the step response of any stable system with a transfer function of the form

$$\frac{C(s)}{R(s)} = \frac{\prod_{i=1}^{n} \tau_{zi}s + 1}{\prod_{j=1}^{m} \tau_{pj}s + 1} \qquad m > n, \text{ all } \tau > 0,\ \tau_{p1} \gg \text{all other } \tau\text{'s}$$

$$(12.1.32)$$

is approximately $1 - e^{-t/\tau_{p1}}$.

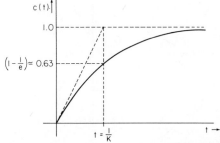

FIG. 6. Step response of first-order system.

The approximate result given above holds even if some of the singularities occur in complex conjugate pairs providing they are located at much greater distances from the origin in the s plane than the dominant pole. However, if the real part of the complex pair is not significantly more negative than the location of the dominant pole, small-amplitude high-frequency oscillations may persist after the dominant transient is over. In cases where the two lowest frequency singularities consist of a complex conjugate pair of poles, a second approximation is useful.

Example 12.3. Consider the system shown in Fig. 5 with $G(s) = K/s(\tau s + 1)$ and closed-loop gain

$$\frac{C(s)}{R(s)} = \frac{1}{(\tau/K)s^2 + s/K + 1}. \tag{12.1.33}$$

If $4\tau K > 1$, the transfer function can be put in the standard form

$$\frac{C(s)}{R(s)} = \frac{1}{s^2/\omega_n{}^2 + 2\zeta s/\omega_n + 1} \qquad 0 < \zeta < 1. \tag{12.1.34}$$

The equation parameters ω_n and ζ are called the *natural frequency* (expressed in radians per second) and the *damping ratio* respectively. The physical significance of these quantities is illustrated in the s-plane plot shown in Fig. 7. The situation depicted is called *underdamped* ($\zeta < 1$) in contrast to the *critically damped* case where the two poles coincide on the real axis ($\zeta = 1$) and the *overdamped* case where the denominator is factorable into two terms with real coefficients ($\zeta > 1$). The step response for this system, determined using the last relationship given in Table I, is

$$c(t) = 1 - \frac{1}{\sqrt{1 - \zeta^2}} e^{-\zeta \omega_n t} \sin(\sqrt{1 - \zeta^2}\, \omega_n t + \phi)$$

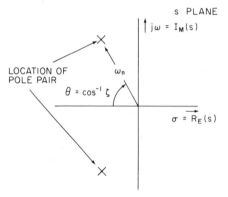

FIG. 7. S-plane plot of complex-pole pair.

where

$$\phi = \tan^{-1} \frac{\sqrt{1 - \zeta^2}}{\zeta}. \tag{12.1.35}$$

Figure 8 shows $c(t)$ for various values of ζ as a function of normalized time $\omega_n t$. These step responses approximate the behavior of any system where one pair of complex poles is located much closer to the origin than all other singularities.

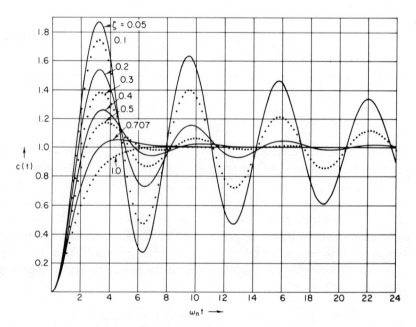

FIG. 8. Step response of second-order system.

12.1.3.2. Error Coefficients. In most feedback systems the desired error, or difference between the input and feedback signals, is zero. For a system not subject to external disturbances, it is possible to evaluate the error as a function of time in terms of the input and its time derivatives by an *error series* of the form

$$e(t) = e_0 r(t) + e_1 \frac{dr(t)}{dt} + e_2 \frac{d^2 r(t)}{dt^2} + \cdots + e_k \frac{d^k r(t)}{dt^k} + \cdots .$$

$$(12.1.36)$$

The complete expansion is valid at all times when the input and its derivatives are continuous. In cases of practical interest, the first few terms of the series yield an acceptable approximation at times which are not close to input-signal discontinuities. This expansion is a particularly useful representation when systems are excited by inputs which do not have simple Laplace transforms.

The coefficients of the series, called *error coefficients*, can be determined from the relationship

$$e_k = \frac{1}{k!} \frac{d^k}{ds^k} \left[\frac{E(s)}{R(s)} \right]_{s=0} .$$

$$(12.1.37)$$

The coefficients can also be obtained by dividing $E(s)$ by $R(s)$ to obtain a series in ascending powers of s. The coefficient of the s^k term is e_k.

An approximate upper bound on the magnitude of the remainder, if the series is terminated after m terms, can be obtained as

$$\left| e(t) - \sum_{k=1}^{m} e_k \frac{d^k r(t)}{dt^k} \right| \leq \left| \frac{d^{m+1} r(t)}{dt^{m+1}} \right|_{max} | e_{m+1} | .$$

$$(12.1.38)$$

One important result of error coefficient analysis applies to systems excited by *singularity functions* $\delta_{-n}(t)$, where

$$\frac{d^n \delta_{-n}(t)}{dt^n} = \delta_0(t).$$

$$(12.1.39)$$

If a system includes a multiple order pole of the form $1/s^k$ in its forward path gain, then the steady-state error in response to the singularity function $\delta_{-n}(t)$ applied as a system input will be zero for $n \leq k$.

12.1.3.3. Frequency Response. The *frequency response* of an element or system is a measure of its steady-state performance under conditions of sinusoidal excitation. Under steady-state conditions, the output of a linear system excited with a sinusoid at a frequency ω (expressed in radians per second) is purely sinusoidal at frequency ω. The frequency response is expressed as a *gain* or *magnitude* $M(\omega)$ which is the ratio of the amplitudes of the output and input sinusoids and a phase angle $\phi(\omega)$ which is the relative angle between the output and input sinusoids. The phase angle is positive if the output leads the input. The two components which complete the frequency response of a system with a transfer function $F(s)$ are given by

$$M(\omega) = |F(j\omega)|, \tag{12.1.40a}$$

$$\phi(\omega) = \angle F(j\omega) = \tan^{-1} \frac{\text{Im}[F(j\omega)]}{\text{Re}[F(j\omega)]}. \tag{12.1.40b}$$

It is frequently necessary to determine the frequency response of a system with a transfer function which is a ratio of polynomials in s. One possible method is to evaluate the frequency response by substituting $j\omega$ for s at all frequencies of interest, but this method is cumbersome, particularly for high-order polynomials. An alternative approach is to present the information concerning the frequency response graphically, as described below.

The transfer function is first factored so that both the numerator and denominator consist of products of first- and second-order terms with real coefficients. The function can then be written in the general form

$$F(s) = \frac{K}{s^n} \left[\prod_{\substack{\text{first-order} \\ \text{zeros}}} (\tau_h s + 1) \right] \left[\prod_{\substack{\text{complex-} \\ \text{zero pairs}}} \left(\frac{s^2}{\omega_{ni}^2} + \frac{2\zeta_i s}{\omega_{ni}} + 1 \right) \right]$$

$$\times \left[\prod_{\substack{\text{first-order} \\ \text{poles}}} \frac{1}{(\tau_j s + 1)} \right] \left[\prod_{\substack{\text{complex-} \\ \text{pole pairs}}} \left(\frac{s^2}{\omega_{nk}^2} + \frac{2\zeta_k s}{\omega_{nk}} + 1 \right)^{-1} \right]. \tag{12.1.41}$$

While several methods such as Lin's method[9] are available for factoring polynomials, this operation is tedious unless machine computation is employed, particularly when the order of the polynomial is large. Fortunately, in many cases of interest the polynomials are either of low order or are available from the system equations in factored form.

Since $F(j\omega)$ is a function of a complex variable, the angle $\phi(\omega)$ is the sum of the angles of the constituent terms. Similarly, the magnitude

[9] S. N. Lin, A method of successive approximations of evaluating the real and complex roots of cubic and higher-order equations. *J. Math. Phys.* **20**, 231 (1941).

$M(\omega)$ is a product of the magnitudes of the components. Furthermore, if the magnitudes of the components are expressed in logarithmic units [normally the decibel scale is used, the magnitude in decibels being given as $20 \log_{10}(\text{magnitude})$], the log of M is given by the sum of the logs corresponding to the individual components.

Plotting is simplified by recognizing that only four types of terms are possible in the representation of Eq. (12.1.41):

(a) constants, K;
(b) single- or multiple-order integrations or differentiations $1/s^m$, where m can be either positive (integrations) or negative (differentiations);
(c) first-order terms of the form $\tau s + 1$ or their reciprocal;
(d) complex conjugate pairs of the form $s^2/\omega_n^2 + 2\zeta s/\omega_n + 1$ or their reciprocal.

It is particularly convenient to represent each of these possible terms as a plot of M (expressed in decibels) and ϕ (expressed in degrees) as a function of ω (expressed in radians per second) plotted on logarithmic coordinates. The magnitude and angle of any rational function can then be determined by adding the magnitudes and angles of its components. This representation of the frequency response of a system or element is called a *Bode plot*.[6]

The magnitude of a constant term K is given by $20 \log_{10} K$, while the angle is either 0 or 180° depending on whether the sign of K is positive or negative. Both quantities are frequency independent.

The magnitude for a term of the form $1/s^n$ is $20 \log_{10}(1/\omega^n)$, a function that passes through 0 dB at $\omega = 1$ and has a slope equal to $-20\,n$ dB per *decade* (factor of 10) in frequency. The angle associated with this function is $-n(90°)$, independent of frequency.

The magnitude and phase for a first-order pole $1/(\tau s + 1)$ as a function of normalized frequency $\omega\tau$ are shown in Fig. 9. An essential feature of the magnitude function is that it can be approximated by two straight lines, one lying along the $M = 0$ dB line and the other with a slope of -20 dB/decade, which intersect at $\omega = 1/\tau$. (This frequency is called the *corner frequency*.) The maximum departure of the actual curves from the *asymptotic representation* is 3 dB and this maximum occurs at the corner frequency. The magnitude and phase for a first-order zero are obtained by inverting the curves shown for the pole, so that the magnitude approaches an asymptotic slope of $+20$ dB/decade beyond the corner frequency, while the phase changes from 0 to $+90°$.

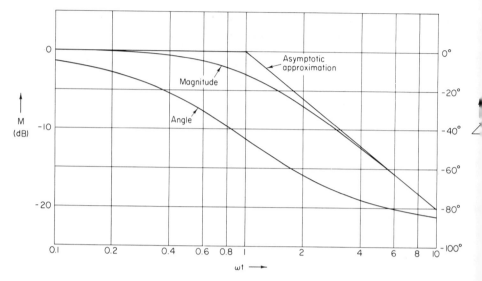

FIG. 9. Magnitude and phase for first-order pole $1/(\tau j\omega + 1)$.

The magnitude and phase curves for a complex conjugate pole pair $(s^2/\omega_n^2 + 2\zeta s/\omega_n + 1)^{-1}$ are shown in Fig. 10 as a parametric family plotted vs. normalized frequency ω/ω_n. Note that the asymptotic approximation to the magnitude is reasonably accurate providing the damping ratio exceeds 0.25. The corresponding curves for a complex conjugate zero are obtained by inverting the curves shown in Fig. 10.

The value associated with the use of Bode plots stems in large part from the ease with which the plot for a complex system can be obtained by combining the magnitude and phase functions corresponding to first- and second-order terms located at appropriate frequencies. In practice, the asymptotic magnitude curve is usually sketched by drawing a series of intersecting straight lines with appropriate slope changes at the intersections. Corrections to the asymptotic curve can be added in the vicinity of singularities if necessary. The phase angle of a composite function can be obtained by either drawing the phase of each component at the correct frequencies and adding the terms graphically with a pair of dividers, or by adding the contributions determined from the curves shown in Figs. 9 and 10 at each frequency of interest and plotting the sum directly.

The information contained in a Bode plot can also be presented as a *gain-phase* plot, which is a more convenient representation for some operations. Rectangular coordinates are used, with the ordinate repre-

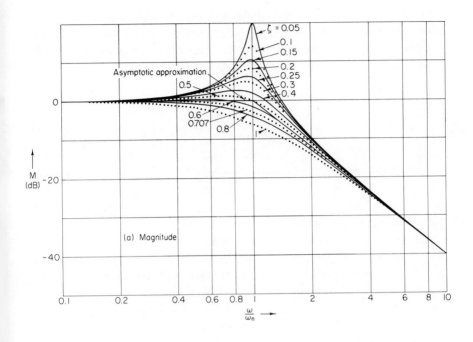

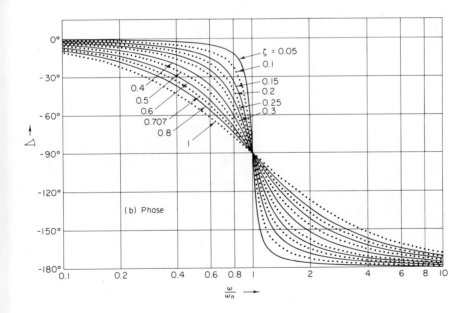

FIG. 10. Magnitude and phase for complex-pole pair $[(2\zeta j\omega/\omega_n) + (1 - (\omega^2/\omega_n^2))]^{-1}$.

senting the magnitude in decibels and the abscissa representing the phase angle in degrees. Frequency expressed in radians per second is a parameter along the gain-phase curve. Gain-phase plots are frequently obtained by transferring data from a Bode plot.

Example 12.4. The transfer function

$$F(s) = \frac{100(0.1s + 1)}{s^2(s^2/10^4 + s/250 + 1)} \tag{12.1.42}$$

is to be presented in Bode and gain-phase form. There are four components of $F(s)$: the constant 100, a second-order integration, a first-order zero with a corner frequency of 10 rad/sec, and a complex pole pair with $\omega_n = 100$ rad/sec and $\zeta = 0.2$. The magnitude of the term $100/s^2$ plots as a straight line with a slope of -40 dB/decade and a magnitude of 40 dB at $\omega = 1$, while the angle associated with this term is $-180°$ at all frequencies. The contributions of the first-order zero and second-order pole, determined from Figs. 9 and 10, are added at frequencies of 10 and 100 rad/sec respectively. The resultant Bode plot for $F(s)$ is shown in Fig. 11. This transfer function is replotted on gain-phase coordinates in Fig. 12.

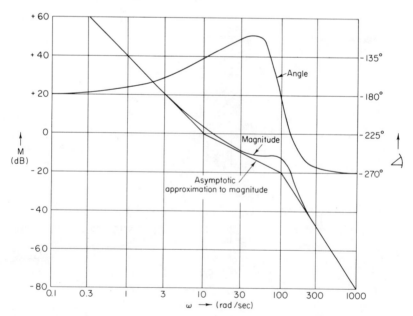

FIG. 11. Bode plot of $[100(0.1s + 1)]/s^2[(s^2/10^4) + (s/250) + 1]$.

12.1.3.4. Relationships between Transient Response and Frequency Response. It is evident that both the impulse response of a linear system, $\mathscr{L}^{-1}[C(s)/R(s)]$, and its frequency response, $C(j\omega)/R(j\omega)$, completely specify the system. Similarly, the transient response to any input can be determined from the frequency response. In many cases experimental measurements on a closed-loop system are most easily made by applying

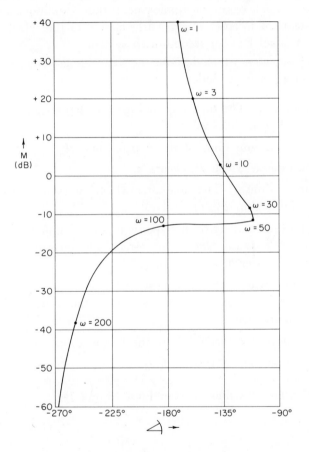

FIG. 12. Gain-phase plot of $[100(0.1s + 1)]/s^2[(s^2/10^4) + (s/250) + 1]$.

a transient input. It may be desirable to predict certain characteristics of the frequency response from the measured transient response. Since the measured transient response does not provide an equation for system output as a function of time, the conversion cannot be accomplished by Laplace techniques. This section lists several approximate relationships between transient response and frequency response which can be used to estimate one performance measure from the other. The approximations are based on properties of first- and second-order systems.

It is assumed that the system under study has unity feedback and has very high open-loop gain at zero frequency so that the closed-loop gain is one at zero frequency. It is also assumed that the lowest frequency singularity is a pole or complex-pole pair. If these conditions are satisfied, many systems of interest adhere fairly closely to the relationships developed for either first- or second-order systems.

Useful parameters which describe the step response and the frequency response of a system include:

(a) Risetime t_r. The time required for the step response to go from 10 to 90% of final value.

(b) The maximum value of the step response P_0.

(c) The time at which P_0 occurs, t_p.

(d) Settling time t_s. The time after which the system step response remains within 2% of final value.

(e) The steady-state error in response to a unit ramp $1/K_v$, where K_v is the *velocity constant* of the system. The quantity $1/K_v$ is equal to e_1 in the error series (Eq. (12.1.36)).

(f) The bandwidth in radians per second ω_h or hertz f_h ($f_h = \omega_h/(2\pi)$). The frequency at which the response of the system is 3 dB below its low-frequency value.

(g) The maximum magnitude of the frequency response M_p.

(h) The frequency at which M_p occurs, ω_p.

These definitions are illustrated in Figs. 13 and 14.

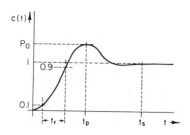

FIG. 13. Step response.

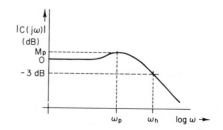

FIG. 14. Frequency reponse.

For a first-order system with $C(s)/R(s) = 1/(\tau s + 1)$, the relationships are:

$$t_r = 2.2\tau = 2.2/\omega_h = 0.35/f_h \qquad (12.1.43)$$

$$P_0 = M_p = 1 \qquad (12.1.44)$$

$$t_p = \infty \qquad (12.1.45)$$

$$t_s = 4\tau \qquad (12.1.46)$$

$$\omega_p = 0 \qquad (12.1.47)$$

$$1/K_v = \tau. \qquad (12.1.48)$$

For a second-order system with

$$\frac{C(s)}{R(s)} = \left(\frac{s^2}{\omega_n^2} + \frac{2\zeta s}{\omega_n} + 1\right)^{-1} \qquad 0 < \zeta < 1$$

the relationships are:

$$t_r \simeq 2.2/\omega_h = 0.35/f_h \qquad (12.1.49)$$

$$P_0 = 1 + \exp\left[\frac{-\pi\zeta}{\sqrt{1-\zeta^2}}\right] \qquad (12.1.50)$$

$$t_p = \frac{\pi}{\omega_n\sqrt{1-\zeta^2}} \qquad (12.1.51)$$

$$t_s \simeq 4/\zeta\omega_n \qquad (12.1.52)$$

$$\frac{1}{K_v} = 2\zeta/\omega_n \qquad (12.1.53)$$

$$M_p = \frac{1}{2\zeta\sqrt{1-\zeta^2}} \qquad (12.1.54)$$

$$\omega_p = \omega_n\sqrt{1-2\zeta^2} \qquad (12.1.55)$$

$$\omega_h = \omega_n[1 - 2\zeta^2 + \sqrt{2 - 4\zeta^2 + 4\zeta^4}]^{1/2}. \qquad (12.1.56)$$

If a system step response or frequency response is similar to that of an approximating system (see Figs. 6, 8, 9, and 10), measurements of t_r, P_0, and t_p permit estimation of ω_h, ω_p, and M_p or vice versa. The steady-state error in response to a unit ramp can be estimated from either set of measurements.

12.1.4. Stability

12.1.4.1. Introduction. As mentioned earlier, major effort in the design of feedback systems is devoted to maintaining stability while simultaneously obtaining sufficiently high open-loop gain to reduce the effects of parameter variations and disturbances to acceptable levels. A stable system is defined as one which produces a bounded output for a bounded input. Thus stability implies that

$$\int_{-\infty}^{\infty} |c(t)| \, dt < \infty \qquad\qquad (12.1.57)$$

for any input such that

$$\int_{-\infty}^{\infty} |r(t)| \, dt < \infty. \qquad\qquad (12.1.58)$$

The stability of a linear feedback system is determined completely by its characteristic equation (the denominator of its transfer function). In particular, a system is unstable if one or more zeros of the characteristic equation (poles of the transfer function) have positive real parts or equivalently lie in the right half of the s plane, since Table I shows that right-half-plane poles contribute terms to the system response which grow exponentially with time. The definition of stability [Eq. (12.1.57)] shows that systems with purely imaginary poles or poles at the origin are also unstable, since these poles contribute constants or constant-amplitude sinusoids to the output.

For some applications it is sufficient to know whether a system is stable or not. In most cases, however, more quantitative information, such as the amount of overshoot in response to a step or the change in a particular parameter which will cause instability, is required.

12.1.4.2. The Routh Criterion.[6] Stability is insured if all zeros of the characteristic equation have negative real parts. The Routh test is a method which determines the number of zeros of an algebraic polynomial with real coefficients which have positive real parts. An advantage is that it is not necessary to factor the polynomial to apply the test.

The test is described for a polynomial of the form

$$a_0 s^n + a_1 s^{n-1} + a_2 s^{n-2} + \cdots + a_{n-1} s + a_n. \qquad (12.1.59)$$

A necessary but not sufficient condition for all the zeros of polynomial (12.1.59) to have negative real parts is that all the a's be present and that they all have the same sign. If this necessary condition is satisfied, an array of numbers is generated from the a's as follows. (This example is for n even. For n odd, a_n terminates the second row.)

$$
\begin{array}{cccccc}
a_0 & a_2 & a_4 \cdots & a_{n-2} & a_n \\[2mm]
a_1 & a_3 & a_5 \cdots & a_{n-1} & 0 \\[2mm]
\dfrac{a_1 a_2 - a_0 a_3}{a_1} = b_1 & \dfrac{a_1 a_4 - a_0 a_5}{a_1} = b_2 & \cdots & \dfrac{a_1 a_n - a_0 \cdot 0}{a_1} = b_{n/2} & 0 \\[4mm]
\dfrac{b_1 a_3 - a_1 b_2}{b_1} & \dfrac{b_1 a_5 - a_1 b_3}{b_1} & \cdots & 0 & 0 \\[4mm]
\cdot & \cdot & \cdots & 0 & 0 \\
\cdot & \cdot & \cdots & \cdot & \cdot \\
\cdot & \cdot & \cdots & \cdot & \cdot \\
0 & 0 & \cdots & \cdot & \cdot \\
0 & 0 & \cdots & 0 & 0
\end{array}
$$

$$(12.1.60)$$

As the array develops, progressively more elements of each row become zero, until only the first element of the $(n+1)$th row is nonzero. The total number of sign changes in the first column is then equal to the number of zeros of the original polynomial which lie in the right half-plane.

Example 12.5. Use the Routh test to determine if any zeros of the polynomial $s^3 + 8s^2 + 6s + 260$ have positive real parts. Since all coefficients are present and positive, the first test does not indicate any right-half-plane zeros. The array is

$$
\begin{array}{cc}
1 & 6 \\[2mm]
8 & 260 \\[2mm]
\text{sign change} \rightarrow \dfrac{8 \times 6 - 260}{8} = -26.5 & 0 \qquad (12.1.61) \\[4mm]
\text{sign change} \rightarrow \dfrac{-26.5 \times 260 - 0 \times 8}{-26.5} = 260 & 0.
\end{array}
$$

The two sign changes in the first column indicate two right-half-plane zeros. This is verified by factoring the original polynomial, since

$$s^3 + 8s^2 + 6s + 260 = (s - 1 + j5)(s - 1 - j5)(s + 10). \qquad (12.1.62)$$

While the Routh test does not give any quantitative design information, it is still possible to use it as a design tool in certain cases.

Example 12.6. A phase-shift oscillator can be constructed by applying sufficient negative feedback around an amplifier which contains three or more poles. If the amplifier has three coincident poles, the system can be represented as shown in Fig. 15. It is possible to determine the value of K required to sustain oscillation as follows.

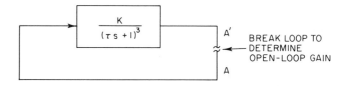

FIG. 15. Phase-shift oscillator.

It is first necessary to determine the characteristic equation of the system. Since the oscillator has no input, no transfer function is evident. The discussion associated with Eq. (12.1.2) illustrated the general result that the characteristic equation of any system is proportional to its open-loop gain plus one. The loop is broken at the indicated point to find the open-loop gain $-A'(s)/A(s)$. The characteristic equation is

$$1 - \frac{A'(s)}{A(s)} = 1 + \frac{K}{(\tau s + 1)^3}. \qquad (12.1.63)$$

Recognizing that the zeros of Eq. (12.1.63) are the same as those of $(\tau s + 1)^3 + K$ allows us to write the Routh array as

$$
\begin{array}{cc}
\tau^3 & 3\tau \\
3\tau^2 & 1 + K \\
(8 - K)\tau^3 & 0 \\
1 + K & 0
\end{array}
\qquad (12.1.64)
$$

Assuming τ is positive, there are roots with positive real parts for

$$K < -1 \quad \text{(one root)} \qquad (12.1.65a)$$

and for

$$K > 8 \quad \text{(two roots)}. \qquad (12.1.65b)$$

Laplace analysis indicates that generation of a constant amplitude sinusoidal oscillation requires a pole pair on the imaginary axis. In practice, a complex-pole pair is located slightly to the right of the imaginary axis. An intentionally introduced nonlinearity can then be used to limit the amplitude of the oscillation (see Example 12.16). Thus, a practical oscillator circuit is obtained with $K > 8$.

Two kinds of difficulties can occur when applying the Routh test. It is possible that the first element in one row of this array is zero, with the rest of the row elements nonzero. In this case the original polynomial is multiplied by $(s + a)$, where a is any positive real number, and the test is repeated.

The second possibility is that an entire row becomes zero. This condition indicates that there are a pair of roots on the imaginary axis, a pair of real roots located symmetrically with respect to the origin, or both kinds of pairs in the original polynomial. The terms in the row above the all-zero row are used as coefficients of an equation in even powers of s called the *auxiliary equation*. The zeros of this equation are the pairs mentioned above. The auxiliary equation can be differentiated with respect to s, and the resultant coefficients are used in place of the all-zero row.

Example 12.7. With K equal to 8 in Example 12.6, both elements in the third row of array (12.1.64) are zero. The corresponding auxiliary equation is obtained by forming a polynomial in even powers of s using the elements of the second row as coefficients. The resultant equation is

$$A(s) = 3\tau^2 s^2 + 9. \qquad (12.1.66)$$

The zeros of this auxiliary equation are at $s = \pm j \sqrt{3}/\tau$, indicating oscillation at a frequency $\sqrt{3}/\tau$ rad/sec. Differentiating the auxiliary equation yields

$$dA(s)/ds = 6\tau^2 s. \qquad (12.1.67)$$

The coefficient $6\tau^2$ can be used as the first element of the third row of array (12.1.64) for $K = 8$, indicating no roots in the right half-plane for this value of K.

12.1.4.3. The Nyquist Stability Criterion.[5] The Nyquist stability test provides an alternative to the Routh criterion for determining the stability of feedback control systems. In addition to absolute stability information, application of the Nyquist test provides a measure of how close a particular system is to instability, and thus is a valuable design tool. Further-more, the Nyquist test can be applied to experimentally determined data obtained by measuring the open-loop frequency response of a linear system. This contrasts with the Routh criterion, which requires that the characteristic equation be expressable as a polynomial in s.

A system used to illustrate the Nyquist test is shown in Fig. 16. The closed-loop gain for this system is

$$\frac{C(s)}{R(s)} = \frac{G(s)}{1 + G(s)H(s)}. \tag{12.1.68}$$

The system is unstable if any zeros of $1 + G(s)H(s)$ (or equivalently -1's of $G(s)H(s)$) occur for values of s with positive real parts. The number of zeros of $1 + G(s)H(s)$ with positive real parts is determined as follows.

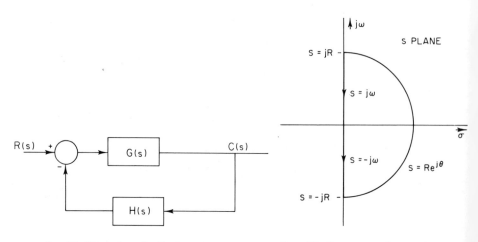

FIG. 16. Single-loop feedback system.   FIG. 17. Contour used to evaluate open-loop gain for Nyquist test.

The open-loop gain $G(s)H(s)$ is evaluated as it takes on values along the contour shown in Fig. 17. This contour includes a portion of the imaginary axis and is closed with a semicircle of radius R. The function is evaluated for the limiting case $R \to \infty$ and the resulting locus is plotted in a GH plane as shown in Fig. 18. The resultant plot will always be symmetrical about the real axis.

The Nyquist criterion can be stated as

$$Z = N + P \tag{12.1.69}$$

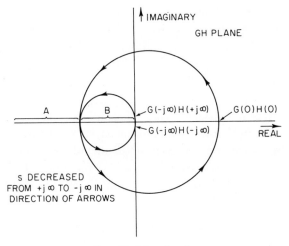

FIG. 18. Nyquist plot.

where:

Z is the number of zeros of $1 + G(s)H(s)$ in the right half of the s plane;

N is the number of encirclements of the -1 point by the GH-plane plot, and is positive if the -1 point is to the left as the plot is traversed in the direction of the arrows, or, equivalently, if the -1 point is encircled in a counterclockwise direction;

P is the number of poles of $G(s)H(s)$ encircled by the contour of Fig. 17, i.e., the number of right-half-plane poles of $G(s)H(s)$.

For stability it is necessary to have no right-half-plane zeros of $1 + G(s)H(s)$, requiring

$$N = -P. \tag{12.1.70}$$

In the example of Fig. 18, the value of N is zero if the -1 point lies somewhere along segment A, while $N = +2$ if the -1 point is on segment B. It is concluded that the system plotted in Fig. 18 will be stable if the -1 point lies on segment A provided $G(s)H(s)$ has no right-half-plane poles. Since P must be greater than or equal to zero, the system is always unstable if the -1 point is on segment B.

Example 12.8. A feedback system of the form illustrated in Fig. 16 is constructed with an open-loop gain

$$G(s)H(s) = \frac{K(0.1s + 1)}{s(s - 1)}. \tag{12.1.71}$$

The values of K for which the system is stable are to be determined. Notice that the factor K simply scales the plot in the GH plane. Thus it is convenient to plot $G(s)H(s)/K$ and determine the encirclements of the $-1/K$ point by this function. This operation is equivalent to determining the encirclements of the -1 point by the function $G(s)H(s)$.

Since there is a pole of $G(s)H(s)$ at the origin, the contour illustrated in Fig. 17 bisects one system pole. This difficulty can be resolved by either shifting the pole at the origin to the left or right by an infinitesimal amount, or by modifying the contour to include a semicircle with an infinitesimally small radius around the pole at the origin. If the alteration used moves the pole inside the contour, it increases the value of P by one.

One contour modification which can be used and which does not increase P is shown in Fig. 19. The singularities of $G(s)H(s)$ are also indicated in this figure. The plot of $G(s)H(s)/K$ evaluated along this contour is shown in Fig. 20. If K is between 0^+ and 10, the $-1/K$ point lies on segment A and is encircled once, so that $N = +1$ for these values of K. If K is between 10 and $+\infty$, the $-1/K$ point is on B. There is one *clockwise* encirclement of a point in this region so $N = -1$. If the $-1/K$ point is on C, implying a negative value for K, $N = 0$. The open-loop transfer function has one pole inside the contour of Fig. 19, or $P = 1$. Equation (12.1.70) shows that the system is stable only for $N = -1$, or for values of K between $+10$ and $+\infty$. The reader may find it helpful to verify these results by means of the Routh criterion.

For a stable system, the proximity of the GH plot to the -1 point is an indication of the relative system stability. If the plot touches the -1 point, it implies that for some frequency $G(j\omega)H(j\omega) = -1$, or that

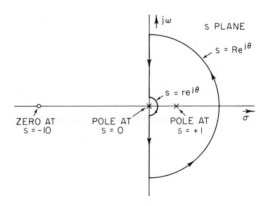

FIG. 19. Contour for Example 12.8.

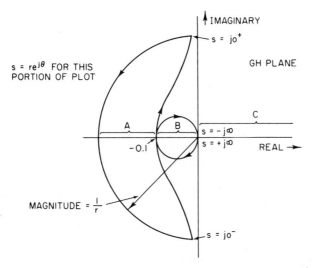

FIG. 20. Evaluation of $G(s)H(s)/K$ along contour shown in Fig. 19.

the system can sustain constant-amplitude oscillations at the frequency ω. Similarly, if $G(j\omega)H(j\omega)$ is close to -1, the characteristic equation of the system transfer function approaches zero at this frequency. Hence, sinusoidal inputs at the frequency ω tend to be accentuated by the system, indicating the existence of a complex pair of poles with a small damping ratio in the closed-loop transfer function of the system.

12.1.4.4. Root-Locus Techniques.[2,10] Root-locus techniques offer a method for determining the roots or zeros of a characteristic equation of the form

$$1 + G(s)H(s) = 1 + K\,\frac{P(s)}{Q(s)} \qquad (12.1.72)$$

where $P(s)$ and $Q(s)$ are polynomials in s. Results are presented as a *root-locus diagram*, which is a plot of the location in the s plane of the zeros of the characteristic equation, or equivalently the poles of the closed-loop transfer function, as a function of K. Since these singularities correspond directly to terms in the closed-loop transient response of a control system, the root-locus diagram provides an excellent method for comparing the behavior of several systems or for evaluating the effects of gain changes on a particular system.

[10] W. R. Evans, Control system synthesis by root-locus method. *Trans. AIEE* **69**, Part I, 66 (1950).

The condition for a point s_1 in the s plane to be a root of the character-istic equation and thus lie on the locus is that

$$K \frac{P(s_1)}{Q(s_1)} + 1 = 0. \tag{12.1.73}$$

The equality of Eq. (12.1.73) requires that simultaneously

$$| P(s_1)/Q(s_1) | = 1/K \tag{12.1.74a}$$

and

$$\measuredangle (P(s_1)/Q(s_1)) = (2n + 1)\,180° \tag{12.1.74b}$$

where n is an integer.

These relationships can be used to develop the following rules which are helpful in establishing the root locus:

1. The number of branches of the root-locus diagram is equal to the number of poles of the open-loop transfer function. Each branch starts at an open-loop pole for $K = 0$ and terminates either on an open-loop zero or at infinity for $K = \infty$.

2. Branches of the diagram lie on those portions of the real axis which are to the left of an odd number of real-axis poles and zeros of the open-loop transfer function.

3. The root locus is symmetrical with respect to the real axis.

4. For large values of K, $P\text{-}Z$ branches approach infinity, where P and Z are the number of poles and zeros of the open-loop transfer function respectively. These branches approach asymptotes which make an angle $(2n + 1)\,180°/(P - Z)$ with the real axis. The asymptotes all inter-sect the real axis at the same point. The location of this point is given by

$$\frac{\sum \text{real parts of poles of } G(s)H(s) - \sum \text{real parts of zeros of } G(s)H(s)}{P - Z}.$$

5. Near a complex open-loop pole, the angle of a branch with respect to the pole is $180° + \sum \measuredangle z - \sum \measuredangle p$, where $\sum \measuredangle z$ is the sum of the angles of vectors drawn from all zeros of $G(s)H(s)$ to the complex pole and $\sum \measuredangle p$ is the sum of the angles of vectors drawn from all other poles to the pole in question. The angle a branch makes with a complex open-loop zero is

$$180° - \sum \measuredangle z + \sum \measuredangle p.$$

6. The points where branches leave or join the real axis are given by the roots of the equation

$$\frac{d}{ds}\left(\frac{P(s)}{Q(s)}\right) = 0.$$

7. The gain K corresponding to a point s_1 on the locus is determined from

$$K = |\,Q(s_1)/P(s_1)\,|.$$

The use of these rules as aids to establishing the root locus is illustrated in the following examples.

Example 12.9. The root-locus diagram for a system with

$$G(s)H(s) = \frac{K}{s(\tau s + 1)} \tag{12.1.75}$$

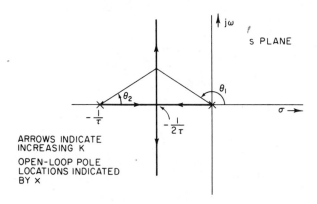

FIG. 21. Root locus for $G(s)H(s) = K/s(\tau s + 1)$.

is shown in Fig. 21. Rule 2 indicates that part of the locus lies between the two open-loop poles on the real axis. Rule 4 shows that the two asymptotes make an angle of $+90°$ and $+270°$ (or $-90°$) with the real axis, and intersect the real axis at a distance $-1/2\tau$ from the origin. For the second-order system illustrated, the poles are located on the asymptotes when they are not on the real axis. This behavior is easily verified, since the geometry of Fig. 21 insures that the sum of θ_1 and θ_2 is $180°$ for any point on the vertical line which bisects the open-loop pole locations. Thus the angle of $G(s)H(s)$ is $-180°$ at any point on this line. Rule 6 verifies that the branches leave the real axis at $s = -1/2\tau$, since

$$\frac{d}{ds}[G(s)H(s)] = \frac{-K(2\tau s + 1)}{[s(\tau s + 1)]^2} = 0 \qquad (12.1.76)$$

for $s = -1/2\tau$.

Example 12.10. Figure 22 shows the root-locus diagram for

$$G(s)H(s) = \frac{K}{(s + 1)(s^2 + 4s + 8)}. \qquad (12.1.77)$$

Rule 4 establishes the asymptotes, while Rule 5 is used to determine the locus near the complex poles. The value of K for which the complex poles cross into the right half-plane and the frequency at which they cross

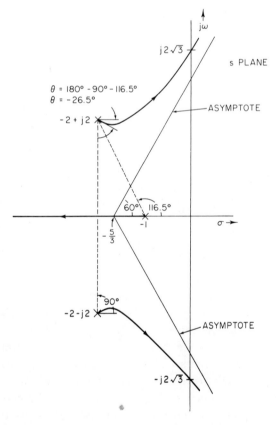

FIG. 22. Root locus for $G(s)H(s) = K/(s + 1)(s^2 + 4s + 8)$.

the imaginary axis can be determined by Routh's criterion. The characteristic equation for the system described by Eq. (12.1.77) is

$$(s + 1)(s^2 + 4s + 8) + K = s^3 + 5s^2 + 12s + 8 + K. \qquad (12.1.78)$$

The corresponding Routh array is

$$\begin{matrix} 1 & 12 \\ 5 & 8 + K \end{matrix} \qquad (12.1.79)$$

Two poles are in the right half-plane for $K > 52$. The auxiliary equation for $K = 52$ is

$$A(s) = 5s^2 + 60 = 0 \qquad \text{for } s = \pm j2 \sqrt{3}, \qquad (12.1.80)$$

indicating that the locus crosses the imaginary axis at $s = \pm j2 \sqrt{3}$.

Example 12.11. The root-locus diagram for a system with

$$G(s)H(s) = \frac{K(s + 2)}{s(s + 1)} \qquad (12.1.81)$$

is shown in Fig. 23. Rule 2 indicates that branches are on the real axis between the two open-loop poles and to the left of the open-loop zero. The points of departure from and reentry to the real axis are obtained

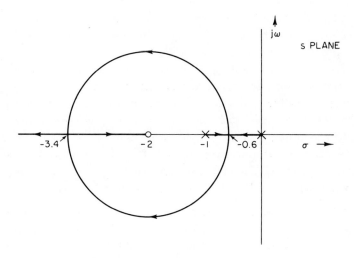

Fig. 23. Root locus for $G(s)H(s) = K(s + 2)/s(s + 1)$.

by solving

$$\frac{d}{ds}\left[\frac{K(s+2)}{s(s+1)}\right] = \frac{-K(s^2+4s+2)}{[s(s+1)]^2} = 0 \qquad (12.1.82)$$

yielding $s = -2 \pm \sqrt{2}$.

12.1.5. Compensation

12.1.5.1. Objectives. The starting point for the design of a feedback control system is normally a set of design specifications or objectives. The system designer has available a wide variety of components to use as elements of the complete control system. In spite of this availability, the range of certain performance indices of components is limited by current technology. For example, the maximum unloaded acceleration of a motor is equal to its maximum torque divided by its inertia. This ratio cannot be selected at will, since material limitations constrain its value. Similarly, it is not currently possible to obtain transistors which provide power gain at frequencies in excess of 10^{10} Hz.

On the other hand, the designer can often include components over which he has relatively greater parameter choice in a system. For example, electronic amplifiers or signal conditioners with essentially any desired gain and transfer function can normally be included in the design for a servomechanism because of the vastly greater frequency response of electronic components compared with electromechanical elements. Also, the various node-to-ground capacitances in a feedback amplifier, while bounded on the lower end by stray capacitances constrained by construction technique, can be increased if such increases seem appropriate.

The process of modifying or adding components to a control system in an attempt to favorably alter certain performance characteristics is called *compensation*. Successful compensation usually reduces to a trial and error procedure, with the experience of the designer often playing a major role in the eventual outcome. The process is normally accomplished by assuming a given type of compensation and evaluating system performance to see if objectives are met. If performance is not adequate, alternate methods of compensation are tried until objectives are either met or it becomes evident that they cannot be achieved.

12.1.5.2. Closed-Loop Performance in Terms of Open-Loop Parameters. A major objective of compensation is to alter the transfer function of either the forward gain path $G(s)$ or the feedback element $H(s)$ in order to improve system characteristics. Alternatively a second feedback path can be formed to modify the system. These two types of modifications are called *series compensation* and *feedback compensation* respectively. Since the effects of either type of compensation on the open-loop transfer function $G(s)H(s)$ are easily determined, techniques which permit approximation of the effects of certain open-loop characteristics on closed-loop performance lend insight to the process of compensation.

The open-loop and closed-loop transfer functions for a system of the type illustrated in Fig. 16 are exactly related by

$$\frac{C(s)}{R(s)} = \frac{G(s)}{1 + G(s)H(s)}.$$ (12.1.83)

The closed-loop gain of a stable system can be approximated for extreme values of $G(j\omega)H(j\omega)$ as

$$\frac{C(j\omega)}{R(j\omega)} \simeq \frac{1}{H(j\omega)} \qquad |G(j\omega)H(j\omega)| \gg 1$$ (12.1.84a)

$$\frac{C(j\omega)}{R(j\omega)} \simeq G(j\omega) \qquad |G(j\omega)H(j\omega)| \ll 1.$$ (12.1.84b)

The closed-loop gain corresponding to an open-loop gain magnitude closer to unity can be determined with the aid of the *Nichols chart* shown in Fig. 24. This chart relates $G/(1 + G)$ to G, where G is any complex number. While this form seems intended for use with unity feedback systems, the chart can be used for any closed-loop system by observing that

$$\frac{C(j\omega)}{R(j\omega)} = \frac{G(j\omega)}{1 + G(j\omega)H(j\omega)} = \frac{1}{H(j\omega)} \frac{G(j\omega)H(j\omega)}{1 + G(j\omega)H(j\omega)}.$$ (12.1.85)

The complete closed-loop frequency response can be determined by multiplying the factor $G(j\omega)H(j\omega)/(1 + G(j\omega)H(j\omega))$ determined via a Nichols chart by $1/H(j\omega)$ using Bode methods.

One quantity of particular interest is the *peak magnitude* M_p (see Fig. 14) equal to the maximum value of $|C(j\omega)/R(j\omega)|$. A large value for M_p indicates a relatively less stable control system, since it shows that there is some frequency for which the characteristic equation approaches zero. Control systems are frequently designed to have an M_p of approximately 3 dB, representing a compromise between the greater relative stability implied by a lower value of M_p and the advantages of a higher

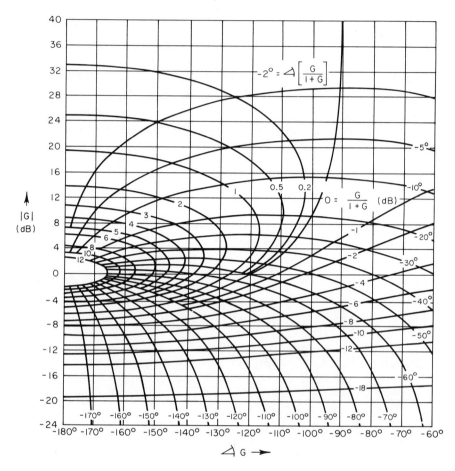

$$-2° = \angle\left[\frac{G}{1+G}\right]$$

$$0 = \frac{G}{1+G} \quad (dB)$$

$|G|$ (dB)

$\angle$ G $\rightarrow$

FIG. 24. Nichols chart.

open-loop gain which often leads to increased M_p. The gain setting which results in a specified M_p for unity-feedback systems is easily determined using the Nichols chart.

Example 12.12. A control system is constructed with

$$G(s) = \frac{K}{s(0.1s + 1)} \qquad H(s) = 1.$$

The relationship between K and M_p for this system is determined by plotting $G(j\omega)/K$ on gain-phase coordinates using the same scale as the Nichols chart. If $G(j\omega)/K$ is plotted on tracing paper, this plot can be lined up with the Nichols chart and slid up or down to illustrate the effects of different gains. The corresponding closed-loop gain as a function of frequency can be determined for any particular value of K. Furthermore, the highest value magnitude curve touched by the $G(j\omega)$ plot corresponds to M_p. Figure 25 shows $G(j\omega)$ plotted on a Nichols chart for values of K equal to 10.5, 16, and 38. The corresponding values of M_p are 1, 3, and 6 dB, respectively.

The relative stability of a feedback system and many important characteristics of its closed-loop response are largely determined by the behavior of its open-loop gain at frequencies where this gain magnitude is close to one. Therefore the characteristics of $G(j\omega)H(j\omega)$ defined below are useful for estimating closed-loop response, predicting the effects of compensation, and comparing the performance of various systems. It is assumed in these definitions that the open-loop gain magnitude is unity at only one frequency and that the open-loop phase is $-180°$ at only one frequency. The *crossover frequency* ω_c is the frequency at which the magnitude of the open-loop gain is unity (0 dB). An increase in the

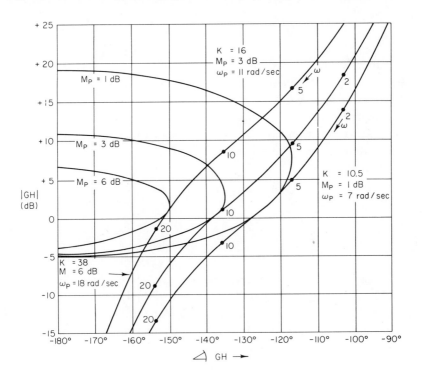

FIG. 25. Gain determination for Example 12.12.

value of ω_c generally speeds up the system transient response and increases system bandwidth. The *phase margin* of a system is $180°$ plus the angle of $G(j\omega)H(j\omega)$ evaluated at the frequency where $|G(j\omega)H(j\omega)| = 1$. Systems with phase margins from 30 to $60°$ usually yield acceptable performance. Lower values indicate reduced stability, while the gain of systems with higher values can often be increased to speed up response. The gain margin of a system is $1/|G(j\omega)H(j\omega)|$ evaluated at the frequency where $\angle G(j\omega)H(j\omega) = -180°$. This quantity is a measure of the gain increase necessary to produce instability. Values in excess of approximately 2.5 (8 dB) are generally satisfactory.

Several approximations interrelate the parameters defined above and two others defined in Section 12.1.3.4:

$$\text{phase margin} \simeq \sin^{-1}(1/M_{\text{p}}) \qquad (12.1.86)$$

$$0.6/\omega_c < t_{\text{r}} < 2.2/\omega_c. \qquad (12.1.87)$$

The shorter values of risetime correspond to lower values of phase margin.

$$t_{\text{s}} > 4/\omega_c. \qquad (12.1.88)$$

The limit is approached only for systems with large phase margins.

A higher-order system with a phase margin of 30 to $60°$ can often be approximated by a second-order system for purposes of estimating frequency response in the vicinity of ω_c and transient response. Damping ratio and natural frequency for the approximating second-order system are determined by

$$\zeta = \text{phase margin}/100 \qquad (12.1.89)$$

and

$$\omega_{\text{n}} = \frac{\omega_c}{\sqrt{1 - 2(\text{phase margin}/100)^2}}. \qquad (12.1.90)$$

It is also possible to determine the first nonzero error coefficient (Section 12.1.3.2) for a unity-feedback system directly from the open-loop gain. If the magnitude $1/G(j\omega)H(j\omega)$ is equal to K/ω^n for sufficiently small values of ω, the first $n - 1$ error coefficients are zero, and e_n is equal to $1/K$. This relationship can be used to predict the effect of compensation on the steady-state error in response to singularity function excitation.

12.1.5.3. Series Compensation. Feedback control systems are series compensated by adjusting the transfer function of the forward gain element until it becomes possible to meet specifications. The resultant forward path can be represented as the cascade connection of a compensating element and the fixed element as shown in Fig. 26. The transfer functions of these two components are called $G_c(s)$ and $G_f(s)$ respectively.

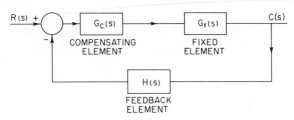

FIG. 26. System with series compensation.

The simplest of series compensations involves the selection of a frequency-independent gain for G_c. The gain is usually adjusted for a specified M_p as explained in the preceding section. If specifications cannot be met with this type of compensation, more elaborate compensating transfer functions can be selected. A mathematically attractive choice uses a $G_c(s)$ which includes a term equal to $1/G_f(s)$. If $G_f(s)$ is rational and *minimum phase* (no singularities in the right half-plane), it is possible to obtain any desired forward path transfer function by this cancellation of the singularities of $G_f(s)$. Unfortunately, this technique is of limited value for physical systems, since the canceling transfer function invariably requires a gain which increases without bound at high frequencies, and such transfer functions result in intolerable amplification of the noise which always accompanies signals.

Two types of networks are normally used for compensation, and these types can either be used separately or can be combined in one system. The *lead network*, realized as shown in Fig. 27, provides the transfer function

$$\frac{E_o(s)}{E_i(s)} = \frac{1}{\alpha}\left[\frac{\alpha\tau s + 1}{\tau s + 1}\right] \qquad (12.1.91)$$

where

$$\alpha = (R_1 + R_2)/R_2 \quad \text{and} \quad \tau = R_1 R_2 C/(R_1 + R_2).$$

As the name implies, this network provides positive phase shift at all

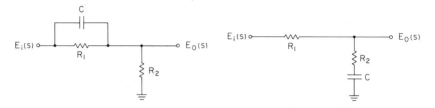

FIG. 27. Lead network.   FIG. 28. Lag network.

frequencies. It is normally used to cancel the negative phase-shift characteristic of most fixed elements so that the crossover frequency can be increased to improve parameters such as risetime.

The *lag network* shown in Fig. 28 has the transfer function

$$\frac{E_o(s)}{E_i(s)} = \frac{\tau s + 1}{\alpha \tau s + 1} \qquad (12.1.92)$$

where

$$\alpha = (R_1 + R_2)/R_2 \quad \text{and} \quad \tau = R_2 C.$$

This network, which provides negative phase shift at all frequencies, is normally located well below crossover so that the system gain can be increased in order to reduce low-frequency errors.

The maximum magnitude of the phase angle associated with either of these functions is

$$\phi_m = \sin^{-1}\left[\frac{\alpha - 1}{\alpha + 1}\right]. \qquad (12.1.93)$$

The maximum phase-angle magnitude occurs at the geometric mean of the frequencies of the two singularities, and this frequency is called ω_m:

$$\omega_m = 1/\tau\sqrt{\alpha}. \qquad (12.1.94)$$

The gain of either network at ω_m is $1/\sqrt{\alpha}$.

The ratio of the frequencies of the two singularities, α, is normally selected to be between 5 and 20. Smaller values do not produce sufficient modification of the transfer function, while larger values greatly accentuate noise in the case of a lead network and may result in poor recovery from overloads or saturation in the case of a lag network.

The magnitude and phase for lead networks with $\alpha = 5$, 10, and 20 are plotted as a function of normalized frequency in Fig. 29. Lag network characteristics are obtained by replacing ϕ by $-\phi$, inverting

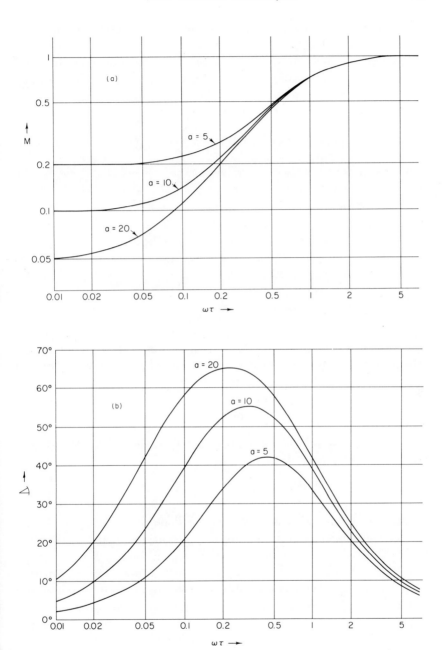

FIG. 29. Magnitude and phase for lead network $(\alpha\tau j\omega + 1)/\alpha(\tau j\omega + 1)$: (a) magnitude; (b) phase.

the magnitude curve about the 0 dB line, and shifting the inverted curve down by $20 \log_{10} \alpha$.

The location of a lead network is selected on the basis of the required phase margin of the system. For a network with a specified α, τ is often chosen so that the crossover frequency of the system is equal to ω_m, with the result that

$$\text{phase margin} = 180° + \phi_m + \angle G_f(j\omega_m). \quad (12.1.95)$$

It is evident that it is necessary to simultaneously adjust the gain of the system to satisfy these two constraints.

In order to locate a lag network, the system is first compensated with a compensating element transfer function equal to K, and K is selected to meet the phase-margin requirement. Assume that this value of K produces a crossover frequency ω_c. The final compensating transfer function is then chosen as

$$G_c(s) = \frac{\alpha K(10s/\omega_c + 1)}{(10\alpha/\omega_c)s + 1}. \quad (12.1.96)$$

This choice of gain and network location insures that the system crossover frequency remains unchanged and that the maximum decrease in phase margin introduced by the compensating network is approximately 5°. The open-loop gain of the system at frequencies small compared to $\omega_c/10\alpha$ is increased by a factor of α.

Example 12.13. Consider the control system shown in Fig. 26 with $H(s) = 1$ and

$$G_f(s) = \frac{1}{s(0.1s + 1)(0.01s + 1)}. \quad (12.1.97)$$

This transfer function is shown in the Bode plot of Fig. 30. If frequency-independent gain is used to compensate the system, a G_c equal to 21 dB raises the magnitude curve by this amount yielding a phase margin of 45° and a crossover frequency of 9 rad/sec. The gain margin is 18 dB for this compensation. The steady-state error in response to a unit ramp applied to the system will be 0.09, since the open-loop gain magnitude is equal to $11/\omega$ at low frequencies. This quantity is numerically equal to the e_1 term of the error series (Section 12.1.3.2) or the reciprocal of the velocity constant, $1/K_v$ (Section 12.1.3.4).

If the system is compensated with a lead network with $\alpha = 10$, 45° of phase margin is possible with a crossover frequency equal to 40 rad/sec,

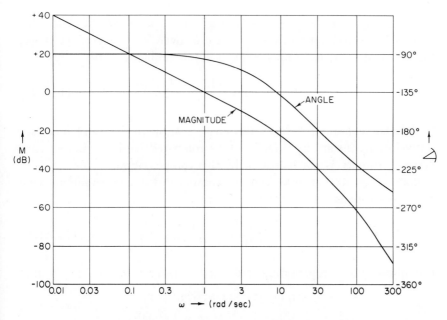

FIG. 30. Uncompensated open-loop Bode plot for Example 12.13.

since the angle of $G_f(j40)$ is $-190°$ and ϕ_m for the lead network is $+55°$. (See Eqs. (12.1.93) and (12.1.95).) The lead network should be selected with $\tau = 0.0078$ sec so that ω_m is 40 rad/sec. It is also necessary to include a gain equal to 500 (54 dB) to raise the crossover frequency to 40 rad/sec, since the magnitude of $G_f(j40) = -44$ dB and the gain of the lead network is $1/\sqrt{10}$ or -10 dB at 40 rad/sec.

The above discussion shows that the transfer function of the required compensating element is

$$G_c(s) = 500\left[\frac{0.078s + 1}{10(0.0078s + 1)}\right]. \tag{12.1.98}$$

The increase in crossover frequency to 40 rad/sec combined within a phase margin equal to that of the system compensated with frequency-independent gain indicates that the risetime and settling time for the lead compensated system should be approximately 4.5 times shorter than corresponding quantities for the first system. The velocity constant is increased to 50 compared with 11 for the gain-compensated system. The price paid for these performance improvements is one of greater noise sentitivity because of the lead network and the improved high-frequency response of the closed-loop system.

The system can also be compensated with a lag network. Since this type of compensation will result in a crossover frequency of 9 rad/sec, the required $G_c(s)$ is (from Eq. (12.1.96) assuming $\alpha = 10$)

$$G_c(s) = 110 \frac{(1.1s + 1)}{(11s + 1)}. \tag{12.1.99}$$

This compensation reduces the phase margin to $40°$, and increases the velocity constant by a factor of 10 compared with the first system. Noise sensitivity is not increased by this compensation.

Both types of networks can be combined for compensation. One possible compensating transfer function is

$$G_c(s) = \frac{500(0.25s + 1)(0.078s + 1)}{(2.5s + 1)(0.0078s + 1)}. \tag{12.1.100}$$

Combining the compensating and fixed transfer functions shows that

$$G_c(s)G_f(s) = \frac{500(0.25s + 1)(0.078s + 1)}{s(2.5s + 1)(0.1s + 1)(0.01s + 1)(0.0078s + 1)}. \tag{12.1.101}$$

This transfer function is plotted in Fig. 31. The crossover frequency

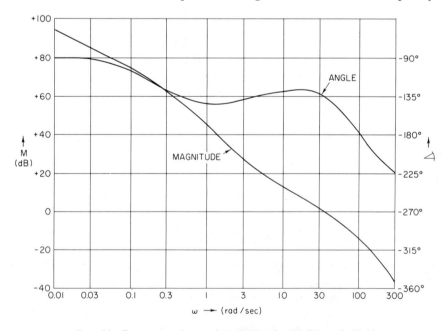

FIG. 31. Compensated open-loop Bode plot for Example 12.13.

is 35 rad/sec and the phase margin is 45°. The gain margin is 14 dB, and the velocity constant is increased to 500.

The phase margin can be increased to more than 50° and the gain margin increased to 20 dB by reducing the constant multiplicative term in Eq. (12.1.101) to 250. The crossover frequency is lowered to 20 rad/sec and the velocity constant is halved by this modification.

12.1.5.4. Feedback Compensation. Figure 32 shows a feedback system which includes a *minor* or *inner* loop consisting of elements $G_2(s)$, $G_f(s)$, and $H_c(s)$. The closed-loop transfer function for this system is

$$\frac{C(s)}{R(s)} = \frac{\dfrac{G_1(s)G_2(s)G_f(s)}{1 + G_2(s)G_f(s)H_c(s)}}{1 + \dfrac{G_1(s)G_2(s)G_f(s)H(s)}{1 + G_2(s)G_f(s)H_c(s)}}. \qquad (12.1.102)$$

If this transfer function is compared with the transfer function for the series-compensated system shown in Fig. 26,

$$\frac{C(s)}{R(s)} = \frac{G_c(s)G_f(s)}{1 + G_c(s)G_f(s)H(s)}, \qquad (12.1.103)$$

it is evident that selecting

$$G_c(s) = \frac{G_1(s)G_2(s)}{1 + G_2(s)G_f(s)H_c(s)} \qquad (12.1.104)$$

makes the two functions equal.

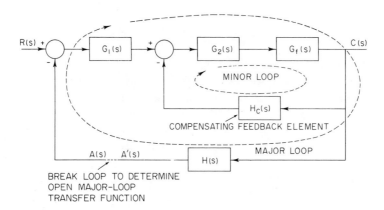

FIG. 32. System illustrating feedback compensation.

Furthermore, the degrees of freedom represented by the $G_1(s)$ and $G_2(s)$ are not required to force equivalence, and these elements need differ from unity only if required by other system considerations such as maintaining linearity or insensitivity to variations in system parameters. The above analysis indicates that theoretically a system can be compensated equally well by either series compensation or by feedback compensation applied around an inner loop. In practice, feedback compensation is often preferable, since it frequently leads to systems with less sensitivity to component variations and to noise.

An appropriate transfer function for feedback compensation can be selected by first determining series compensation which will meet specifications and then converting to equivalent feedback compensation. Alternatively, feedback compensation can be chosen to approximate a desired inner-loop transfer function as illustrated by the following example.

Example 12.14. The open-loop transfer function for the system of Example 12.13 using lead and lag network compensation is

$$G_c(s)G_f(s) = \frac{500(0.25s + 1)(0.078s + 1)}{s(2.5s + 1)(0.1s + 1)(0.01s + 1)(0.0078s + 1)}. \quad (12.1.105)$$

Virtually no change in any important characteristics such as crossover frequency, phase margin, or steady-state error in response to a ramp is introduced by shifting the two zeros by approximately 20% in opposite directions with a resultant

$$G_c(s)G_f(s) = \frac{500(0.2s + 1)(0.1s + 1)}{s(2.5s + 1)(0.1s + 1)(0.01s + 1)(0.0078s + 1)}$$

$$= \frac{500(0.2s + 1)}{s(2.5s + 1)(0.01s + 1)(0.0078s + 1)}. \quad (12.1.106)$$

This transfer function can be approximated by an appropriate choice of parameters in the system topology shown in Fig. 32 as follows. For correspondence with Example 12.13, it is necessary to have $H(s) = 1$ and

$$G_f(s) = \frac{1}{s(0.1s + 1)(0.01s + 1)}. \quad (12.1.107)$$

Evaluating the open-major-loop transfer function yields

$$-\frac{A'(s)}{A(s)} = \frac{G_1(s)G_2(s)G_f(s)}{1 + G_2(s)G_f(s)H_c(s)}. \quad (12.1.108)$$

Note that

$$-\frac{A'(j\omega)}{A(j\omega)} \simeq G_1(j\omega)G_2(j\omega)G_f(j\omega), \qquad |\,G_2(j\omega)G_f(j\omega)H_c(j\omega)\,| \ll 1 \tag{12.1.109a}$$

and

$$-\frac{A'(j\omega)}{A(j\omega)} \simeq \frac{G_1(j\omega)}{H_c(j\omega)}, \qquad |\,G_2(j\omega)G_f(j\omega)H_c(j\omega)\,| \gg 1. \tag{12.1.109b}$$

The transfer function $G_1(s)$ is arbitrarily chosen equal to 1. Since the desired open-loop transfer function is equal to $500/s$ for small values of s (Eq. (12.1.106)) and since $G_f(s)$ is equal to $1/s$ for small values of s, a value of 500 is selected for G_2. At frequencies between 0.4 and 100 rad/sec, the desired function is approximately $200(0.2j\omega + 1)/(j\omega)^2$. This transfer function can be obtained at frequencies where the gain of the inner loop is large compared to unity if

$$H_c(s) = \frac{s^2}{200(0.2s + 1)}. \tag{12.1.110}$$

While this feedback transfer function may seem unrealistic because of its unlimited high-frequency gain, it can be approximated by a *tachometer* (see Section 12.2.2.1) followed by a resistor–capacitor high-pass network in some servomechanisms, and by a resistor–capacitor T network in certain feedback amplifiers. The high-frequency gain is acceptable because of the lower noise sensitivity of systems compensated by internal feedback.

Substituting the values developed above into Eq. (12.1.108) yields

$$-\frac{A'(s)}{A(s)} = \frac{\dfrac{500}{s(0.1s+1)(0.01s+1)}}{1 + \dfrac{2.5s}{(0.2s+1)(0.1s+1)(0.01s+1)}}$$

$$= \frac{500(0.2s+1)}{s(2.8s+1)\left[\dfrac{s^2}{(120)^2} + \dfrac{2(0.5)}{120}s + 1\right]}. \tag{12.1.111}$$

If this open-major-loop transfer function is compared with the transfer function originally selected because it closely approximates the results obtained by series compensation (Eq. (12.1.106)), the difference is a shift of approximately 10% in the location of the low-frequency pole and replacement of two poles at $s = -100$ and $s = -128$ by a complex pair with $\omega_n = 120$ and $\zeta = 0.5$. This initially selected feedback compensation therefore lowers crossover frequency by about 10% and increases phase margin by nearly 20° compared with the series-compensated system. In order to increase the crossover frequency and to reduce the phase margin (since 65° of phase margin results in relatively sluggish transient response), the feedback element is modified so that

$$H_c(s) = \frac{s^2}{600(0.07s + 1)}. \qquad (12.1.112)$$

With this value for $H_c(s)$ combined with the previous values for $G_1(s)$, $G_2(s)$, and $G_f(s)$, the open-major-loop transfer function is

$$-\frac{A'(s)}{A(s)} = \frac{500(0.07s + 1)}{s(s + 1)\left[\dfrac{s^2}{(115)^2} + \dfrac{2(0.5)s}{115} + 1\right]}. \qquad (12.1.113)$$

This function is plotted in Fig. 33. The Bode plot shows a crossover frequency of 35 rad/sec, a phase margin of 50°, and a gain margin of

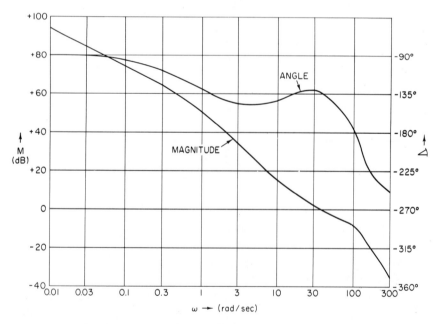

FIG. 33. Open-major-loop Bode plot for Example 12.14.

10 dB. It is concluded that the performance of this feedback-compensated system should be similar to that of the series-compensated system plotted in Fig. 31. One advantage of the feedback-compensated system is that the steady-state error in response to a ramp input should be reached in a shorter time period since the low-frequency pole is located at 1 rad/sec for this system compared with 0.4 rad/sec for the series-compensated system. A partial fraction expansion of the closed-loop response to ramp excitation shows that the location of the low-frequency pole determines the time required to reach the steady-state error.

12.1.6. Nonlinear Systems[11,12]

12.1.6.1. Linearization. The successful application of the analytic techniques introduced in earlier sections is predicated on the assumption that the system in question can be described by a linear differential equation. While the computational advantages of the assumption of linearity are obvious, virtually all real systems are nonlinear when examined in sufficient detail.

One approximate method which is particularly useful when the range of a variable is restricted to small variations about an operating point is called the *tangent approximation* or *linearization*. This approximation is based on the use of a Taylor series estimation of the function of interest. In general, it is assumed that the variable y is related to n variables x_j as

$$y = F(x_1, x_2, \ldots, x_n) \qquad (12.1.114)$$

where F is any function. The function y can be expanded in a power series for deviations $\Delta x_1, \Delta x_2, \ldots, \Delta x_n$ about an operating point x_{10}, $x_{20}, \ldots, x_{n0}$. For some restricted range surrounding the operating point, the first-order approximation will be sufficient:

$$\Delta y = y - y_0 \simeq \sum_{j=1}^{n} \Delta x_j \left. \frac{\partial F}{\partial x_j} \right|_{x_{10}, x_{20}, \ldots, x_{n0}} \qquad (12.1.115)$$

where $y_0 = F(x_{10}, x_{20}, \ldots, x_{n0})$.

Equation (12.1.115) is used to write linear system equations which relate incremental variables and which approximate the actual nonlinear system equations over some restricted range of operation.

[11] W. J. Cunningham, "Introduction to Nonlinear Analysis." McGraw-Hill, New York, 1958.
[12] G. J. Thaler and M. P. Pastel, "Analysis and Design of Nonlinear Feedback Control Systems." McGraw-Hill, New York, 1962.

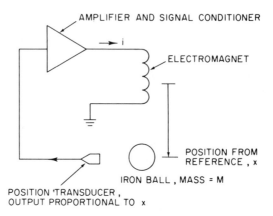

AMPLIFIER AND SIGNAL CONDITIONER

ELECTROMAGNET

POSITION FROM REFERENCE , x

IRON BALL , MASS = M

POSITION TRANSDUCER ,
OUTPUT PROPORTIONAL TO x

FIG. 34. System used to suspend ball in magnetic field.

Example 12.15. Figure 34 diagrams a system which can be used to suspend a ball in a field generated by an electromagnet. Consideration is limited to motion of the ball in the x direction.

The downward force produced by gravity must be cancelled by the upward magnetic force to suspend the ball. It is clear that stabilization with constant current is impossible, since while a value of x where there is no force on the ball exists, a small deviation from this position changes the magnetic force in such a way as to accelerate the ball further from equilibrium. This effect can be cancelled by appropriately controlling the magnet current as a function of measured ball position.

For certain geometries and with appropriate choice of the reference position for x, the magnetic force F_e directed in the positive x direction is approximated as

$$F_e = - Ci^2/x^2 \qquad (12.1.116)$$

where C is a constant.

Assuming incremental changes Δi and Δx about operating-point values i_0 and x_0,

$$F_e = - \frac{Ci_0^2}{x_0^2} - \frac{2Ci_0}{x_0^2} \Delta i + \frac{2Ci_0^2}{x_0^3} \Delta x$$
$$+ \text{ higher-order terms.} \qquad (12.1.117)$$

The equation of motion of the ball is

$$M \frac{d^2x}{dt^2} = Mg + F_e \qquad (12.1.118)$$

where g is the acceleration of gravity.

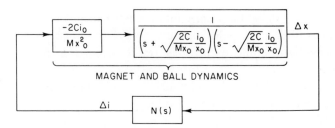

$$\boxed{\dfrac{-2Ci_0}{Mx_0^2}} \quad \boxed{\dfrac{1}{\left(s + \sqrt{\dfrac{2C}{Mx_0}\dfrac{i_0}{x_0}}\right)\left(s - \sqrt{\dfrac{2C}{Mx_0}\dfrac{i_0}{x_0}}\right)}} \quad \Delta x$$

MAGNET AND BALL DYNAMICS

$\Delta i \quad \boxed{N(s)}$

FIG. 35. Linearized block diagram for System of Fig. 34.

The equilibrium values x_0 and i_0 are selected so that

$$Mg - Ci_0^2/x_0^2 = 0. \qquad (12.1.119)$$

Introducing this value and noting that the derivative of an incremental variable is the same as the derivative of the total variable results in the linearized equation

$$M\frac{d^2\,\Delta x}{dt^2} - \frac{2Ci_0^2}{x_0^3}\,\Delta x = -\frac{2Ci_0}{x_0^2}\,\Delta i. \qquad (12.1.120)$$

This equation can be transformed and manipulated to yield the block diagram shown in Fig. 35. The top two blocks represent the relationship between Δi and Δx, while the bottom block represents the amplifier which can be used to control Δi as a linear function of Δx. The transfer function of this amplifier may in general be frequency dependent, and is denoted by $N(s)$. The open-loop transfer function includes two poles located symmetrically with respect to the origin on the real axis which arise from ball-magnet dynamics. If $N(s)$ is frequency independent and equal to N_0, the root-locus diagram for the system (Fig. 36) shows that

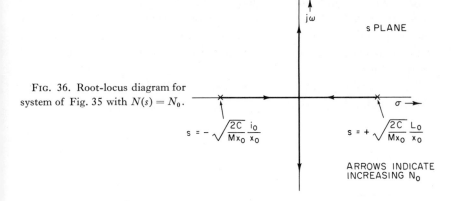

FIG. 36. Root-locus diagram for system of Fig. 35 with $N(s) = N_0$.

s PLANE

$$s = -\sqrt{\frac{2C}{Mx_0}\frac{i_0}{x_0}} \qquad s = +\sqrt{\frac{2C}{Mx_0}\frac{i_0}{x_0}}$$

ARROWS INDICATE INCREASING N_0

it is not possible to get both poles into the left half-plane. For sufficiently large values of N_0, the poles are both on the imaginary axis, indicating that constant-amplitude oscillations result when x is perturbed from equilibrium.

The system can be stabilized by including a lead network in the amplifier. The root behavior for one possible lead-network location is shown in Fig. 37.

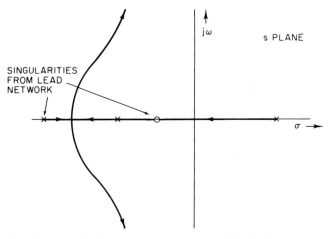

FIG. 37. Stabilization of system of Fig. 35 with a lead network.

12.1.6.2. Describing Functions. Describing-function techniques provide a method for the analysis of nonlinear systems which is similar to the frequency-domain methods used for linear systems. While it is possible to determine the closed-loop frequency response of some nonlinear feedback systems by means of describing functions, the discussion of this section will be limited to determining if steady-state oscillations or *limit cycles* are possible for a particular nonlinear system, and the frequency and amplitude of such oscillations if they exist. The methods can be applied to experimentally determined data as well as data available in analytic form.

The describing function for a nonlinear element is determined by making its input e_I a sinusoid, or

$$e_I = E \sin \omega t. \qquad (12.1.121)$$

If the nonlinearity does not *rectify* the input (provide a dc output) and does not introduce subharmonics, the output e_O can be expanded in a

Fourier series of the form

$$e_0 = F_1(E, \omega) \sin[\omega t + \theta_1(E, \omega)]$$
$$+ F_2(E, \omega) \sin[2\omega t + \theta_2(E, \omega)] + \cdots. \quad (12.1.122)$$

The describing function for the nonlinear element is defined as

$$G_D(E, \omega) = |F_1(E, \omega)/E| \measuredangle \theta_1(E, \omega). \quad (12.1.123)$$

Table II lists describing functions for several common nonlinearities. Since all of the nonlinearities shown are frequency independent, the describing functions are dependent only on the amplitude of the input signals. While this restriction is not necessary, it does apply to many nonlinearities of practical interest and does significantly simplify both the derivation and the use of a describing function.

In order to analyze a system by the describing-function method, it is necessary to first arrange it in a form similar to that shown in Fig. 38. Since the intent of the analysis is to determine if steady-state oscillations exist, the location of input and output points is unimportant. Similarly, several linear elements may be present. The important feature is that all linear elements can be combined into one transfer function which appears in a loop with a single nonlinear element.

The describing-function approximation states that oscillations may be possible if there are values of E and ω such that

$$G_1(j\omega)G_D(E, \omega) = -1 \quad (12.1.124a)$$

or

$$G_1(j\omega) = -1/G_D(E, \omega). \quad (12.1.124b)$$

Furthermore, describing-function analysis can be used to determine if stable-amplitude oscillations exist at points where Eq. (12.1.124) is satisfied and if the system oscillates spontaneously or must be triggered into oscillation by applying certain signal levels. When oscillations exist, the frequency of oscillation and the amplitude at the input to the nonlinear element are the values which satisfy Eq. (12.1.124).

The basic approximation of describing-function analysis is now evident. It is assumed that under conditions of steady-state oscillation, the input to the nonlinear element consists of a single-frequency sinusoid. While this assumption is certainly not exactly satisfied because the nonlinear element generates harmonics which propagate around the loop, it is

TABLE II. Describing Functions

Nonlinearity (Input $= e_1 = E \sin \omega t$)	Describing function (all are frequency independent)
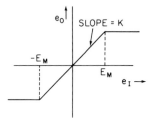	$G_D(E) = K \angle 0°, \ E \leq E_M$ $G_D(E) = \dfrac{2K}{\pi}$ $\times (\sin^{-1} R + R \sqrt{1 - R^2}) \angle 0°,$ $E > E_M$ where $R = E_M/E$
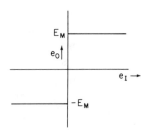	$G_D(E) = \dfrac{4E_M}{\pi E} \angle °0$
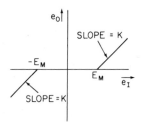	$G_D(E) = 0 \angle 0° \ E \leq E_M$ $G_D(E) = \left[K - \dfrac{2K}{\pi} \right.$ $\left. \times (\sin^{-1} R + R \sqrt{1 - R^2}) \right] \angle 0°,$ $E > E_M$ where $R = E_M/E$
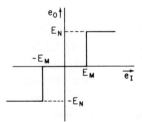	$G_D(E) = 0 \angle 0°, \ E \leq E_M$ $G_D(E) = \dfrac{4E_N}{\pi E} \sqrt{1 - R^2} \angle 0°,$ $E > E_M$ where $R = E_M/E$

TABLE II (*continued*)

Nonlinearity (Input $= e_{\mathrm{I}} = E \sin \omega t$)	Describing function (all are frequency independent)
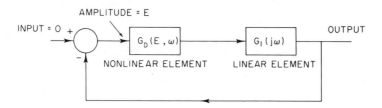	E must exceed E_{M} or a dc term results $$G_{\mathrm{D}}(E) = \frac{4E_{\mathrm{N}}}{\pi E} \angle -\sin^{-1} R$$ where $R = E_{\mathrm{M}}/E$

often valid for two reasons. First, most nonlinearities generate harmonics with amplitudes which are small compared to the fundamental. Second, since most linear elements are low-pass in nature, harmonics are attenuated to a greater degree than the fundamental by the linear elements. The second reason indicates that the approximation is better for higher order systems.

The existence of the relationship indicated in Eq. (12.1.124) is often determined graphically. The transfer function of the linear element is plotted in gain-phase form. The function $-1/G_{\mathrm{D}}(E, \omega)$ is also plotted on the same graph. If G_{D} is frequency independent, $-1/G_{\mathrm{D}}(E)$ is a single curve with E a parameter along the curve. The oscillation condition is satisfied if an intersection of the two curves exists. The frequency can be determined from the $G_{1}(j\omega)$ curve, while amplitude is determined from the $-1/G_{\mathrm{D}}(E)$ curve. If the nonlinearity is frequency dependent, a family of curves $-1/G_{\mathrm{D}}(E, \omega_{1})$, $-1/G_{\mathrm{D}}(E, \omega_{2})$, ... are plotted. The oscillation condition is satisfied if the $-1/G_{\mathrm{D}}(E, \omega_{i})$ curve intersects the point $G_{1}(j\omega_{i})$.

FIG. 38. System in form for describing-function analysis.

It is also necessary to insure that the oscillation predicted by the intersection is of stable amplitude. In order to do this, it is assumed that the amplitude E increases slightly, and the point corresponding to the perturbed value of E is found on the $-1/G_D(E, \omega)$ curve. If this point lies to the left of the $G_1(j\omega)$ curve, it implies that the system poles lie in the left half-plane for an increased value of E, tending to restore the amplitude to its original value. Alternatively, if the perturbed point lies to the right of the $G_1(j\omega)$ curve, a growing amplitude oscillation results from the perturbation and a limit cycle with parameters predicted by the intersection is not possible.

Example 12.16. The gain necessary to produce stable-amplitude oscillations in a phase-shift oscillator was determined in Example 12.6. A more practical phase-shift oscillator, which provides sufficient gain to insure oscillation and a limiter to control the amplitude of the oscillation, is shown in Fig. 39. Table II shows that the describing function for the limiter is

$$G_D(E) = \frac{2}{\pi}\left[\sin^{-1}\frac{1}{E} + \frac{1}{E}\sqrt{1 - \frac{1}{E^2}}\right] \angle 0° \qquad \text{for } E > 1, \tag{12.1.125a}$$

$$G_D(E) = 1 \angle 0° \qquad \text{for } E < 1. \tag{12.1.125b}$$

The conditions for oscillation can be determined directly from Eq. (12.1.124a). Since the angle of $G_D(E)$ is 0°, oscillation is possible only at $\omega = \sqrt{3}$, the frequency at which the phase shift of the linear ele-

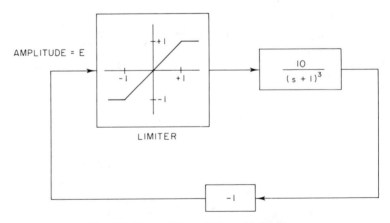

FIG. 39. Phase-shift oscillator with limiting.

ments is $-180°$. In order to satisfy the magnitude condition,

$$| G_D(E) | = |j\sqrt{3} + 1|^3/10 = 0.8. \qquad (12.1.126)$$

The magnitude of the signal at the input to the nonlinearity, E, is equal to 1.45 and is most easily determined by trial and error from Eq. (12.1.125a). It can be shown that this value of E produces a limit cycle.

The validity of the describing-function assumption concerning the purity of the signal at the input of the nonlinear element can also be demonstrated for this example. If a sinusoid is applied to the limiter, only odd harmonics are present in its output, and the amplitudes of higher harmonics decrease monotonically. The ratio of the magnitude of the third harmonic to that of the fundamental is 0.14. Similarly, the linear elements attenuate the third harmonic by a factor of 18 greater than the fundamental. It is therefore estimated that the ratio of third harmonic to fundamental is approximately 0.008 at the input to the nonlinear element.

Example 12.17. A circuit used in several commercial electronic function generators is shown in idealized form in Fig. 40. It can be shown by direct evaluation that the signal at the input to the nonlinear element is a 2-V peak-to-peak triangle wave with a period of 4 sec and that the

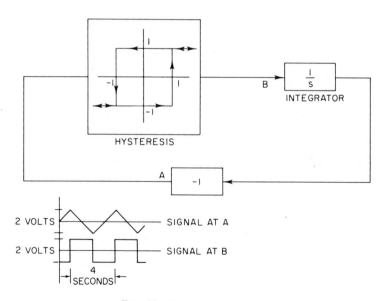

FIG. 40. Function generator.

signal at the output of the nonlinear element is a 2-V peak-to-peak square wave at the same frequency, with zero crossing displaced from those of the triangle wave by one second. The ratio of third harmonic to fundamental is $\frac{1}{9}$ at the input to the nonlinearity, a considerably higher value than in the preceding example.

Table II shows that the describing function for this example is

$$G_D(E) = \frac{4}{\pi E} \angle - \sin^{-1}\frac{1}{E}, \qquad E \geq 1. \qquad (12.1.127)$$

This function is plotted along with the transfer function of the linear element in Fig. 41. This curve predicts a limit cycle with a period of approximately 5 sec and a peak-to-peak amplitude at the input to the nonlinear element $2E$ equal to 2. The correspondence between these values and those of the exact solution is excellent considering the actual nature of the signals involved.

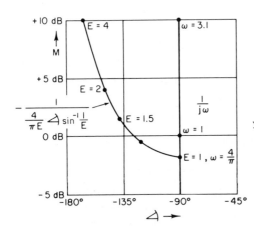

FIG. 41. Describing-function analysis of function generator.

Example 12.18. The final example of the use of describing functions involves a negative feedback loop which combines a limiter as shown in Fig. 39 with linear elements which have a transfer function

$$G_1(s) = \frac{3 \times 10^3 (0.1s + 1)^2}{s(s + 1)^2 (0.01s + 1)^2}. \qquad (12.1.128)$$

This type of transfer function can result if a system is designed to have a large velocity constant combined with a relatively low crossover frequency.

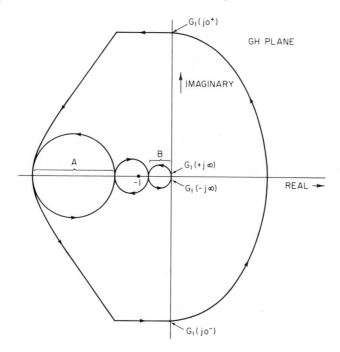

FIG. 42. Nyquist plot for conditionally-stable system of Example 12.18.

Essential features of the Nyquist plot for this transfer function are shown in Fig. 42, and this plot shows that the system is stable. However, if the gain decreases so the -1 point is on segment A or increases so that it is on segment B, the system becomes unstable. Since the limiter can decrease loop gain in a describing function sense, the system can become unstable.

Figure 43 shows $G_1(j\omega)$ and $-1/G_D(E)$ in gain-phase form. The lower of the two intersections does not represent a stable-amplitude limit cycle, since an increase in the amplitude into the nonlinear element causes the $-1/G_D$ curve to lie to the right of the $G_1(j\omega)$ curve, indicating that the amplitude increases further. A limit cycle at a frequency of 1.3 rad/sec is possible with $E = 1300$ since the top intersection depicts a stable-amplitude oscillation.

Systems of this general type, called *conditionally stable* systems, exhibit interesting behavior. The system is stable as long as the signal amplitude at the input to the limiting element remains small. However, if this signal level becomes large, either because of large inputs or because of momentary transients which may exist during turn-on or at other times, the

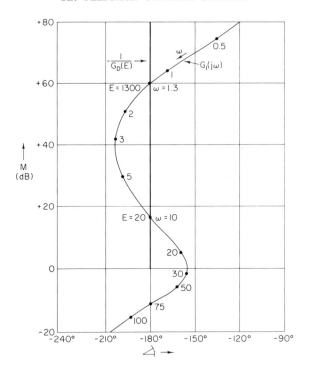

FIG. 43. Describing-function analysis of conditionally-stable system of Example 12.18.

system may become unstable and maintain a limit cycle. Some form of nonlinear compensation which detects the presence of an oscillation or a large signal and acts to modify the transfer function of the linear elements can be used to insure unconditional stability.

12.2. Servomechanisms

12.2.1. Introduction

A feedback control system with at least one variable mechanical quantity is called a *servomechanism*. Examples range in size and complexity from an automatic system which regulates film exposure in a camera to rolling operation controls which maintain the thickness of sheet steel.

The techniques required for the analysis of servomechanisms have already been described in Chapter 12.1. The purpose of this chapter is to indicate some of the components used to realize servomechanisms and to describe two types of systems of particular interest because of their frequent use.

12.2.2. Components[13]

12.2.2.1. Motors. Motors are devices which convert power supplied in some other form to mechanical power. High-power systems often use hydraulic or pneumatic motors. Many control systems use electric servo-motors because of the ease of control and application of such devices. This section describes two types of frequently used electric motors.

Permanent-magnet dc motors include a winding which generates a controllable magnetic field. Force is produced by the interaction of the electrically-generated field with the field of a permanent magnet. The windings are usually placed on the moving portion of the motor called the *armature* or *rotor*, with the permanent magnets attached to the *stator* or stationary portion, although designs with the windings stationary are available. The stator is usually located outside the rotor, but designs with the rotor external exist. Power levels from milliwatts to several kilowatts are practical, and electrical-to-mechanical conversion efficiencies above 70% can be obtained with some designs.

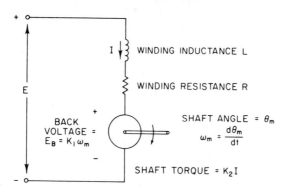

FIG. 44. Model for a dc motor.

A model for a dc motor is shown in Fig. 44. The input voltage is applied to a winding which has a series resistance and inductance. Mechanical energy conversion is represented by two relationships: the back voltage is

$$E_{\mathrm{B}} = K_1 \omega_m \qquad (12.2.1)$$

where ω_m is the angular velocity of the armature expressed in radians

[13] J. E. Gibson and F. B. Tuteur, "Control System Components." McGraw-Hill, New York, 1958.

per second, and the torque is

$$T = K_2 I. \qquad (12.2.2)$$

Combining Eqs. (12.2.1) and (12.2.2) yields

$$E_\mathrm{B} I = K_1 \omega_m T / K_2. \qquad (12.2.3)$$

However, the motor model implies that all losses are in the resistor. Therefore, the power applied to the back-voltage generator must equal the mechanical power supplied by the motor. Equation (12.2.3) thus indicates that K_1 and K_2 are equal if the units used to measure electrical and mechanical power are identical.

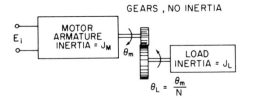

FIG. 45. Motor with inertia load.

Figure 45 shows a dc motor driving a load consisting of an inertia only. The load is geared to the motor through a gear reduction of $N{:}1$. Assuming the motor is modeled as shown in Fig. 44 with $K_1 = K_2 = K$, the relationship between $\theta_\mathrm{L}(s)$ and $E_\mathrm{i}(s)$ can be determined from the block diagram shown in Fig. 46. The transfer function relating θ_L to E_i is

$$\frac{\theta_\mathrm{L}(s)}{E_\mathrm{i}(s)} = \left(NKs \left\{ \left[\frac{s(J_m + J_\mathrm{L}/N^2)(Ls + R)}{K^2} \right] + 1 \right\} \right)^{-1}. \qquad (12.2.4)$$

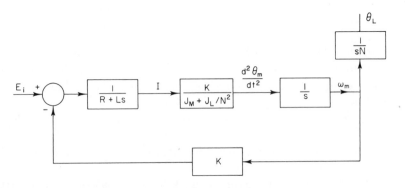

FIG. 46. Block diagram for motor with inertia load.

Parameter values are usually such that

$$\frac{R(J_m + J_{\mathrm{L}}/N^2)}{K^2} \gg \frac{L}{R}. \tag{12.2.5}$$

As a result of this inequality

$$\frac{\theta_{\mathrm{L}}(s)}{E_i(s)} \simeq \left\{ NKs\left[\frac{R(J_m + J_{\mathrm{L}}/N^2)}{K^2} s + 1\right]\left(\frac{L}{R} s + 1\right)\right\}^{-1}$$

$$\triangleq \{NKs(\tau_m s + 1)(\tau_e s + 1)\}^{-1}. \tag{12.2.6}$$

In contrast to the dynamic situation described above, the relationship between input voltage, motor velocity, and a torque applied to the shaft, when all three quantities are time invariant, can be presented as a family of torque-speed curves (Fig. 47). All curves are parallel and have a slope $-F_m$, where F_m is equal to K^2/R and is called the *viscous damping coefficient*. The time constant τ_m is obtained by dividing the total inertia of the armature and the load (properly scaled if gearing is used) by F_m.

A dc motor can also be used as a *tachometer* or velocity-measuring device, since the model of Fig. 44 shows that the open-circuit voltage of the motor will be proportional to armature velocity. Tachometers can be used for feedback compensation in servomechanisms (see Example 12.14).

Another type of motor frequently used in servomechanisms is the *two-phase servomotor*.[13] The stator for a motor of this type consists of two

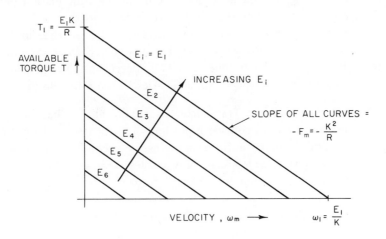

FIG. 47. Torque-speed curves for a dc motor.

windings with their magnetic axes at right angles. A rotating magnetic field with controllable amplitude and direction of rotation is established by applying a fixed amplitude sinusoid (usually at either 60 or 400 Hz) to the *reference winding*, and a variable amplitude sinusoid which is 90° out-of-phase with the reference signal to the *control winding*. Torque is generated in the rotor (which is made of conducting material) because of a magnetic field which results from currents induced in this member.

The torque-speed curves for a two-phase servomotor are similar to those of a dc motor (Fig. 47). The torque for $\omega_m = 0$ is proportional to control voltage, but the slopes are more negative for higher applied voltages, with the result that most of the curves terminate at speeds only slightly below *synchronous speed*.

The efficiency of the two-phase servomotor is relatively low because of design compromises required to obtain linear torque-speed curves. As a result, these motors are normally used only in low-power applications.

It is interesting to note that any load acceleration requirements can theoretically be met if a large number of any type of motor are available. Assume that the motor can provide a maximum acceleration α_m with no external load applied to the motor. This quantity is called the *theoretical acceleration* of the motor and is equal to its maximum torque divided by its rotor inertia. The acceleration imparted to an inertialess load can be increased to $N\alpha_m$ by gearing the load to the motor with a step-up gear ratio equal to N. Furthermore, if a large number of motor shafts are connected together, the unloaded acceleration of the combination (after gearing) remains $N\alpha_m$, but the inertia attributable to the load can be made a small fraction of the armature inertia so that accelerations arbitrarily close to unloaded values can be obtained. Similar arguments apply to situations which combine velocity and acceleration requirements and constrain the torque-speed characteristics of the available motor.

12.2.2.2. Angular Position Transducers.

Virtually all feedback control systems use *transducers* to convert input and output variables from their original physical form to a form more easily used. Transducers are available to convert a wide range of variables such as light intensity, temperature, or acceleration to electrical form.

The input and output variables for servomechanisms are frequently the angular position of shafts. While potentiometers can be used to detect shaft angle, the limited travel of such devices prohibits their use in systems where unlimited rotation is required.

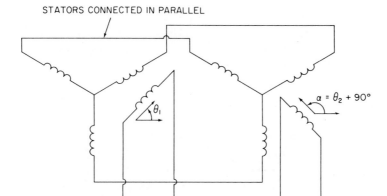

STATORS CONNECTED IN PARALLEL

$\alpha = \theta_2 + 90°$

e_O

EXCITATION
E sin ωt

FIG. 48. Synchro pair.

A large class of devices which use variable magnetic coupling to determine angular position are available. The *synchro* is one commonly used device type.[13] A synchro consists of a stator having three coils, normally Y connected, with their magnetic axes forming mutual angles of 120°. The rotor is a single coil which can be rotated relative to the stator. Brushes and slip rings are used to achieve unlimited rotation. If two synchros are connected as shown in Fig. 48, the variables are related as

$$e_O = KE \sin \omega t \cos(\theta_1 - \alpha). \qquad (12.2.7)$$

Since under normal conditions of operation the angular shaft positions differ by approximately 90°, it is convenient to introduce a new variable $\theta_2 = \alpha - 90°$. Substituting this relationship into Eq. (12.2.7) yields

$$e_O = KE \sin \omega t \sin(\theta_1 - \theta_2) \simeq KE \sin \omega t[\theta_1 - \theta_2] \qquad (12.2.8)$$

for small angular differences.

Thus the output signal is an alternating voltage at the carrier frequency with an amplitude or envelope proportional to the difference between the two shaft angles. A change in sign of the difference results in a change of the phase of the carrier. If the output voltage is processed with a *phase-sensitive demodulator*, which produces a voltage proportional to the envelope with polarity determined by phase angle, a dc voltage proportional to the angular difference is obtained. A pair of synchros can therefore be used to perform the comparison or subtraction of signals represented by a summing element in a block diagram as well as to measure angles.

12.2.2.3. Amplifiers and Signal Conditioners. Two distinct types of amplifiers are frequently used in servomechanisms. One of these is a signal conditioner which may be used to provide gain, addition, or subtraction of several signals, or to provide a transfer function required for compensation. These signal-conditioning functions are frequently performed with inexpensive integrated-circuit operational amplifiers. (See Section 12.3.2).

Power amplifiers are used to drive the output member of a servomechanism. These amplifiers may be separate or they may be incorporated into the output stages of a signal conditioning amplifier.

Obtaining high output power from an electronic amplifier was more difficult before the advent of modern semiconductor devices. One alternative to electronic amplification which was used extensively was the *Ward–Leonard* system.[13] This type of amplifier consists of an externally excited dc generator which is driven at constant velocity. The signal to be amplified is applied to the field winding of the generator, and an open-circuit voltage proportional to this current results in the armature winding. The armature output is used to drive a dc motor. The *amplidyne* is a related type of rotary power amplifier.[13] A two-stage, self-compensated generator design is used to obtain higher power gains than are possible with the Ward–Leonard system.

Modern servomechanisms usually use semiconductor power amplifiers. Designs which supply upwards of 10 kW are available. Popular types include linear amplifiers which use a class-B connection of power transistors and switching amplifiers which minimize power dissipation in the load handling transistors by operating them only in the saturated or cut-off regions (see Chapters 5.2 and 6.7).

12.2.3. Carrier Systems

In a *carrier system* the signal information is the amplitude (*or envelope*) and phase of a single-frequency ac voltage called the carrier. If the signal is $f(t)$, the corresponding variable in a carrier system is $f(t)E_c \sin \omega_c t$, where E_c and ω_c are the amplitude and frequency of the carrier. This type of signal is an example of *suppressed-carrier modulation*, since if $f(t)$ is a sinusoidal signal at frequency ω_s,

$$E_s \sin \omega_s t \, E_c \sin \omega_c t = \frac{E_s E_c}{2} \left[\cos(\omega_c - \omega_s)t - \cos(\omega_c + \omega_s)t \right]. \quad (12.2.9)$$

The two-phase servomotor and the synchro are components which use or generate suppressed-carrier signals. Most transducers which use magnetic coupling to detect relative position or angular orientation also produce this type of signal.

Entire systems can be constructed where all signals are carrier signals, and such systems were frequently used for low-power servomechanisms before the availability of semiconductor dc amplifiers. These systems are frequently analyzed by ignoring the carrier and considering an equivalent dc system where carrier signals are replaced by their envelopes. This approximation yields excellent agreement with actual behavior providing that all signal frequencies and the crossover frequency of the system are low compared to the carrier frequency. The modulation process introduces a time delay, and the negative phase shift associated with this time delay causes carrier systems to become progressively less stable as the crossover frequency approaches the carrier frequency. It is normally necessary to limit the system crossover frequency to about one tenth of the carrier frequency to maintain acceptable performance.

The networks often used for series compensation are significantly more difficult to design for carrier than for dc systems. While it is possible to design various networks with singularities close to the imaginary axis in the s plane which apply lag or lead functions to the signal-frequency components, these networks are extremely sensitive to component variations or carrier-frequency shifts. Providing effective, stable, series compensation normally requires detecting the envelope with a phase-sensitive demodulator and operating on the detected signal itself with a network. The network output signal can be remodulated if required to drive a component which uses a carrier signal.

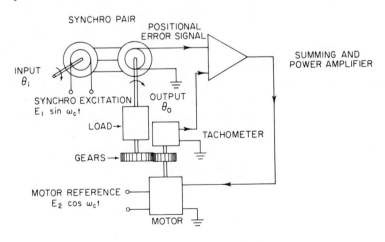

FIG. 49. Carrier servomechanism.

A representative carrier system is shown in Fig. 49. The objective of this system is to cause the position of an output shaft to follow the position setting of an input shaft. An error signal proportional to the angular error is generated by the synchro pair. This signal is amplified and applied to a two-phase servomotor. The tachometer can be used to provide feedback compensation as described in Section 12.1.5.4.

12.2.4. Relay Servomechanisms

In systems where relatively low performance is acceptable, cost can be reduced if power amplification is provided by a relay. One possible topology for a system of this type is shown in Fig. 50. If the relay switching time is short compared to the response time of the system, the relay can be characterized by its static input-output relationships. Typical relay characteristics are shown in Fig. 51.

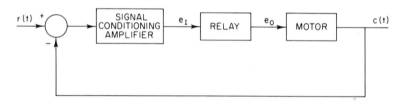

FIG. 50. Relay servomechanism.

Describing-function analysis (Section 12.1.6.2) is used to determine if a relay servomechanism is stable or the signal amplitudes if a limit cycle exists. The complete transient response of the system can be determined by solving the linear equations which apply as long as the relay remains in one state. Furthermore, the time at which the relay switches can be determined from its characteristics and the signal at its input. Thus a complete transient response can be obtained by solving a sequence of linear problems, with the values of system variables at a relay switching time used as initial conditions for the following time interval. Digital computation facilitates these calculations.

The use of a relay with characteristics similar to those shown in Fig. 51a and 51b normally results in a limit cycle. The limit cycle can often be eliminated by using a relay characterized by Figs. 51c or 51d, but systems with these characteristics often have a dead zone or range of error between input and output for which the output does not change.

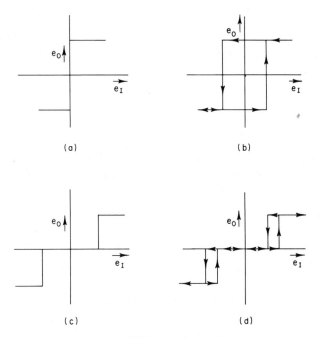

(a) (b) (c) (d)

FIG. 51. Relay characteristics.

12.3. Electronic Feedback Systems

12.3.1. Introduction

Feedback systems where all variables are electrical quantities are also analyzed by means of the methods introduced in Chapter 12.1. There are, however, certain characteristics of electronic feedback systems which differ from the corresponding characteristics of servomechanisms. The more important of these differences are described in the following paragraphs.

The open-loop transfer function of a servomechanism often includes a pole at the origin because of the integration which relates the output angle of a motor to its input voltage. This pole is not present in electronic systems. In many electronic systems, the open-loop gain has a very low-frequency pole and very high (but finite) dc gain, and in most practical situations these two features combine to give performance equivalent to that which would be obtained with an integration.

The frequency response of electronic systems can be many orders of magnitude greater than that of servomechanisms. While the crossover frequency of mechanical systems is typically of the order of 10 Hz and seldom exceeds 1 kHz, feedback amplifiers with crossover frequencies in excess of 100 MHz have been built. The wide bandwidth of many electronic systems gives the designer relatively less freedom in compensation techniques for electronic systems, since it is often impossible to include elements in the loop which provide significant gain at the crossover frequency.

The purpose of this chapter is to introduce two types of electronic feedback systems, the operational amplifier and the voltage regulator, and briefly indicate how the general techniques are applied to these systems.

12.3.2. Operational Amplifiers[14,15]

An *operational amplifier* is a high-gain direct-coupled amplifier, generally with a differential input and single-ended output, intended to be used with external feedback networks. If the amplifier characteristics are satisfactory, the closed-loop transfer function of the amplifier with feedback is determined primarily by the stable and well-known values of the passive feedback components.

The name is derived from original applications in analog computation, where these amplifiers are used to perform certain mathematical operations such as summation and integration. Present applications extend far beyond the original ones, and design improvements coupled with the availability of low-cost integrated circuits have led to the use of the operational amplifier as a general-purpose analog data processing element.

The symbol for an operational amplifier is shown in Fig. 52, where the notation implies that

$$E_o(s) = A(s)[E_1(s) - E_2(s)]. \qquad (12.3.1)$$

NONINVERTING INPUT

$E_1(s)$ $A(s)$ $E_0(s)$ FIG. 52. Symbol for an operational amplifier.

$E_2(s)$

INVERTING INPUT

[14] G. A. Korn and T. M. Korn, "Electronic Analog and Hybrid Computers." McGraw-Hill, New York, 1964.

[15] Applications Manual for Computing Amplifiers. Philbrick Res. Inc., Dedham, Massachusetts.

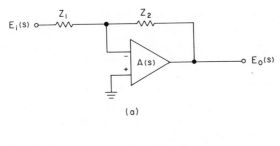

(a)

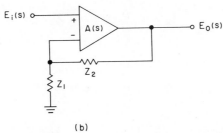

(b)

FIG. 53. Operational-amplifier connections: (a) inverting, (b) noninverting.

All voltages are assumed measured with respect to the ground used for the amplifier power supply. The power supply connections are normally not shown.

Operational amplifiers are frequently used as illustrated in Fig. 53. The following assumptions, which are usually valid if the proper amplifier has been selected, are introduced to facilitate analysis of these two connections.

1. The current required at either input terminal is negligibly small.
2. The offset referred to the input of the amplifier, defined as the voltage which must be applied between the two inputs to make the output voltage zero, is negligibly small.
3. The feedback network is not loaded significantly by the input impedance of the amplifier.
4. The gain of the amplifier remains $A(s)$ when the amplifier is loaded with the feedback network and any other elements which may be connected to its output.

With these simplifications the closed-loop gain for the inverting amplifier (Fig. 53a) is

$$\frac{E_o(s)}{E_i(s)} = \frac{-A(s)Z_2/(Z_1 + Z_2)}{1 + A(s)Z_1/(Z_1 + Z_2)} \qquad (12.3.2)$$

while that of the noninverting amplifier is

$$\frac{E_o(s)}{E_i(s)} = \frac{A(s)}{1 + A(s)Z_1/(Z_1 + Z_2)}. \qquad (12.3.3)$$

These expressions indicate that the open-loop gain for either connection is $A(s)Z_1/(Z_1 + Z_2)$. If the input signal contains only frequencies for which the open-loop gain is large, the gain expressions for the amplifiers with feedback reduce to

$$\frac{E_o(s)}{E_i(s)} \simeq \frac{-Z_2}{Z_1} \qquad (12.3.4)$$

for the inverting connection and

$$\frac{E_o(s)}{E_i(s)} \simeq \frac{Z_1 + Z_2}{Z_1} \qquad (12.3.5)$$

for the noninverting connection.

Operational amplifiers are normally used in situations where the actual closed-loop gain is closely approximated by the ideal value of Eq. (12.3.4) or Eq. (12.3.5). The ideal gain for a more complex operational-amplifier connection can be determined from Kirchhoff's laws by assuming that the current required at either input terminal and the voltage between the two terminals are negligibly small. This analytic approach is illustrated for the connections shown in Figs. 54 and 55. In the case of the summing amplifier, Kirchhoff's current law at the node including the inverting input terminal requires that

$$I_1(s) + I_2(s) + I_f(s) = 0 \qquad (12.3.6)$$

or

$$E_o(s) = -\frac{Z_f E_1(s)}{Z_1} - \frac{Z_f E_2(s)}{Z_2}. \qquad (12.3.7)$$

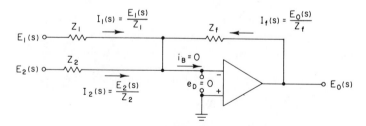

FIG. 54. Summing amplifier.

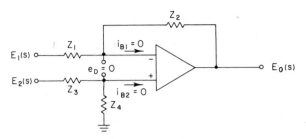

FIG. 55. Difference amplifier.

Similarly, the ideal gain of the difference amplifier can be determined by noting that

$$\frac{E_1(s)Z_2}{Z_1 + Z_2} + \frac{E_0(s)Z_1}{Z_1 + Z_2} = \frac{E_2(s)Z_4}{Z_3 + Z_4} \qquad (12.3.8)$$

or

$$E_0(s) = -\frac{Z_2}{Z_1}E_1(s) + \left[\frac{Z_1 + Z_2}{Z_1}\right]\left[\frac{Z_4}{Z_3 + Z_4}\right]E_2(s). \qquad (12.3.9)$$

While the ideal closed-loop gain of an operational amplifier is determined only by passive feedback components, it is clear that the stability of the amplifier and the frequency range for which the ideal gain approximates the actual is determined by $A(s)$ and the characteristics of the feedback network used in conjunction with the amplifier. Many amplifiers are designed to have one dominant open-loop pole at some low frequency, and no other singularities in the amplifier gain at frequencies where the magnitude of $A(j\omega)$ exceeds unity. Since the low-frequency gain is typically in excess of 10^4, $A(s) \approx K/s$ over some wide range of frequency, where K is the ratio of the dc gain to the time constant of the dominant pole. This type of amplifier gain may compromise performance in certain connections, but does insure stability for most feedback networks and loads of practical interest.

Greater flexibility is possible with an amplifier which can be compensated by the user. The basic topology for an amplifier which can be readily compensated and which is representative of several modern integrated-circuit designs is shown in simplified form in Fig. 56.[16] The input stage of this amplifier consists of a differential-amplifier pair (Q_1 and Q_2) which insures low voltage drift and the differential inputs required for noninverting operation. Transistor Q_3 is loaded with the very

[16] R. J. Widler, A New Monolithic Operational Amplifier Design. Nat. Semicond. Tech. Paper TP-2 (1968).

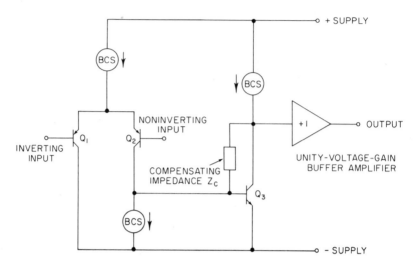

FIG. 56. Basic operational-amplifier topology. BCS: Bias Current Sources.

high incremental resistance of a current source to provide high voltage
gain. The high-impedance node which includes the collector of Q_3 is
buffered so that loads applied to the amplifier do not reduce its voltage
gain.

This approach has at least two advantages compared with earlier
designs which used more than two voltage amplifying stages. First, the
number of open-loop poles increases with the number of transistors used
in the amplifying path, so that stabilization becomes progressively more
difficult for designs with more stages. Second, the amplifier can be easily
and precisely compensated by means of a compensating impedance (Z_c
in Fig. 56) which provides feedback around the high-gain portion of the
circuit. The techniques described in Section 12.1.5.4 can be used to
show that the open-loop gain of an amplifier compensated this way is
equal to KZ_c over a wide range of frequency. The constant K is dependent
on the transconductance of the input transistor pair and can be determined
from the manufacturer's data.

A fundamental limitation to this type of compensation is that the gain
of the amplifier with compensation must be less than that of the uncompensated amplifier at most frequencies. Exceptions can occur only
near the frequency where the gain around the inner loop is unity because
of possible resonance in this loop. As a result of this limitation, the
maximum dc gain and the unity-gain frequency of the amplifier cannot
be improved by compensation.

Figure 57 shows four types of networks frequently used to compensate amplifiers and also shows the corresponding open-loop asymptotic magnitude plots. The limitations introduced at high and low frequencies by the uncompensated amplifier characteristics are also indicated in this figure.

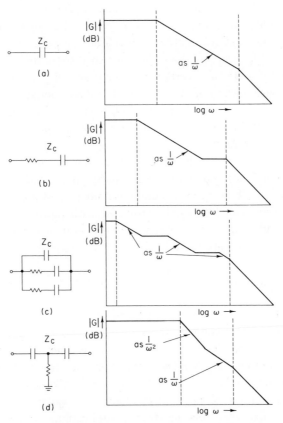

FIG. 57. Representative compensation networks and resultant transfer functions. Gain outside of dashed lines is not controlled by compensation used.

The use of a capacitor for Z_c results in an amplifier open-loop gain which approximates a single integration over a wide range of frequency. If additional open-loop poles are not introduced by the feedback network or by loading, a system compensated in this way can provide highly stable performance since phase margins approaching 90° are possible. The amplifier gain at intermediate frequency can be increased by decreasing the capacitor size so that a relatively constant crossover frequency can be maintained as the attenuation of the feedback network used around the amplifier is increased. This conservative compensation is the most frequently used type.

If feedback networks or loads which introduce one additional pole are anticipated, a series *RC* network can be used to produce a zero in the amplifier transfer function. The zero can cancel the additional pole so that single pole open-loop response is obtained from the over-all loop.

Several paralleled *RC* sections provide an amplifier transfer function which falls off less rapidly than $1/\omega$. The corresponding phase remains at approximately $-45°$ for certain combinations of element values. This insures a phase margin of close to $45°$ if one additional open-loop pole is introduced at an arbitrary frequency. This type of compensation is useful, for example, when a number of different sized capacitive loads may be applied to the amplifier.

An *RC* T network can be used to obtain an initial $1/\omega^2$ gain decrease followed by a $1/\omega$ region to reduce negative phase shift near the cross-over frequency. This type of compensation provides higher open-loop gain at intermediate frequency compared with more conservative compensation, but amplifiers compensated this way may become unstable for relatively small capacitive loads.

12.3.3. Voltage Regulators

12.3.3.1. Introduction. Many types of systems require precisely controlled or regulated voltages for satisfactory operation. Regulators are designed to prevent changes in output voltage when the current supplied by the regulator changes (*load regulation*) or when the unregulated voltage supplied to the regulator changes (*line* or *input regulation*). While regulating methods which do not use feedback are available and do provide adequate regulation in less demanding applications, accurate regulation, particularly at high-power levels, usually involves feedback.

Feedback regulators can be classified as either *dissipative* or *nondissipative*. Dissipative regulators include an active element either between the unregulated supply voltage and the regulated output (*series regulator*) or in parallel with the output (*shunt regulator*). A significant fraction of the total power from the unregulated source is often dissipated in these elements in order to achieve regulation.

The active element in a nondissipative regulator is operated as a switch to minimize its power losses. The switched signal is filtered to provide a regulated output. Circuits of this type can achieve efficiencies in excess of 90% and are used when the total system power requirements must be minimized or when it would be difficult to transfer heat away from a dissipative element.

12.3.3.2. Dissipative Regulators. Excellent regulators can be constructed using an operational amplifier and a *Zener diode* to provide a reference voltage. A representative design is shown in Fig. 58. Assuming that the voltages applied to the two inputs of the amplifier are equal, the output voltage is

$$V_O = \frac{R_1 + R_2}{R_1} V_Z. \tag{12.3.10}$$

The current through the Zener diode is

$$I_Z = \frac{V_O - V_Z}{R} = \frac{R_2 V_Z}{R_1 R_3}. \tag{12.3.11}$$

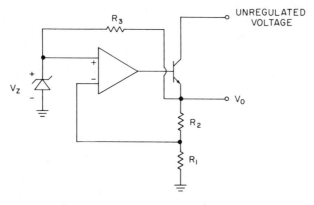

FIG. 58. Series regulator.

Since the Zener current is derived from its own well-regulated voltage, voltage changes attributable to the Zener series resistance are reduced.

The emitter follower shown in the diagram, which effectively provides a higher current output stage for the amplifier, can be eliminated when the required output current is within the capacity of the operational amplifier. In this case, the output transistor of the operational amplifier functions as the series element between the unregulated voltage supplied to the amplifier and the output voltage.

Alternate designs where the amplifier, possibly with a high-power external output stage, shunts the output voltage to provide regulation are also possible.

Because capacitive loads are anticipated for voltage regulators, compensation such as that illustrated in Figs. 57b or 57c is often used to maintain stability with these loads.

Several integrated-circuit designs for regulators are available which include both the Zener diode and the amplifier.[17] Voltage is adjusted by means of an external feedback network, and provision for adding external transistors to increase output current capability is included.

A slightly modified dissipative series regulator can be used to illustrate an important feature of all control systems where the intent is to regulate or keep an output variable constant rather than to have it follow a rapidly changing input signal.

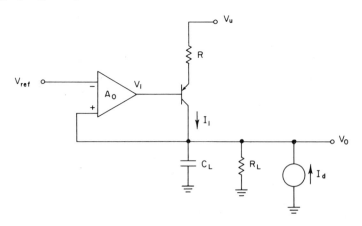

FIG. 59. Series regulator for Example 12.19.

Example 12.19. Figure 59 illustrates a series regulator which combines a grounded emitter output transistor with an operational amplifier. The load for the regulator consists of a parallel RC network, and it is assumed that the amplifier gain is frequency independent and equal to A_o to well beyond the system crossover frequency so that the load network determines system dynamics. There are two sources of disturbance, one variation in the unregulated input voltage V_u and the other changes in load current represented by the I_d generator. Note that the transistor provides an inversion in this connection so that negative feedback is achieved by applying the feedback signal to the noninverting input of the amplifier.

If it is assumed that the current into the noninverting input of the

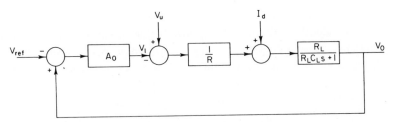

FIG. 60. Block diagram for regulator of Fig. 59.

operational amplifier is negligible, that the base-to-emitter voltage of the transistor is small compared to the voltage across R, and that the common-base current gain of the transistor is one, the regulator is represented by the block diagram shown in Fig. 60. Reducing this block diagram shows that

$$V_0 = \frac{V_{\text{ref}}}{\left(1 + \frac{R}{A_0 R_L}\right) + \frac{RC_L s}{A_0}} + \frac{V_u}{\left(A_0 + \frac{R}{R_L}\right) + RC_L s}$$

$$+ \frac{I_d}{\left(\frac{A_0}{R} + \frac{1}{R_L}\right) + C_L s}. \qquad (12.3.12)$$

The first term on the right-hand side of the Eq. (12.3.12) relates V_0 to the reference, and is approximately one for large values of open-loop gain $A_0 R_L/R$ and a time-invariant reference. The second and third terms represent disturbances attributable to the unregulated voltage and load current respectively. The essential feature of both of these terms is that the output perturbation for low-frequency disturbances can be limited to any desired level by the appropriate choice of A_0 and A_0/R, and that as the frequency of the disturbing input increases, the effect of the disturbance decreases. Thus at some sufficiently high disturbing frequency, the output remains constant because of filtering provided by the load, and the loop does not need to respond to these high-frequency disturbances.

The result is a general one. If a regulator is built, adding inertia to the output of a mechanical system or capacitance to the output of an electrical system will improve its rejection of high-frequency disturbances. This conclusion is directly opposite that which applies to systems intended to track dynamic inputs, where lowering the frequency of the output pole deteriorates performance.

Controlling the loop dynamics of a regulator with its output-member pole also makes the system stability insensitive to variations in the load. Such changes are anticipated as the resistive or capacitive load applied to a voltage regulator is altered. Figure 61 shows the effects of variations in R_L and C_L on the asymptotic open-loop Bode plot. The maximum crossover frequency can be limited to insure that it is lower than the frequency of any singularities associated with the amplifier by limiting the minimum value of C_L. The phase margin of the system shown exceeds 90° for any R_L–C_L combination.

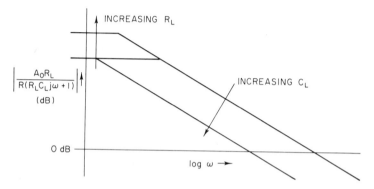

FIG. 61. Effects of changing load parameters for regulator of Fig. 59.

Certain additional features are frequently included in voltage regulators. The output current can be sensed and used to provide an override signal which limits this current to a value which protects the regulating element or the load. Alternatively, current values can be combined with voltage values to derive a feedback signal, and in this case the output resistance of the regulator can be precisely controlled.

Provision for *remote sensing* is also included in certain regulator designs. The voltage feedback signal is obtained at the load rather than at the output terminals of the supply so that the effects of voltage drops along the leads connecting the regulator to its load are eliminated by feedback.

12.3.3.3. Nondissipative Regulators. The power lost in the series- or shunt-regulating element of a dissipative regulator can be saved by operating the power handling element in a switching mode and filtering the switched signal to obtain the regulated output voltage. One configuration is shown in Fig. 62. This connection is called an up-regulator since the output-voltage magnitude must exceed that of the input voltage.

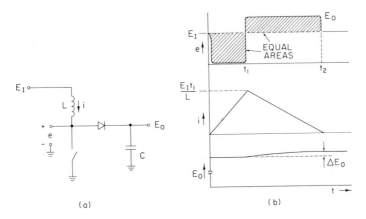

FIG. 62. Up-regulator: (a) circuit, (b) waveforms. $E_O > E_I > 0$; $t_2 = t_1 E_O/(E_O - E_I)$; $\Delta E_O = E_I^2 t_1^2/2LCE_O$.

Operation is explained with the aid of the included waveforms. The switch is closed at $t = 0$, and the inductor current increases linearly for $t > 0$. At time t_1, when the energy stored in the inductor is

$$W = \frac{1}{2} L[i(t_1)]^2 = \frac{E_I^2 t_1^2}{2L} \qquad (12.3.13)$$

the switch opens. Since the inductor current cannot change instantaneously, the current continues flowing through the diode into capacitor C. The current decreases to zero in a time determined by the relative values of E_I and E_O. During this time the energy stored in the inductor is transferred to the capacitor. Assuming that the capacitor is large enough so that the incremental voltage change ΔE_O is small compared to E_O, the magnitude of the change is given by

$$\Delta E_O = \frac{W}{CE_O} = \frac{E_I^2(t_1)^2}{2LCE_O}. \qquad (12.3.14)$$

The frequency at which the process of storing energy in the inductor and transferring it to the output capacitor is repeated can be changed to change the power being transferred from the input to the output of the circuit. The magnitude of the output voltage ripple or increment on each cycle of operation can be limited by proper choice of operating parameters as indicated by Eq. (12.3.14).

Two other configurations which can be used for energy transfer are shown in Fig. 63. Figure 63a shows a circuit for which the output voltage must be more negative than the input. In contrast to the circuit shown

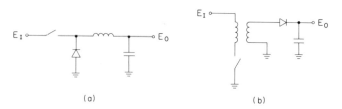

FIG. 63. Possible switching-regulator circuits.

in Fig. 62, the output voltage for this circuit increases when either the switch or the diode conducts. The circuit shown in Fig. 63b temporarily stores energy in the magnetic field of a transformer. The transformer allows the output voltage to be higher or lower than the input, isolates output from input, and can provide multiple output voltages if more than one transformer secondary winding, diode, and capacitor are used.

Any of these switching circuits can be controlled in one of two modes in order to regulate its power transfer level and thus output voltage for a given load. One possibility is to operate the switch for a fixed time on each cycle, thus fixing the energy transferred, and vary frequency to change power. The second possibility is to keep the frequency of operation constant and change the switch ON time to change the energy transferred on each cycle of operation.

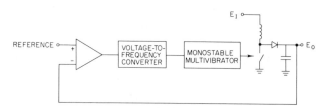

FIG. 64. Control system for high-efficiency regulator.

Figure 64 illustrates a feedback system which regulates output by controlling frequency. The difference between the output voltage and a reference is amplified and used to control a voltage-to-frequency converter in such a way that the output frequency of the converter increases for low output voltage. This signal triggers a monostable multivibrator which provides the pulse used to operate the switch.

12.3.4. Current Regulators

The discussion of Section 12.3.3 outlined the operation of regulators intended to supply a fixed output voltage. In certain applications, constant current rather than constant voltage is required to power a system. The current regulator which is used to meet this requirement can be realized in several ways.

Most voltage regulators, particularly those intended for general laboratory use, include some form of current limiting to prevent the destruction of the regulator in the event of excessive loading. For example, the maximum output current which can be obtained from the series regulator shown in Fig. 59 can be constrained by designing the amplifier so that the maximum voltage it can produce across the resistor R is $V_{\rm L}$. This limit constrains emitter current (and therefore output current) to a maximum value of $V_{\rm L}/R$. If either $V_{\rm L}$ or R can be adjusted, the limiting-current level can be controlled. This feature is useful for protecting low-power experimental circuits from being destroyed by the power supply while being tested. Although the current regulation provided by the limiter described above is not particularly accurate, this type of protected supply can be used as a crude current regulator simply by operating it in its current-limited region.

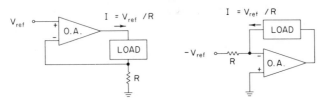

FIG. 65. Current-regulator configurations.

More sophisticated current regulators use feedback to establish precisely the output current. Two possible topologies which use an operational amplifier as a current regulator are shown in Fig. 65. In both of these connections, the load current is identical to the current through the resistor, which in turn is equal to $V_{\rm ref}/R$. Adjustment of the resistor value or the reference voltage changes the output-current level. Several commercially available supplies allow the feedback connection to be changed to provide either voltage or current regulation.

High-efficiency current regulators can be designed by using a switching connection similar to that described in Section 12.3.3.3 in a current-controlling feedback configuration.

13. NOISE IN ELECTRONIC DEVICES*

13.1. Introduction

The purpose of this part is to present a discussion of those aspects of noise in electronic devices which are most important in choosing devices as detectors and input stages of sensitive measurement systems. The discussion of the details of the physical origin of noise sources in the various equivalent circuits will be kept to a minimum. It is intended that this part complement the discussion of amplifier design found in Part 6. *Details of definitions of common terms used in the characterization of noise in devices and circuits are summarized in the Appendix.* A brief description of noise in visible and infrared photodetectors is included. It is assumed that the reader is familiar with the physical operation of these devices.

13.2. Sources of Noise

It is useful to have some physical understanding of the origin of the noise sources in the device equivalent circuits. It is common practice to refer to thermal noise and shot noise. These terms will be used frequently in this Part. However, a somewhat more precise picture is presented by use of the terms *electronic noise* and *modulation noise.*[1]

Electronic noise has been introduced to describe the purely statistical fluctuations inherent in the operation of a device. A useful physical picture of such sources is stimulated by the names *transition noise* and *transport noise.*[2] Thermal noise and generation–recombination (g-r) noise are examples of transition noise. G-r noise results from statistical fluctuations in the numbers of carriers making transitions between energy levels

---

[1] R. L. Petritz, *Proc. IRE* **40**, 1440 (1952).

[2] K. M. van Vliet and J. R. Fassett, *in* "Fluctuation Phenomena in Solids" (R. E. Burgess, ed.), p. 267. Academic Press, New York, 1965.

---

* Part 13 is by E. R. Chenette and K. M. van Vliet.

as, for example, between impurity levels and the valence or conduction bands in a semiconductor. Thermal noise in a conductor is caused by intraband transitions which result in the scattering of carriers. *Transport noise* results because the carrier transport mechanism, whether drift or diffusion, is a random process. Shot noise in a temperature-limited vacuum diode is a common example.

Modulation noise results from fluctuations which control the average characteristics of a device. The best example here is probably flicker noise in vacuum tubes and transistors.

We now present a brief description of the most important sources of electronic and modulation noise. Those desiring more detail are referred to the references cited and to pertinent texts.[3-9]

13.2.1. Thermal Noise

The most important point about *thermal noise* is its universality. It is characteristic of all dissipative media.[10] It is found in all practical electronic devices and circuits.

According to Nyquist's theorem,[11] the noise power available in a narrow frequency interval df from any resistive element at a uniform temperature of $T°K$ is

$$dP = kTp(f)\,df \tag{13.2.1}$$

where $p(f) = (hf/kT)[1 + (\exp(hf/kT) - 1)^{-1}]$ arises because of the quantization of energy levels. The frequency dependence of $p(f)$ becomes important at frequencies above those corresponding to the far infrared but is of little significance at lower frequencies. For most of this review $p(f) = 1$. Oliver has presented an interesting and readable

[3] A. van der Ziel, "Noise; Sources, Characterization, Measurement." Prentice-Hall, Englewood Cliffs, New Jersey, 1970.

[4] D. K. C. MacDonald, "Noise and Fluctuations." Wiley, New York, 1962.

[5] W. R. Bennett, "Electrical Noise." McGraw-Hill, New York, 1960.

[6] A. van der Ziel, "Noise." Prentice-Hall, Englewood Cliffs, New Jersey, 1954.

[7] A. van der Ziel, "Fluctuation Phenomena in Semiconductors." Butterworths, London and Washington, D.C., 1959.

[8] R. E. Burgess (ed.), "Fluctuation Phenomena in Solids." Academic Press, New York, 1965.

[9] *Proc. Conf. Phys. Aspects Noise Electron. Devices, Univ. Nottingham, Sept. 11-13, 1968.* Peter Peregrinus, Stevenage, Hertfordshire.

[10] H. B. Callen and T. A. Welton, *Phys. Rev.* **83**, 34 (1951).

[11] H. Nyquist, *Phys. Rev.* **32**, 110 (1928).

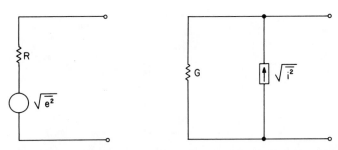

FIG. 1. Circuit representations of Nyquist's theorem. The magnitudes of the noise generators are given by $\overline{e^2} = 4kTRdf$ and $\overline{i^2} = 4kTGdf$, where $G = 1/R$.

discussion of the relationship between thermal noise and quantum fluctuations.[12]

Either of the circuits shown in Fig. 1 can be used to represent the thermal noise of a resistance. The spectral density of the series noise emf of Fig. 1a is

$$S_v(f) = 4kTRp(f) \qquad (13.2.2a)$$

and that of the parallel noise current generator in Fig. 1b is

$$S_i(f) = 4kTGp(f) \qquad (13.2.2b)$$

where $G = 1/R$.

The validity of Nyquist's theorem was well-established by Johnson's work.[13]

13.2.2. Shot Noise

Schottky's theorem[14] is concerned with fluctuations in the current which arise because of statistical fluctuations in the number of carriers being emitted across a potential barrier. The classic example is emission from the cathode of a temperature-limited vacuum diode. If the emission of each charge carrier is an independent random event and if each carrier carries an electronic charge of magnitude $q = 1.6 \times 10^{-19}$ C, the noise may be represented as shown in Fig. 2. The spectral density of the noise current generator is

$$S_i(f) = 2qI \qquad (13.2.3)$$

where I is the total dc current resulting from the emission.

[12] B. M. Oliver, *Proc. IEEE* **53**, 436 (1965).
[13] J. B. Johnson, *Phys. Rev.* **32**, 97 (1928).
[14] W. Schottky, *Ann. Phys.* [4] **57**, 541 (1918).

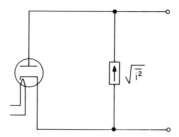

FIG. 2. Circuit representation of Schottky's theorem. Fluctuations in the anode current are represented by the noise current generator with $\overline{i^2} = 2qIdf$, where I is the dc anode current.

According to Eq. (13.2.3) the spectrum is flat. This cannot be the case since $\int_0^\infty S(f) \, df$ must converge. In the high-frequency limit, the spectrum decreases as the collision time of carriers gives rise to correlation effects. In most practical devices, transit time and parasitic effects modify the observed noise at frequencies far below this limit.

13.2.3. Generation–Recombination Noise

G-r noise occurs in semiconducting materials because of the random generation and recombination of carriers. If the number of electrons in a semiconductor sample is fluctuating about the average number n_0 and if the average lifetime of added carriers is τ, it can be shown that the spectral density of the current fluctuations in a current carrying sample is

$$S_i(f) = 4(I^2/n_0)\varkappa\tau/(1 + \omega^2\tau^2). \tag{13.2.4}$$

Here I is the average or dc current carried by the n_0 electrons. The factor $\varkappa$ arises from the statistics of carriers in semiconductors.[15] For an n-type semiconductor, $\varkappa$ can be considered to correspond to the probability that a carrier is bound to an ionized donor. In nearly intrinsic material it can be shown that

$$\varkappa = (b + 1)^2 n_0^2 p_0/[(n_0 + p_0)(bn_0 + p_0)^2] \tag{13.2.5}$$

where $b = \mu_n/\mu_p$ is the ratio of the mobilities of electrons and holes, n_0 is the average number of electrons, and p_0 is the average number of holes.

A general discussion of g-r noise and of the interaction of g-r fluctuations with transport fluctuations is given by van Vliet and Fassett.[2]

[15] R. E. Burgess, *Physica* **20**, 1007 (1954).

Noise spectra of the form of Eq. (13.2.4) are frequently found in photodetectors. It is not uncommon to observe spectra which show the effect of several different time constants. G-r noise is also an important source of noise in some types of transistors, as will be discussed below.

13.2.4. Flicker Noise

Flicker noise is the name which has been applied to the usual excess noise (i.e., noise above that which can readily be attributed to thermal, shot, or g-r noise sources) found at low frequencies in almost all current carrying devices. The predominant characteristic of this noise is the $1/f$ frequency dependence of its spectral density.

The causes of flicker noise are not well known; its magnitude cannot be predicted with any useful accuracy.

A mathematical model frequently presented to "explain flicker noise" is one with a superposition of spectra such as given by Eq. (13.2.4) with a wide distribution of time constants.[8] Many ingenious models have been presented to make this idea reasonable.

A model which seems to be valid for an important part of the $1/f$ noise in semiconductor devices attributes the noise to modulation of the surface recombination velocity.[16,17] There are devices in which other mechanisms are at work as well.[18-20] Hooge believes that $1/f$ noise in many materials is a volume effect.[21] A poor metal-semiconductor contact or a conducting channel paralleling the base-collector junction may provide substantial $1/f$ noise. The noise of any current carrying resistor, other than one of continuous metal, may be a serious source of $1/f$ noise in an otherwise low-noise circuit. Controls with metal-carbon contacts are best avoided for the lowest noise applications. For that matter, even the best of sockets and connectors can become anomalous sources of excess noise—with a $1/f$ spectrum—after only a little wear.

A practical solution to the $1/f$ noise problem for many applications has been made available by the recent development of junction field-effect transistors which show very little excess noise. Figure 3 shows a comparison of the low-frequency noise performance of several devices:

[16] W. H. Fonger, *in* Transistors I, p. 239. RCA Lab., Princeton, New Jersey, 1956.
[17] T. B. Watkins, *Proc. Phys. Soc.* [1] **73**, 59 (1959).
[18] I. D. Gutkov, *Radiotekh. Electron.* **12**, 946 (1967).
[19] S. T. Hsu, D. J. Fitzgerald, and A. S. Grove, *Appl. Phys. Lett.* **12**, 287 (1968).
[20] O. Mueller, *ISSCC Digest Tech. Papers* **8**, 68 (1965).
[21] F. N. Hooge, *Phys. Lett.* **29A**, 139 (1969).

a gold-doped JFET, a typical JFET, a high-performance low-noise vacuum tube, and a very low-noise JFET. This lowest noise JFET is typical of those currently available from a large number of manufacturers. The data shown were obtained by Lukaszek and Wong in the Department of Electrical Engineering at the University of Florida. Work is continuing to determine the basic reason for the dramatic improvements which have apparently resulted from skillful application of advanced fabrication technology.

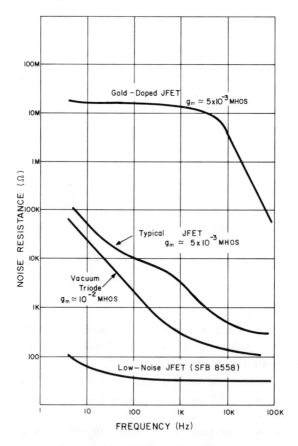

FIG. 3. Equivalent noise resistance (see Section 13.A.2.1 for the definition) of several different devices as a function of frequency. The main purpose of this figure is to show the very low magnitude of the excess low-frequency noise for the SFB 8558 (a developmental unit provided by Texas Instruments). This JFET is typical of those that are available as a result of recent advances in device fabrication technology. The data of the vacuum tube are the noise performance of the best low-noise vacuum tube currently available in the noise laboratory at the University of Florida.

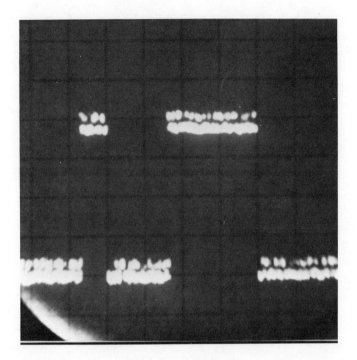

FIG. 4. Collector waveform of a bipolar transistor showing burst noise. Two different levels of bursts (of the order of 0.3 and 4 divisions) with different time constants can be seen. Normal shot and flicker noise waveform appear to be superimposed on the bursts.

13.2.5. Burst Noise

Burst noise is another source of excess noise at low frequencies. It is found in a large number of different types of devices.[22-25] The name burst noise was applied because the current through a device, when viewed on an oscilloscope, shows large pulses or bursts. Another name commonly used is *popcorn noise* because of the characteristic sound when this noise occurs at the input of an amplifier.

Figure 4 shows the waveform of the collector current of a silicon planar transistor with burst noise. The current fluctuates, apparently at

[22] W. H. Card and P. K. Chaudhari, *Proc. IEEE* **53**, 652 (1965).
[23] G. Giralt, J. C. Martin, and F. X. Mateu-Perez, *C.R. Acad. Sci. Paris* **261**, 5350 (1965).
[24] R. C. Jaeger and A. J. Brodersen, *IEEE J. Solid-State Circuits* **SC-5**, 63 (1970).
[25] S. T. Hsu and R. J. Whittier, *Solid-State Electron.* **12**, 867 (1969).

a random rate, between the two well-defined levels. The waveform which results closely approximates an ideal random telegraph signal; the spectral density of the resulting noise reflects this (Fig. 5).

This burst noise is found in a large portion of modern silicon planar transistors. Its effect can be minimized by careful circuit design; low burst noise units may be hand-selected.

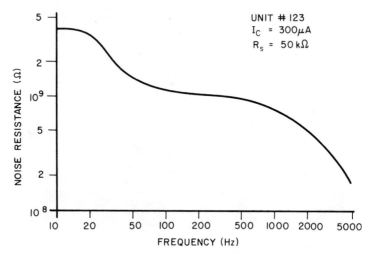

FIG. 5. Low-frequency noise spectrum of a bipolar transistor with burst noise. The spectrum shows clearly the effect of the two time constants occurring in the collector waveform of Fig. 4.

A model suggested for burst noise in bipolar transistors involves the modulation of the barrier potential of a junction by the trapping of carriers at impurity centers.[26] It has been shown that the presence of burst noise in diodes can be increased by heavy doping with metallic impurities.

13.2.6. Other Noise Sources

The list of topics discussed briefly above is not complete. In the discussion which follows it will also be useful to refer to avalanche break-down noise,[27,28] induced grid (or base or gate) noise, partition noise, and secondary emission noise.[3]

[26] S. T. Hsu, R. J. Whittier, and C. A. Mead, *Solid-State Electron.* **13**, 1055 (1972).
[27] K. S. Champlin. *J. Appl. Phys.* **30**, 1039 (1959).
[28] M. E. Hines, *IEEE Trans. Electron Devices* **ED-13**, 169 (1966).

13.3. Device Noise Models

The purpose of this chapter is to present noise models of several different devices and to demonstrate the use of these models in predicting the noise performance of an amplifier or detector.

13.3.1. Noise in Junction Diodes

The equivalent circuit shown in Fig. 6 has been shown to describe the noise performance of a junction diode over a wide range of operating conditions. The model has been derived by various workers using different techniques ranging from corpuscular arguments to elegant collective derivations.[1,29-34] The equivalent circuit approach of Polder and Baelde[33] displays the underlying assumptions very clearly, and has the additional merit that it can be extended to multijunction devices in a straightforward manner.

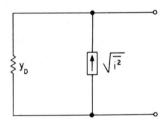

FIG. 6. Noise equivalent circuit for a junction diode. The spectral density of the noise current generator connected in parallel with the diode admittance is given by Eq. (13.3.1).

The spectral density of the noise current generator in Fig. 6 is given by the expression

$$S_i(f) = 4kT \, \mathrm{Re}(y) - 2qI/m. \qquad (13.3.1)$$

Here $\mathrm{Re}(y)$ is the real part of the junction admittance and m is a factor which depends on the ratio of diffusion current to g-r current. $1 \leq m \leq 2$;

[29] B. Schneider and M. J. O. Strutt, *Proc. IRE* **48**, 1731 (1960).
[30] L. J. Giacoletto, *in* Transistors I, p. 296. RCA Lab., Princeton, New Jersey, 1956.
[31] A. van der Ziel, *Proc. IRE* **43**, 1639 (1955).
[32] M. Solow, Ph.D. Thesis, Catholic Univ. of Amer., Washington, D.C., 1957.
[33] D. Polder and A. Baelde, *Solid-State Electron.* **6**, 103 (1963).
[34] K. M. van Vliet, *Solid State Electron.* **15**, 1033 (1972).

$m = 1$ for an ideal diffusion diode and $m = 2$ if all the diode current results from g-r processes in the junction transition region.

Expressions for the noise of diodes operating at high injection levels have been worked out only recently.[34-37] Only a collective carrier model can be employed, since the injection of carriers into the minority carrier region is no longer a sum of independent events. In particular, correlations occur as a consequence of the ambipolar transport, necessary to keep the bulk regions quasineutral, and as a consequence of the carrier-induced field in these regions. In addition, the mode of recombination (generally, via Shockley–Read levels) affects the results. Min et al.[36] obtained the following result, which replaces Eq. (13.3.1):

$$S_i(f) = \frac{b+1}{b} \left\{ 2kTg_j(\omega) - qI \left[1 - \Psi \left(\frac{1-x}{1+x} \right)^2 \frac{3g_j(0)}{2g_j(\omega) + g_j(0)} \right] \right\}.$$

(13.3.1a)

Here b is the mobility ratio (μ_n/μ_p for p^+n diodes and μ_p/μ_n for n^+p diodes), $g_j(\omega) = \mathrm{Re}\, y_j$ is the conductance of the junction transition region, $g_j(0)$ is the same for zero frequency; x reflects the recombination process, being 1 for band–band recombination or symmetric Shockley–Read levels (equal capture cross sections for electrons and holes) and 0 for strongly asymmetric Shockley–Read levels; Ψ relates to the junction geometry and lies between $\frac{1}{3}$ (long diode) and $\frac{1}{6}$ (short diode). Under the most common conditions $x = 0$ and $\Psi = \frac{1}{3}$, the low-frequency noise becomes $\frac{4}{3}[(b+1)/b]qI$, which means a reduction for p^+n junctions ($\approx \frac{5}{6}$ for silicon) and an enhancement for n^+p junctions ($\approx 3\frac{1}{3}$ for silicon) compared to low injection noise. Sometimes the above noise is further altered by the thermal noise and the current noise from the conductivity-modulated bulk impedances.[37]

Real diodes show $1/f$ noise at low frequencies. Burst noise has also been observed with both forward and reverse bias applied. Zener breakdown noise and avalanche multiplication noise become important as relatively large reverse bias is applied.

This model also holds for hot-carrier or Schottky-barrier metal-semiconductor junctions. Figure 7 shows a comparison of the low-frequency

[35] M. L. Tarng and K. M. van Vliet, Solid State Electron. 15, 1055 (1972).

[36] H. S. Min, K. M. van Vliet, and A. van der Ziel, Phys. Status Solidi (a) 10, 605 (1972); also 13, 701 (1973).

[37] H. S. Min and K. M. van Vliet, Phys. Status Solidi (a) 11, 653 (1972).

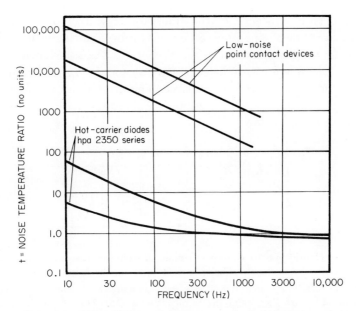

FIG. 7. Comparison of the noise performance of selected low-noise point contact diodes and planar hot-carrier (metal-semiconductor) diodes. The difference in $1/f$ noise probably results from improved surface characteristics of the planar device (courtesy of Hewlett Packard Associates). For the definition of noise temperature, see Section 13.A.2.2.

noise performance of planar hot-carrier diodes with typical low-noise point contact junctions.

13.3.2. Noise in Bipolar Transistors

For many applications, the noise performance of a bipolar transistor amplifier is described with good precision by the equivalent circuit shown in Fig. 8. The spectral densities of i_{Bn} and i_{Cn} as well as the noise emf e_{bn} are well known. They are

$$S_{i_{Bn}}(f) \simeq 2qI_B$$
$$S_{i_{Cn}}(f) \simeq 2qI_C \qquad (13.3.2)$$
$$S_{e_{bn}}(f) \simeq 4kTr_x.$$

For a wide range of operating conditions, there is very little correlation between these generators so that we can assume $\overline{i_{Bn}^{*}i_{Cn}} = 0$ and $\overline{e_{bn}^{*}i_{Bn}} = 0$. Here I_B and I_C are the dc base and emitter bias currents and r_x is the extrinsic base resistance of the transistor (essentially r_b'). These generators are shown superimposed on a hybrid-pi small signal equivalent circuit for a bipolar transistor (see Section 6.2.4). Also shown in Fig. 8 are two excess noise current generators i_{fn} and i_{bn}; i_{fn} is a flicker noise source and i_{bn} is a burst noise source. The magnitudes of these sources cannot be predicted with precision. These excess noise sources will be discussed briefly below.

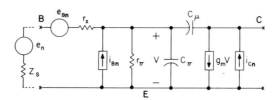

FIG. 8a. Hybrid-pi noise equivalent circuit for a bipolar transistor when excess low-frequency noise is negligible. The spectral densities of the noise generators are given by Eq. (13.3.2). The source resistance R_s and its thermal noise emf e_n ($\overline{e_n{}^2} = 4kTR_s df$) are included because they are needed to calculate Eq. (13.3.3).

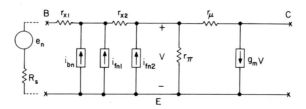

FIG 8b. Hybrid-pi noise equivalent circuit for a bipolar transistor when excess low-frequency noise dominates the performance. The magnitudes of these generators must be determined experimentally. i_{bn} represents burst noise, i_{fn1} represents flicker noise generated at the surface, i_{fn2} represents flicker noise generated "under the emitter." The magnitudes—and existence—of these sources varies greatly from unit to unit.

It should be clear that one of the first requirements for a good noise equivalent circuit is an adequate small signal equivalent circuit. The hybrid-pi has been shown to be accurate for frequencies ranging as high as $0.3f_\alpha$—where transit time effects begin to become important.

It is a straightforward exercise to modify the hybrid-pi circuit and the noise sources to include the description of transit time effects. However, a more common approach has been to turn to the physical-Tee equivalent circuit shown in Fig. 9.

The basic noise model of the bipolar transistor, as was the case for the junction diode, has been derived by several workers using different

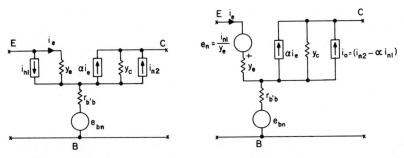

FIG. 9. Physical-Tee noise models for a bipolar transistor. Here $S_{i_{n1}}(f) = 4kT\mathrm{Re}(Y_e)$ $- 2qI_E/m_E$, $S_{i_{n2}}(f) = 2qI_C$, and $S_{i_{n1} * i_{n2}}(f) = 2kT\alpha y_e$. These circuits are equivalent to those of Fig. 8.

techniques.[30,31,33,34,38–41] The model has been well-verified experimentally from very low frequencies (below 10 Hz for selected low-noise units) to the microwave frequency range.[42–49]

Recently, the noise theory has been extended to intermediate and high injection levels.[50] Contrary to the junction diode case, not much change is to be expected, mainly because the base region is much smaller than the ambipolar diffusion length. This explains why experimentally the low injection model has been found to be valid even far outside the proper low injection regime. Some deviations occur, however, in that a correlation between input and output noise generators was observed by Tong and van der Ziel.[48] Presumably, this is caused by the conductivity-modulated series resistance in the base-emitter path. Only very accurate measurements of the noise figure (see below) show this effect.

[38] B. Schneider and M. J. O. Strutt, *Proc. IRE* **47**, 546 (1959).

[39] A. van der Ziel and A. G. T. Becking, *Proc. IRE* **46**, 589 (1958).

[40] P. O. Lauritzen, *IEEE Trans. Electron Devices* **ED-15**, 770 (1968).

[41] J. D. Patterson, *IEEE J. Solid-State Circuits* **SC-4**, 75 (1969).

[42] W. Guggenbuehl and M. J. O. Strutt, *Proc. IRE* **45**, 839 (1957).

[43] G. H. Hanson and A. van der Ziel, *Proc. IRE* **45**, 1538 (1957).

[44] E. R. Chenette and A. van der Ziel, *IEEE Trans. Electron Devices* **ED-9**, 123 (1962).

[45] G. J. Policky and H. F. Cooks, *NEREM Record* **7**, 254 (1965).

[46] H. Fukui, *IEEE Trans. Electron Devices* **ED-13**, 329 (1966).

[47] D. C. Agouridis and A. van der Ziel, *IEEE Trans. Electron Devices* **ED-14**, 808 (1967).

[48] A. H. Tong and A. van der Ziel, *IEEE Trans. Electron Devices* **ED-15**, 307 (1968).

[49] W. Baechtold and M. J. O. Strutt, *IEEE Trans. Microwave Theory Tech.* **MTT-16**, 578 (1968).

[50] H. S. Min and K. M. van Vliet. Solid State Electron. **17**, 285 (1974).

13.3.2.1. Noise Figure of a Bipolar Transistor. Figure 8 can be used to calculate the noise figure (see Section 13.A.2.2 for the definition) of a transistor amplifier operating with a source impedance Z_s. The result is[†]

$$1 + \frac{\overline{e_{bn}^2}}{\overline{e_{sn}^2}} + \frac{(\overline{i_{Bn}^2} + \overline{i_{bn}^2} + \overline{i_{Cn}^2}) \mid Z_s + r_x \mid^2}{\overline{e_{sn}^2}} + \frac{\overline{i_C^2} \mid Z_s + r_x + z_\pi \mid^2}{\overline{e_{sn}^2}}.$$

$$(13.3.3)$$

If the flicker and burst noise sources are negligible, this may be rewritten as

$$F = 1 + \frac{r_x}{R_s} + \frac{g_m}{2R_s}$$

$$\times \left\{ \mid Z_s + r_x \mid^2 \left(\frac{1}{h_{FE}} + \left(\frac{\omega}{\omega_T} \right)^2 \right) + \frac{1}{\beta_0^2} \left| \frac{\beta_0}{g_m} + Z_s + r_x \right|^2 \right\}.$$

$$(13.3.3a)$$

Here $h_{FE} = I_C/I_B$ is the ratio of the dc bias currents and $\beta_0 = I_c/I_b$ is the low-frequency ac gain for the common-emitter circuit.

The device parameters depend on I_C.[51] In particular, $g_m = (q/kT)I_C$ and $\omega_T = g_m/(C_\pi + C_\mu)$. Here $C_\pi = (C_{je} + g_m \tau_f)$, where C_{je} is the transition capacitance of the emitter-base junction and $\tau_f \simeq W^2/2D_b$ is the average base transit time for a carrier. r_x, h_{FE}, and β_0 are also slow functions of I_C. Hence, it is apparent that Eq. (13.3.3a) is a rather complicated function of operating point, frequency, and source impedance.

Figure 10 shows several sets of contours of constant noise figure in the I_C, R_s plane for a transistor designed and manufactured for superior low-frequency noise performance. Figure 11 shows several noise figure spectra for different operating conditions.

It is apparent from the contours that the noise figure has a broad minimum in the I_C, R_s plane. Some insight can be gained by considering that minimum.

[51] P. E. Gray, D. DeWitt, A. R. Boothroyd, and J. F. Gibbons, "Physical Electronics and Models of Transistors," Chapter 7. Wiley, New York, 1964.

[†] The notation used in (13.3.3) is intended to conform to IEEE standards. Lower-case impedance symbols refer to elements in a device equivalent circuit and upper-case impedance symbols refer to circuit elements external to the transistor.

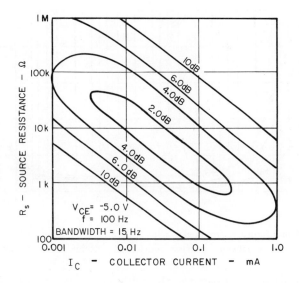

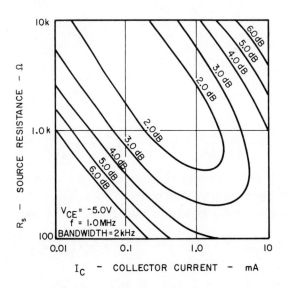

FIG. 10. Contours of constant noise figure in the R_s, I_C plane at two different frequencies for a 2N5138 *pnp* silicon bipolar transistor. The 1.0 MHz contour agrees well with Eq. (13.3.3) and Fig. 8a. The lower frequency contour shows the existence of some excess noise. This is one of the lowest noise and best characterized devices available for relatively low-frequency applications (data courtesy of Fairchild Semiconductor Division).

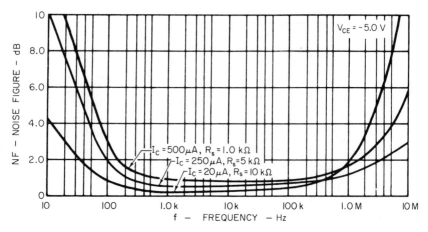

FIG. 11. Noise figure vs. frequency with R_s and I_C as parameters for a typical type 2N5138 low-noise *pnp* silicon transistor (courtesy of Fairchild Semiconductor Division).

For a fixed source resistance, the minimum noise figure is

$$F(\text{min})|_{R_s=\text{const}} = \left(1 + \frac{r_x}{R_s}\right)$$
$$\times \left(1 + \frac{1}{\sqrt{h_{FE}}}\sqrt{(1 + \omega^2 h_{FE}\tau_F{}^2)(1 + \omega^2(R_s + r_x)^2 C_t{}^2)}\right)$$
$$+ \frac{\omega^2 C_t(R_s + r_x)^2 \tau}{R_s} \simeq \left(1 + \frac{r_x}{R_s}\right)\left(1 + \frac{1}{\sqrt{h_{FE}}}\right) \qquad (13.3.4)$$

when

$$I_C = \frac{kT}{q}\frac{\sqrt{h_{FE}}}{(R_s + r_x)}\left[\frac{1 + \omega^2(R_s + r_x)^2 C_t{}^2}{1 + \omega^2 h_{FE}\tau_F{}^2}\right]^{1/2}. \qquad (13.3.5)$$

Here $C_t = C_\pi + C_\mu$ is essentially the sum of the transition capacitances of the emitter and collector junctions.

For a fixed operating point, the minimum noise figure is

$$F(\text{min})|_{I_C=\text{const}} \simeq 1 + \left|\frac{1 + h_{FE}(\omega/\omega_T)^2}{h_{FE}}(1 + 2g_m r_x)\right|^{1/2} \qquad (13.3.6)$$

when

$$R_s = \left|r_x{}^2 + \frac{(h_{FE}/g_m{}^2)(1 + 2g_m r_x)}{1 + h_{FE}(\omega/\omega_T)^2}\right|^{1/2}. \qquad (13.3.7)$$

Both Eqs. (13.3.4) and (13.3.6) go to $F(\text{min}) = (1 + 1/\sqrt{h_{FE}})$ when r_x is small and $\omega \ll \omega_T/\sqrt{h_{FE}}$. This is the "bottom of the valley" in the noise figure contours.

Thus a transistor such as the 2N5138 ($h_{FE} = 225$, $r_x = 50\ \Omega$) can be expected to yield a noise figure of about 1.12 or 1 dB with $R_s = 1000\ \Omega$ and $I_C = 360\ \mu A$ in good agreement with the value shown on the 10 kHz contour.

Good high-frequency noise performance requires a small transit time, small transition capacitance of both emitter and collector junctions, low base resistance, and high dc current gain. Figure 12 shows the noise performance of a transistor designed especially for high-frequency applications.

To this point, no mention has been made of the temperature dependence of the noise performance of a transistor. It is clear that the parameters of the hybrid-pi small signal circuit do depend strongly on temperature. Figure 13 shows the effect of change in temperature on the noise figure of the transistor of Fig. 12. Equation (13.3.3) is accurate in predicting this performance if the temperature-dependent parameters are used. More detailed discussions of the noise performance of transistor amplifiers are available.[52-54]

13.3.2.2. Noise Performance at Low Frequencies.

At low frequencies, the flicker noise and burst noise sources may dominate the noise performance of a bipolar transistor. The equivalent circuit of Fig. 8b can be used to help understand the problem. Two different flicker noise current sources are shown in addition to a burst noise generator. In many transistors, the flicker noise and the burst noise both seem to arise from surface effects; in others, the excess noise sources seem to lie in the bulk material in the active volume of the transistor.[55-57] For a given transistor, measuring the noise performance as a function of source resistance will yield information concerning the location of the noise generators. In general, it appears that both surface and bulk flicker noise sources are required although one or the other may dominate any given unit. Jaeger has used external field plates to control the surface recombination velocity and has been able to demonstrate some control of the surface noise sources.[55]

[52] E. C. Nielsen, *Proc. IRE* **45**, 957 (1957).

[53] E. C. Nielsen, *in* "Amplifier Handbook" (R. F. Shea, ed.), Chapter 7. McGraw-Hill, New York, 1966.

[54] R. D. Thornton, D. DeWitt, E. R. Chenette, and P. E. Gray, "Circuit Limitations of Transistors." Wiley, New York, 1966.

[55] R. C. Jaeger, Ph.D. Thesis, Univ. of Florida, Gainesville, Florida, 1969.

[56] J. F. Gibbons, *IRE Trans. Electron Devices* **ED-9**, 308 (1962).

[57] J. L. Plumb and E. R. Chenette, *IRE Trans. Electron Devices* **ED-10**, 304 (1963).

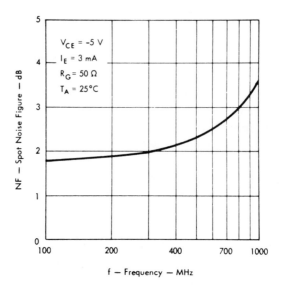

FIG. 12. Noise figure vs. frequency for a type 2N5043 *pnp* germanium planar transistor intended for low-noise applications at frequencies ranging to about 1 GHz (courtesy of Texas Instruments).

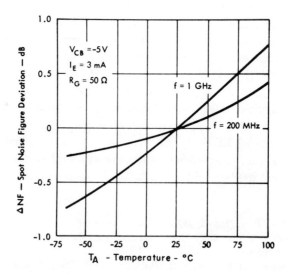

FIG. 13. Effect of temperature on the noise figure of the 2N5043 (courtesy of Texas Instruments).

The minimum noise figure of the circuit of Fig. 8b is

$$F(\min) = 1 + (\overline{i_{bn}^2} + \overline{i_{fn1}^2})2r_{x1} + 2r_{x2}\overline{i_{fn2}^2}$$
$$+ 2[(\overline{i_{bn}^2} + \overline{i_{fn1}^2} + \overline{i_{fn2}^2})(\overline{i_{bn}^2} + \overline{i_{fn1}^2})r_{x1}^2 + \overline{i_{fn2}^2}(r_{x1} + r_{x2})^2]^{1/2}$$

$$(13.3.8)$$

when

$$R_s = \left[\frac{(\overline{i_{bn}^2} + \overline{i_{fn1}^2})r_{x1}^2 + \overline{i_{fn2}^2}(r_{x1} + r_{x2})^2}{\overline{i_{bn}^2} + \overline{i_{fn1}^2} + \overline{i_{fn2}^2}}\right]^{1/2}. \qquad (13.3.9)$$

Thus the minimum noise figure occurs when $R_s = r_{x1}$ if i_{bn} or i_{fn1} dominates the performance and at $R_s = r_{x1} + r_{x2}$ when i_{fn2} dominates. r_{x1} agrees well in value with the resistance of the bulk material between the case and emitter junctions of a planar transistor; r_{x2} agrees well with the value for the small signal base resistance from measurements of input resistance and also agrees well with the value required for good agreement between theory and experiment in the shot noise region.

As mentioned before, these excess noise current generators cannot be predicted with precision. The spectral densities of both i_{fn1} and i_{fn2} are of the general form

$$S_{if} = KI_B^\gamma / f^\alpha \qquad (13.3.10)$$

where γ and α both range from 1 to 2. Hence, the low-frequency noise performance is usually improved by operating at the lowest possible bias currents consistent with other design requirements.

These noise sources provide an interesting limitation to the ultimate stability which can be obtained in a precision direct-coupled amplifier.[58]

13.3.3. Noise in Field Effect Transistors

The noise performance of a field effect transistor, whether it be a junction gate (JFET) or insulated gate (MOSFET) device of either the depletion or enhancement mode, and whether it be p- or n-channel is well represented by the circuit shown in Fig. 14.[59-66] The spectral densi-

[58] A. H. Hoffait and R. D. Thornton, *Proc. IEEE* **52**, 179 (1964).

[59] A. van der Ziel, *Proc. IEEE* **51**, 461 (1963).

[60] A. G. Jordan and N. A. Jordan, *IEEE Trans. Electron Devices* **ED-12**, 148 (1965).

[61] H. Johnson, *in* "Field Effect Transistors" (J. T. Wallmark and H. Johnson, eds.), p. 160. Prentice-Hall, Englewood Cliffs, New Jersey, 1966.

[62] M. Shoji, *IEEE Trans. Electron Devices* **ED-13**, 520 (1966).

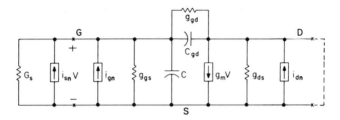

FIG. 14. Noise equivalent circuit for a field effect transistor. The circuit is valid for both junction and insulated gate devices when excess low-frequency noise is negligible.

ties of the two noise current generators have been shown to have the theoretical values

$$S_{ig} = 4kTg_{11}P(V_{GS}, V_{DS}) \qquad (13.3.11)$$

and

$$S_{id} = 4kTg_m(\text{max})\, Q(V_{GS}, V_{DS}).$$

Here $g_{11} = g_{gs} + g_{gd}$ is the total input conductance with the output ac short-circuited and $g_m(\text{max})$ is the maximum transconductance of the FET. $P(V_{GS}, V_{DS})$ and $Q(V_{GS}, V_{DS})$ are both relatively complicated functions of the gate and drain bias voltages. For normal operation $P \geq 1$ for a JFET and $P = \frac{4}{3}$ for a MOSFET; $\frac{1}{2} < Q \leq \frac{2}{3}$ for a JFET while $Q = \frac{2}{3}$ for a MOSFET.

These gate and drain noise current generators are partially correlated since they are caused by the same incremental thermal noise emfs in the conducting channel of the FET. The correlation coefficient, as calculated by van der Ziel, is

$$C = \frac{\overline{i_{gn}i_{dn}^*}}{\overline{i_{gn}^2}\,\overline{i_{dn}^2}} = -jK$$

where $0.395 < |K| < 0.446$ for normal bias conditions.[59]

These are essentially the expressions first derived by van der Ziel for a JFET when the main source of noise is thermal noise of the conducting channel. The best low-noise FETs approach this performance (see

[63] W. C. Bruncke and A. van der Ziel, *IEEE Trans. Electron Devices* **ED-13**, 323 (1966).

[64] F. M. Klaasen and J. Prins, *Philips Res. Rep.* **22**, 504 (1967).

[65] C. T. Sah, S. Y. Wuh, and F. H. Hielscher, *IEEE Trans. Electron Devices* **ED-13**, 410 (1966).

[66] F. M. Klaasen and J. Prins, *IEEE Trans. Electron Devices* **ED-16**, 952 (1969).

Fig. 3). However, there are many units in which various sources of excess noise are very important. The spectrum of the noise resistance of the gold-doped JFET in Fig. 3, for example, shows a large excess g-r noise.

The circuit of Fig. 14 can be used to calculate the noise figure of a typical FET amplifier in the same way as for the bipolar transistor. One does not have as much freedom in choosing the operating point of an FET as in the case of the bipolar transistor. However, it is still worthwhile to consider the optimum noise performance which can be obtained by varying the source impedance. The minimum noise figure is

$$F(\text{min}) = 1 + 2\sqrt{r_n g_n} \tag{13.3.12}$$

when

$$R_s = \sqrt{r_n/g_n} . \tag{13.3.12a}$$

Here it has been assumed that the correlation between i_{gn} and i_{dn} has no significant effect on the magnitude of the minimum noise figure. In that case $g_n = g_{11} \cdot P(V_{GS}, V_{DS})$ and $r_n = (g_m(\text{max})/|y_{tr}|^2) \cdot Q(V_{GS}, V_{DS})$ where y_{tr} is the net transfer admittance of the FET.

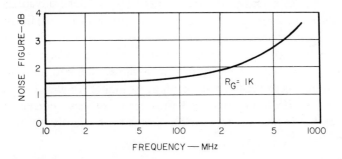

FIG. 15. Noise figure vs. frequency for a 2N5397 JFET intended for low-noise applications at frequencies up to 800 MHz. The performance is in good agreement with the model of Fig. 14 (courtesy of Siliconix).

Of course, the use of this circuit requires accurate small signal characterization of the FET.

Figure 15 shows the noise figure of a type 2N5397 JFET as a function of frequency. This unit is designed particularly for superior noise performance at high frequencies. Its performance is in good agreement with the model.

13.4. Noise in Amplifiers

13.4.1. Comparison of Noise in Devices

The purpose of Table I is to present a compact summary of the optimum noise performance of the various devices. Information on the noise performance of vacuum tubes is included for completeness. Readers interested in the details of the derivations of the expressions for noise in vacuum tubes are referred to van der Ziel.[3]

At this time the highest performance JFETs (such as the SFB 8558 whose noise performance was shown in Fig. 3) are clearly the optimum choice for the input stage of a low-noise, low-frequency amplifier. For low source impedance applications a bipolar transistor with small r_x can provide comparable performance.

13.4.2. Noise in Tuned Amplifiers

The theoretical expressions for the noise figure of the various devices as derived above is most useful for relatively narrow-band amplifiers such as are often required in the input stage of a receiver or other measurement system. In that case, it is often possible to make sensible use of the tuned input circuit as a lossless matching network and take advantage of the optimum noise figure characteristics. Over most of the useful frequency range of a device, optimum performance is obtained simply by adjusting the tuning to provide the desired resistive source impedance.

Figure 16 shows that maximum small signal gain and minimum noise figure do not occur for the same tuning. At higher frequencies (i.e., when transit time effects and induced grid, gate, or base current becomes important) it is possible to adjust a circuit for a slightly improved noise performance.

13.4.3. Noise in Wideband Amplifiers

The noise models presented are valid over a wide frequency range. The expressions for noise figure, equivalent noise resistance, and so on, however, have all been "spot" expressions. They are of little value for amplifiers with relatively wide bandwidths.[67] Figure 17 shows the noise performance of a typical 2N5138 over the band from 0 to 15.7 kHz. Both the low-frequency excess noise and the shot and thermal noise sources are included in this one figure.

[67] J. A. Ekiss and J. W. Halligan, *Proc. IRE* **49**, 1216 (1961).

TABLE I. Comparison of Noise Performance of Devices

	Bipolar transistor	Field effect transistor	Vacuum triode
Equivalent noise resistance $R_n' = \lim\limits_{R_s \to 0} F R_s$	$r_x + kT/2qI_C$	$Q(V_{GS}, V_{DS})/g_m \simeq 0.7/g_m$	$\varepsilon/g_m \simeq 2.5/g_m$
Minimum noise figure	$1 + \dfrac{r_x}{R_s}\left(1 + \dfrac{1}{\sqrt{h_{FE}}}\right)$ when $I_C = \dfrac{kT\sqrt{h_{FE}}}{q(R_s + r_x)}$	$1 + 2\sqrt{G_c(0.7/g_m)}$ [a]	$1 + 2\sqrt{G_c(2.5/g_m)}$ [a]
Equivalent noise conductance $G_n' = \lim\limits_{R_s \to \infty} F/R_s$	$qI_B/2kT$	$G_c + (q/2kT)I_{\text{Gate}}$ [a]	$G_c + (q/2kT)I_{\text{Grid}}$ [a]

[a] Here G_c is the equivalent noise conductance of the input circuit, i.e., external gate or grid bias resistor, plus the gate or grid noise of the device. The $(q/2kT)I$ term accounts for the probably very small shot noise of gate or grid bias current.

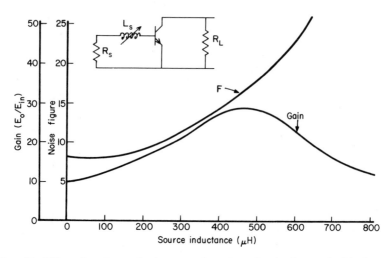

FIG. 16. Effect of tuning at the input on the gain and noise figure of a bipolar transistor amplifier.

The calculations which must be made to predict the total noise performance of a wideband amplifier are straightforward but relatively tedious. Probably the most sensible procedure is to make those calculations for any circuit of interest on a digital computer.

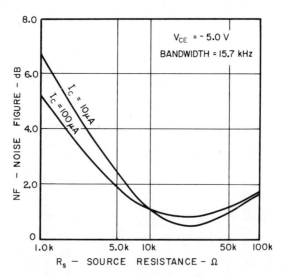

FIG. 17. Wideband (0 to 15.7 kHz) noise figure for a 2N5138 transistor as a function of source resistance with I_C as a parameter (courtesy of Fairchild Semiconductor Division).

13.4.4. Noise in Integrated Amplifiers

Many useful amplifiers are available in silicon monolithic integrated form. The noise processes involved in the devices found in silicon monolithic circuits are essentially those found in discrete devices. The noise performance of an integrated amplifier, whether it be a dc operational amplifier or a chip intended for use as a building block in an rf amplifier, can be predicted with the noise models presented in Chapter 13.3.[68,69]

Unfortunately, the compromises required to build good stable direct-coupled integrated amplifiers have not permitted realization of noise performance which matches that of the best discrete-component amplifiers. For example, the base resistance of the integrated *npn*-planar diffused transistors found in the input stages of such amplifiers as the μA709, LM101, RA909, and others is relatively large. Hence the minimum noise resistance or noise figure is significantly higher than is available with a well-designed amplifier using transistors such as the 2N5138. However, there are many applications where the ultimate noise performance is not required. For those applications, the low cost and small size of these integrated units is most attractive.

13.4.5. Noise in Maser Amplifiers

Maser amplifiers are capable of nearly ideal performance for amplification of microwave signals.[70] These amplifiers employ the interaction of electromagnetic waves with molecular or electronic energy levels in solids or gases. Suppose, for example, we have two energy levels ε_1 and $\varepsilon_2 > \varepsilon_1$ in a solid with $\varepsilon_2 - \varepsilon_1 = h\nu_0$ appropriate for microwave absorption (as, for instance, obtainable in several paramagnetic materials with suitable magnetic fields). In thermal equilibrium, the occupancies N_1 and N_2 of the levels ε_1 and ε_2 are related by the Boltzmann factor and $N_2 < N_1$. When the pump signal is applied to a maser, $N_2 > N_1$. Under this condition, a large stimulated emission caused by transitions $N_2 \to N_1$ may occur when radiation $h\nu_0$ is absorbed and amplification may occur. This is a very low-noise process. The limit for the equivalent noise temperature is found to be $T_{eq} \geq h\nu_0/(k \ln 2)$.

[68] J. R. Maticich, *Proc. IEEE* **53**, 605 (1965).

[69] A. J. Brodersen, E. R. Chenette, and R. C. Jaeger, 1970 *ISSCC Digest of Tech. Papers*, p. 164. (1970).

[70] A. L. Schawlow and C. H. Townes, *Phys. Rev.* **112**, 1940 (1958).

The input noise of a helium-cooled traveling wave maser has been measured to be $= 2.0 \pm 0.5$ °K in good agreement with the theory.[71]

With such sensitivity, it becomes useful to consider the average number of quanta which can be detected in a time given by the reciprocal of the amplifier bandwidth. This can be shown to be

$$N = \frac{P_n}{h\nu \, d\nu} = \frac{1}{2} \coth \frac{h\nu}{2kT_m}. \tag{13.4.1}$$

Here we have used the expression $P_n = \frac{1}{2}h\nu \, d\nu \coth(h\nu/2kT)$ which includes the zero point energy in black body radiation.

Thus, by Eq. (13.4.1) an average of eight quanta can be detected at 10 GHz if $T_m = 4$ °K.

Similar quantum limitations limit the sensitivity of laser amplifiers. Equation (13.4.1) is useful for predicting the limitation of a laser amplifier.[72]

13.5. Noise in Photodetectors

Photodetectors are used for the detection of visible light or of infrared radiation. They can broadly be divided into photoemissive devices, photojunctions or photovoltaic devices, and photoconductive devices (see Chapter 11.1). Basically, each device can be represented by a black box with a light input and an electrical output. In a photoemissive device an input photon flux is converted into an output charge flux, represented by an electrical current generator. In a photojunction the conversion is similar, while in a photovoltaic cell the junction is open-circuited so that the output signal is best represented by a voltage generator. In a photoconductive detector the basic conversion is to a change in conductivity which, in turn, may be represented by an equivalent Norton or Thévénin generator if a standard dc current through the device is specified. The signal black box representations are given in Fig. 18a–d.

The noise in these devices stems from three causes. First, there is the noise associated with the radiation field "seen" by the detector (photo-induced noise). Second, there is noise associated with the incomplete absorption of photons (photon partition noise). Third, there is noise

[71] R. W. DeGrasse and H. E. D. Scovil, *J. Appl. Phys.* **31**, 443 (1960).

[72] See M. Ross, "Laser Receivers: Devices, Techniques, Systems." Wiley, New York, 1966.

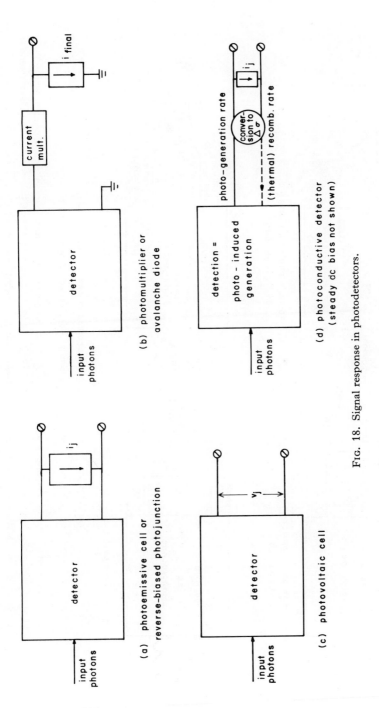

Fig. 18. Signal response in photodetectors.

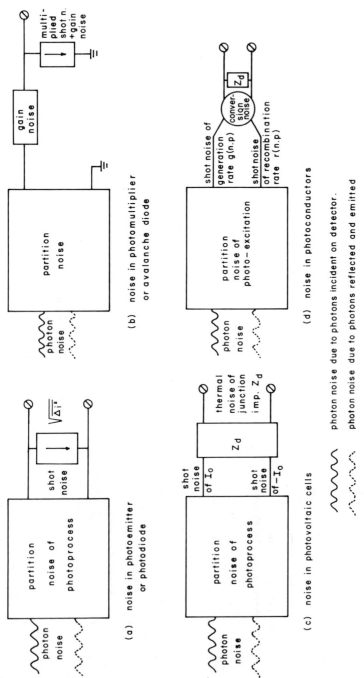

Fig. 19. Noise in detectors. The first square box indicates the basic conversion procedure: photon flux → excited carrier flux. In radiative equilibrium the reverse process contributes an equal part to the noise. In photoconductors, the excitation and de-excitation rates cause rate fluctuations Δn and Δp which are converted into steady-state carrier concentration fluctuations $\overline{\Delta n^2} \sim \overline{\Delta \dot n^2} \tau^2$, where $\tau = \tau(\bar n, \bar p)$ is the appropriate lifetime.

associated with the spontaneous excitation processes which act parallel to the photon absorption and there is noise associated with the spontaneous de-excitation processes in the detector which are responsible for the steady state under which it operates. Similarly as in a transistor, one can either represent all the noise at the input as an equivalent photon source or *noise equivalent power* (NEP), or at the output as an *equivalent electrical generator* $[S_{det}(f)\, \Delta f]^{1/2}$; cf. Fig. 19a–d.

The situation at the input and the output of photodetectors and the relevant concepts will be considered. For more details consult the book by Kruse *et al.*[73] and a recent review paper.[74]

13.5.1. Description of Noise as Viewed at the Input of a Photodetector

In order to judge the performance of a detector, it is assumed that the signal is available as a wave of optical frequency v within a band of frequencies Δv_S, and of information frequency f. Such is the case if an optical wave is perfectly sinusoidally chopped or if its amplitude is sinusoidally modulated by other means, for example, using the electro-optic effect or Faraday rotation.[75]

The signal-to-noise ratio when viewed at the input will be studied. With reference to Figs. 18 and 19, *it is most expedient to compare the signal and the noise in the primary particle flux Φ following photon absorption.* Note that $q\Phi$ stands for the primary photocurrent I_p in photoemissive devices such as the photomultiplier. In reverse-biased photojunctions $q\Phi = q(\Phi_n + \Phi_p)/2$ is the current I_0 resulting from photoelectrons and photoholes created in the junction and "sliding down" the potential hill; in photovoltaic devices the same flux is generated but the carriers are collected at the contacts, setting up a photovoltage. In photoconductors Φ is the generation rate $g(n, p)$ of the liberated carriers, induced by light or when referring to the noise, also by spontaneous parallel processes.

An internal signal-to-noise ratio can be defined as

$$\sigma_{int} = \Phi_S/\Phi_N \qquad \text{with} \quad \Phi_N \equiv \langle \Delta\Phi^2 \rangle^{1/2}. \qquad (13.5.1)$$

For a given signal, the ratio σ_{int} is maximal if Φ_N is minimal. Now, in all cases, Φ_N is given by a shot noise-like formula, i.e., $\Phi_N \sim \sqrt{2\bar{\Phi}_d\, \Delta f}$,

[73] P. W. Kruse, R. D. McGlauchlin, and R. B. McQuistan, "Elements of Infared Technology." Wiley, New York, 1962.

[74] K. M. van Vliet, *Appl. Opt.* **6**, 1145 (1967).

[75] C. C. Robinson, *Appl. Opt.* **3**, 1163 (1964).

where Φ_d is the "dark flux" of the particle current Φ referred to in each of the above cases and Δf is the electrical bandwidth (for the Boson factor see below). Clearly, the flux Φ_d must be kept as low as possible. (In some cases this reasoning is fallacious; however, since the noise is determined by the circuitry, e.g., by the thermal noise of the load impedance, these cases will not be discussed here.)

Physically, the dark flux Φ_d stems from spurious causes (such as leakage currents in photodiodes) or is caused by phonon-excited and (or) photon-excited carrier transitions. At a given temperature T_e (of the environment), the photon rate is fixed by Planck's law. Thus, Φ_d cannot be lower than the rate given by photo-induced transitions. The detector is in this case *photon-limited*. However, as was pointed out before,[74] the temperature T_e may be unnecessarily high; for example, the detector may be exposed to black body radiation of 300°K, while being itself at 77°K. Thus, lowering of T_e by a cooled radiation shield may result in a decrease in Φ_d. The optimum behavior obtainable corresponds to a detector of temperature T_d exposed to a radiation field of the same temperature, which alone accounts for the observed dark rate Φ_d. Such a detector is termed *radiation-limited*.

An *ideal detector* is a detector with unit quantum efficiency, i.e., Φ_S is equal to the signal photon flux, while further the detector has no other noise Φ_N than photo-induced noise. The signal-to-noise ratio at the input of the detector, σ_{inp}, is equal to the above defined σ_{int}, and is, moreover, maximal for a given signal.

More realistically, we consider a *quasi-ideal detector* which (i) has only photo-induced noise *and* partition noise, (ii) a responsive quantum efficiency η_r of unity (i.e., every absorbed photon gives rise to any of the primary events composing the flux Φ) while the incident quantum efficiency η_i may be less than unity, and (iii) which has a constant responsivity in the optical passband. For such a detector we have

$$\Phi_S = \int_{\Delta v_S} \mathsf{j}_S(v, f)\eta_i(v)\, dv \qquad (13.5.2)$$

where $\mathsf{j}(v, f)$ is the ac optical signal. Further for the noise

$$\Phi_N{}^2 = \int_{\Delta v} 2\overline{\mathsf{J}}(v)\eta_i(v)[1 + \eta_i(v)B(f, v)]\, \Delta f \qquad (13.5.3)$$

where B is the Boson factor and $\overline{\mathsf{J}}(v)$ is the photon flux of the background radiation of frequency ν; for thermal radiation fields, both B and $\bar{\mathfrak{J}}(\nu)$ are found in texts on statistical mechanics.[76]

We assume further that $j_S(\nu)/p_S = \mathfrak{J}_S(\nu)/P_S$, where j_S and p_S denote the photon flux at frequency ν and the total power of the ac component of the optical signal, whereas $\mathfrak{J}_S$ and P_S refer mutatis mutandis to the unchopped light signal. This is substituted into the above equations; next we set σ_{int} equal to one and solve for the power $p_S \equiv P_{\text{eq}}(\lambda, f, \Delta f, A)$, where $\lambda = c/\nu$ is the signal wavelength with $\Delta \nu_S \ll \nu$, f is the center electrical frequency of a bandwidth $\Delta f \ll f$, and A is the detector area. For $A = 1$ cm², $\Delta f = 1$ Hz, P_{eq} denotes the spectral specific noise-equivalent-power (SNEP), and its inverse is the spectral photon-limited detectivity $\mathscr{D}_\lambda^*$ or radiation-limited detectivity $\mathscr{D}_\lambda^\dagger$. For $\mathscr{D}_\lambda^*$, it is easily found that

$$\mathscr{D}_\lambda^* = \frac{\int_{\Delta\nu_S}[\mathfrak{J}_S/P_S]\eta_i(\nu)\,d\nu}{\{\int_{\Delta\nu}2\bar{\mathfrak{J}}(T_e;\nu)A^{-1}\eta_i(\nu)[1+\eta_i(\nu)B(T_e;f,\nu)]\,d\nu\}^{1/2}}. \quad (13.5.4)$$

For a near monochromatic signal the numerator equals $\eta_i(\nu)/h\nu$, whereas the denominator is a Bose-type integral over the passband (ν_0, ∞) for a detector with absorption limit ν_0. For $\mathscr{D}_\lambda^\dagger$, replace T_e by T_d. Clearly, the detectivity depends on $\eta_i(\nu)$ and ν_0. The result has been evaluated when $\eta_i(\nu)$ in front of the Bose factor can be replaced by 1. It has been shown before that this is exact for photoconductors. However, in that case, $\mathscr{D}_\lambda^*$ must be divided by $\sqrt{2}$, since it is not realistic to omit the fluctuations associated with the recombination rate $r(n, p)$ which, in the case of radiative equilibrium, contribute again to the black body photon field surrounding the detector. For the case that the signal frequency ν is equal to the peak photoresponse ν_0, Eq. (13.5.4) can be fully computed. Curves for λ varying from 0.6 to 40 μ and $T_e = 300°$K, $200°$K, and $90°$K are given in Fig. 20a and b (after van Vliet[74]).

Lastly, what about a detector which is neither ideal, nor quasi-ideal? First, if the responsive quantum efficiency is not unity, other processes occur parallel to the basic excitations. Second, if the noise Φ_N is not photo-induced for a given environmental temperature T_e, it is necessarily higher. This holds a fortiori for the output noise, and the signal-to-noise ratio in the output σ_{out} is smaller than σ_{int}. The observed detectivity D_λ^* is accordingly less than the computed photon-limited detectivity $\mathscr{D}_\lambda^*$.

---

[76] See e.g., E. Schrödinger, "Statistical Thermodynamics," Chapter 9. Cambridge Univ. Press, London and New York, 1960. For blackbody photons we have

$$J(\nu) = 2\pi A(\nu^2/c^2)/(e^{h\nu/kT} - 1), \qquad B(\nu) = 1/(e^{h\nu/kT} - 1).$$

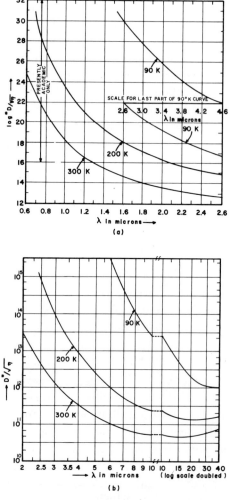

FIG. 20. Computed values of $\mathscr{D}_\lambda$* [after K. M. Van Vliet, *Appl. Opt.* **6**, 1145 (1967)].

It is emphasized that the curves for the latter were universal, i.e., independent of the actual detection mechanism (photovoltaic, photoconductive, etc.). This does not mean that certain forms of detection may not be preferable in order to reach the optimum detectivity.

13.5.2. Description of Noise as Viewed at the Output of a Photodetector

The connection between the output electrical signal and the input optical signal is called the *responsivity* $R(\nu, f)$. Clearly, the output noise and the input photon noise are related in the same fashion, i.e., by $[R(\nu, f)]^2$. Thus, for a quasi-ideal detector the output noise is readily obtained. However, in many cases, the conversion $R(\nu, f)$ adds noise which is essential, viz. in all devices in which multiplication of the primary excitation rate takes place, and also in all photoconductors. Moreover, when viewed at the output, one may observe shot noise-like fluctuations of an electrical current or of the g-r rates and "almost all" references to the excitation process disappear—whether they stem from photons or from other causes. It is therefore more direct to consider the essential noises in the output using standard models such as considered in the other sections of this chapter. In photoconductors the connection between photon fluctuations at the input and the electron fluctuations at the output is far from simple.

Some devices are now discussed very briefly.

13.5.2.1. Photodiodes and Photomultipliers. A *photodiode* is usually limited by circuitry noise because of its high output impedance. In a photomultiplier with n stages of secondary emission multiplication δ there is an extra gain $G = \delta^n$. The output current is then of reasonable magnitude. The responsivity is $q\eta_i G$. The output noise is multiplied shot noise—the Boson factor being usually negligible—plus secondary emission noise from the multiplication process. The output signal-to-noise ratio is

$$\sigma_{\text{out}} = \frac{p_S \eta_i G / h\nu}{\{2qI_a G[(\varkappa - 1)/(\delta - 1)] \, \varDelta f \}^{1/2}} \qquad (13.5.5)$$

where $\varkappa$ is a number close to $\delta + 1$, I_a is the anode current $= I_p G$. Setting $\sigma = 1$ and solving for $1/p_S$, the result

$$D_\lambda^* = \frac{q\eta_i}{h\nu \{2qI_a A^{-1} G^{-1}[(\varkappa - 1)/(\delta - 1)]\}^{1/2}} \qquad (13.5.6)$$

is obtained for the spectral detectivity. The connection with Eq. (13.5.4) is easily traced. The primary current $I_p = q\Phi_d$ (with Φ_d as in Section 13.5.1) is seldom caused by photon excitation of the background, unless the detector is cooled and the photocathode is exposed to a photon field of sufficiently high T_e. The noise factor is defined as the total noise referred back to the input divided by the photon noise of thermal origin. This definition is to some extent analogous to that of Section 13.A.2.2. One finds

$$F = [D_\lambda^* / \mathscr{D}_\lambda^*]^2. \qquad (13.5.7)$$

13.5.2.2. Photojunction Devices. In a reverse-biased photodiode the noise is shot noise-like and the responsivity is $q\eta_i$. It follows that

$$D_\lambda^* = \frac{q\eta_i}{h\nu\{2qI_0A^{-1}(1 + \eta_iB)\}^{1/2}} \qquad (13.5.8)$$

where I_0 is the steady-state diode current. The Boson factor shows up if η_i is large enough. It was recently measured this way using an indium antimonide detector by Kattke at the University of Minnesota, following the correlation method originally used by Hanbury-Brown and Twiss.[77]

In an avalanche diode the sensitivity is increased by a factor $G \sim 10^4$–10^6 due to avalanche ionization in the junction region. However, the output noise becomes proportional to qI_0G^3 (rather than the expected $2qI_0G^2$), so the detectivity deteriorates by a factor $\sim G^{-1/2}$. For details, see McIntyre.[78]

In a photovoltaic cell the shot noise must be doubled because of the two opposing currents I_0 for an open junction. In radiative equilibrium the Boson factor must be omitted. In other cases the situation is more complex, but this will not be considered here. The following equivalent formulas are then valid for the detectivity:

$$D_\lambda^* = \frac{q\eta_i/h\nu}{\{4I_0qA^{-1}\}^{1/2}} = \frac{(\eta_ikT/I_0)A^{1/2}/h\nu}{(4kTR_0)^{1/2}} = \frac{\eta_i/2h\nu}{(p_0L_h/\tau_h)^{1/2}}. \qquad (13.5.9)$$

The first expression follows from Eq. (13.5.8); in the second one we set $R_0 = kT/qI_0$, the numerator being the open voltage response and the denominator being just the thermal noise of the junction impedance; in the last expression, which is useful for comparison with photoconductive detectors, we set $I_0/A = qp_0L_h/\tau_h$, where L_h is the hole diffusion length and τ_h the hole lifetime (electron current is neglected). The noise factor is simply

$$F = \tau_h^*/\tau_h \qquad (13.5.10)$$

where τ_h^* is the radiative lifetime. Thus, if this ratio can be determined, one can simply estimate the detectivity from Fig. 20.

13.5.2.3. Photoconductive Detectors. In recent years most photoconductors show some form of g-r noise. Whereas the noise reflects simple intrinsic transitions in materials like Ge, Si, InSb, $Hg_{1-x}Cd_xTe$, the

[77] R. Hanbury-Brown and R. Q. Twiss, Nature (London) 178, 1046 (1956); also Proc. Roy. Soc. A243, 291 (1957).
[78] R. J. McIntyre, IEEE Trans. ED-13, 164 (1966).

noise spectra are quite complex in CdS, CdSe, GaAs, Si:Au, etc. Rather simple extrinsic noise is sometimes observed for various impurities in Ge or Si. If the g-r noise has only one relaxation mode, the detectivity is of the form

$$D_\lambda{}^* = \frac{\eta_i}{2h\nu} \left[\frac{\alpha}{p_0} \frac{\tau}{d} \right]^{1/2} \qquad (13.5.11)$$

where d is the thickness of the active region of the detector, p_0 is the hole density (or mutatis mutandis the electron density), τ is the transient lifetime, and α is a factor related to $\varkappa$ of Section 13.2.3. Note that the above formula is correct for all frequencies, even though the absolute responsivity and the noise fall off like $(1 + \omega^2\tau^2)^{-1/2}$.

For an extrinsic detector with hydrogen-like acceptor centers which are already largely ionized at $T = 0$ due to counterdoping—as often is the case—$\alpha \approx 1$. For a detector with intrinsic or quasi-intrinsic transitions (via Shockley–Read centers, for example) $\alpha = (n_0 + p_0)/n_0$. If the detector is p-type due to *ionized* impurities (which do not contribute themselves to the noise), $\alpha \approx p_0/n_0 \gg 1$. Thus, as pointed out by Long,[79] a marked increase in the detectivity over that of extrinsic detectors is obtainable providing the other quantities (τ, p_0, and d) are the same.

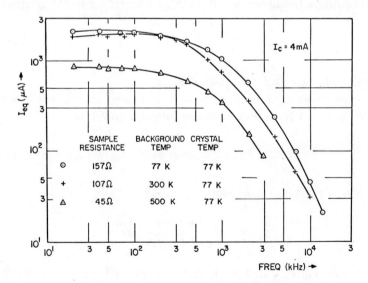

FIG. 21. An example of a noise spectrum observed in mercury-cadmium telluride [after R. N. Sharma and K. M. van Vliet, *Phys. Status Solidi* (a) **1**, 765 (1970)].

[79] D. Long, *Infrared Phys.* **7**, 121 (1967).

A photoconductive detector cooled to 77°K is easily found to be photon-limited when exposed to 300°K background. Detectivities of 10^{13} cm sec$^{-1/2}$ W^{-1} for a detector operating at 2μ are not uncommon. We do not know, however, if a detector exists at 77°K which has been found to be radiation-limited. For example, recent experiments on Hg$_{1-x}$Cd$_x$Te with a peak sensitivity at $15\,\mu$ showed "perfect" g-r noise, but the lifetime was a factor ~ 100 less than the calculated radiative lifetime. A typical noise spectrum of the photocurrent fluctuations for this material is given in Fig. 21.

13.A. Appendix

The purpose of this appendix is to provide a brief summary of some of the techniques and definitions used in Part 13. For a description of the laboratory techniques of noise measurement, see van der Ziel.[3]

13.A.1. Spectral Resolution of a Fluctuating Process

The total noise intensity $\overline{\Delta y^2}$ of a fluctuating process $y(t)$ can be resolved into a frequency spectrum $S_y(f)$ according to the relation

$$\overline{\Delta y^2} = \int_0^\infty S_y(f)\,df \qquad (13.A.1)$$

where $S_y(f)$ is called the spectral density.

The mean-squared fluctuation of a recording instrument observing this $y(t)$, with a frequency response function $G(f)$, is

$$\overline{\theta^2} = \int_0^\infty S_y(f)\,|\,G(f)\,|^2\,df \simeq S_y(f_0)\,|\,G(f_0)\,|^2\,B \qquad (13.A.2)$$

where B is the noise bandwidth, defined by

$$B = \int_0^\infty \left|\frac{G^2(f)}{G^2(f_0)}\right|\,df.$$

The second equality in Eq. (13.A.2) is reasonable only if $S_y(f)$ is relatively constant over the bandwidth B.

13.A.1.1. Calculation of Spectral Density. A random process $y(t)$ can be well-characterized by its correlation function $\phi(\delta) = \overline{y(t)y(t+\delta)}$. Ac-

cording to the Wiener–Khintchine theorem:[80,81]

$$S_y(f) = 4 \int_0^\infty \phi(\delta) \cos(2\pi f \delta) \, d\delta; \qquad (13.A.3)$$

also

$$\phi(\delta) = \int_0^\infty S(f) \cos(2\pi f \delta) \, df. \qquad (13.A.4)$$

If the correlation function is such that there is no correlation, except for extremely short times $\delta \simeq 0$, the process is called *purely random*. Equation (13.A.3) shows that in that case the spectrum is *white* or independent of frequency. If the correlation function is exponential so that $\phi(\delta) = \overline{y^2} \exp(-\delta/\tau)$, the process is called *Markoffian random*.[82] Then the spectrum is

$$S_y(f) = \overline{4y^2}\tau/(1 + \omega^2\tau^2) \qquad (13.A.5)$$

where $\omega = 2\pi f$.

13.A.1.2. Use of the Short Time Average. The average of a fluctuating process $y(t)$ over a period $(t, t + \theta)$ is a new stochastic variable depending on t:

$$y_\theta(t) = \frac{1}{\theta} \int_t^{t+\theta} y(\xi) \, d\xi. \qquad (13.A.6)$$

There are cases where this short time average can be interpreted more easily than the basic process $y(t)$ itself.

The mean-squared fluctuation of the short time average is

$$\overline{\Delta y_\theta^2} = \frac{1}{\theta^2} \overline{\left| \int_t^{t+\theta} \{y(\xi) - \bar{y}\} \, d\xi \right|^2}$$

$$= \frac{1}{\theta^2} \int_0^\theta \int_0^\theta \overline{\Delta y(\xi) \, \Delta y(\eta)} \, d\xi \, d\eta. \qquad (13.A.7)$$

Changing the variables one can show that

$$\overline{\Delta y_\theta^2} = \frac{2}{\theta^2} \int_0^\theta dt' \int_0^{t'} \overline{\Delta y(\xi) \, \Delta y(\xi + \delta)} \, d\delta. \qquad (13.A.8)$$

Combining this with Eq. (13.A.4) yields the relation between $\overline{\Delta y_\theta^2}$

[80] N. Wiener, *Acta Math. Stockholm* **55**, 117 (1930).
[81] A. Khintchine, *Math. Ann.* **109**, 604 (1934).
[82] M. C. Wang and G. E. Uhlenbeck, *Rev. Mod. Phys.* **17**, 323 (1945).

and $S_y(f)$ known as MacDonald's theorem.[83] In particular, if $S_y(f)$ is independent of frequency,

$$S_y(f) = 2\theta \, \overline{\Delta y_0{}^2}. \tag{13.A.9}$$

13.A.1.3. Calculation of Noise Signals. Current and voltage fluctuations in electrical devices and circuits are now considered. From a Fourier analysis of the noise it can readily be seen that *the normal theory of ac network analysis and of device electronics is applicable to noise signals. The noise output of any network can be calculated once the spectral intensities of the various noise sources are known.* It is common practice to represent the noise signals by current and voltage generators and to represent their rms value in a narrow frequency interval by $\sqrt{S_i(f)\,df} = \sqrt{\overline{i^2}}$ and $\sqrt{S_v(f)\,df} = \sqrt{\overline{v^2}}$ respectively.

Many other methods of analyzing fluctuating processes are known. The reader is referred to the text and references cited.[84–86]

13.A.2. Circuit Representation of Noise

13.A.2.1. Noise of Two-Terminal Networks. The noise of a two-terminal linear network or device in a narrow frequency interval can be represented either by a noise emf in series with the impedance of the network or by a noise current generator in parallel with the admittance of the device.

1. *Noise resistance.* The mean-squared value of any noise emf in a narrow frequency interval can be represented by the thermal noise of a fictitious resistor by using Nyquist's theorem formally. The result is

$$R_{eq} = \overline{e_n{}^2}/4kT\,\Delta f \tag{13.A.10}$$

where R_{eq} is called the *equivalent noise resistance.*

2. *Noise conductance.* Nyquist's theorem is also used to define the noise conductance. The mean-squared value of any noise current generator in a narrow frequency interval can be represented by the thermal noise of a fictitious conductance;

$$G_{eq} = \overline{i_n{}^2}/4kT\,\Delta f \tag{13.A.11}$$

where G_{eq} is called the *equivalent noise conductance.*

[83] D. K. C. MacDonald, *Rep. Progr. Phys.* **12**, 56 (1948).
[84] S. Chandrasekhar, *Rev. Mod. Phys.* **15**, 1 (1943).
[85] J. M. W. Milatz, *Ned. Tijdschr. Natuurk.* **8**, 19 (1941).
[86] S. O. Rice, *Bell Syst. Tech. J.* **23**, 282 (1944); **24**, 46 (1945).

3. *Saturated diode current.* Another way of representing the spectral intensity of a noise current generator is to apply Schottky's theorem and equate it to the shot noise of a saturated (or temperature-limited) diode. The result is

$$I_{eq} = \overline{i_n^2}/2q\,\Delta f. \tag{13.A.12}$$

13.A.2.2. Noise in Four-Terminal Networks. The discussion in the text of Part 13 has emphasized physical modeling of the noise sources in the "four-terminal devices" (tubes and transistors) discussed. This is essentially in agreement with the recommendations of the IEEE Subcommittee on Noise.[87] The reader is referred to the report of that subcommittee for a discussion of the representation of noise in linear two-ports for those cases when one cannot make a sensible physical noise model.

There are several derived methods of describing and/or comparing the noise performance of four-terminal networks in common use. Most refer to a single equivalent noise generator—or a parameter which describes its spectral density—at the input of an otherwise "noise-free" network.

1. *Noise factor (noise figure).* The noise factor, at a specified input frequency, is defined as the ratio of (1) the total noise power per unit bandwidth at a corresponding output port when the noise temperature of the input termination is standard (290°K) to (2) that portion of (1) caused by the thermal noise of the input termination.

1a. The noise figure in decibels is simply $F(\text{db}) = 10\log_{10}F$, where F is the noise factor defined above.

2. *Noise temperature.* The effective input noise temperature of a two-port transducer is defined as the temperature which, when the input was connected to a noise-free equivalent of the transducer, would result in the same output noise power as that of the actual transducer connected to a noise-free input termination. The effective noise temperature in degrees Kelvin is related to the noise factor by the expression

$$T_e = 290(F - 1). \tag{13.A.13}$$

3. *Noise resistance.* There are times when it is convenient to refer all of the noise of a transducer to a single noise emf in series with the input termination. This emf will, in general, depend on the impedance of the

[87] H. A. Haus *et al.*, IRE subcommittee 7.9 on noise. *Proc. IRE* **48**, 69 (1960).

input termination. The mean-squared value of this *single* emf can be expressed as an equivalent noise resistance of the two-port:

$$R_n = \overline{e_{\mathrm{eq}}^2}/4kT\,\Delta f. \tag{13.A.14}$$

4. *Noise conductance.* There are also occasions when the most expedient way of representing the noise of a two-port is with a *single* noise current generator in parallel with the admittance of the source termination seen by the transducer. The mean-squared value of this noise current generator can be expressed in terms of *equivalent noise conductance*:

$$G_N = \overline{i_{\mathrm{eq}}^2}/4kT\,\Delta f. \tag{13.A.15}$$

It is also possible to express the spectral density of this generator in terms of equivalent saturated diode current:

$$I_{\mathrm{eq}} = \overline{i_{\mathrm{eq}}^2}/2q\,\Delta f. \tag{13.A.16}$$

5. *Noise measure.* The noise measure of a transducer has been defined by Haus and Adler as[88]

$$M = \frac{F-1}{1 - 1/G_a} \tag{13.A.17}$$

where F is the noise factor as defined above and G_a is the available power gain.

The noise figure of several transducers connected in cascade is given by the Friis formula[89]

$$F_s = F_1 + \frac{F_2 - 1}{G_1} + \frac{F_3 - 1}{G_1 G_2} + \cdots. \tag{13.A.18}$$

If the expression for the noise measure is substituted into this formula, one finds

$$F_s \simeq 1 + M. \tag{13.A.19}$$

Haus and Adler have shown that the optimum noise measure of a device is invariant under reciprocal lossless circuit transformations. Thus, for example, common-base, common-emitter, and common-collector transistor stages can all be optimized to have the same cascaded noise figure.

[88] H. A. Haus and R. B. Adler, "Circuit Theory of Linear Noisy Networks." MIT Press, Cambridge, Massachusetts and Wiley, New York, 1959.

[89] H. T. Friis, *Proc. IRE* **32**, 419 (1944).

AUTHOR INDEX

Numbers in parentheses are footnote numbers. They are inserted to indicate that the reference to an author's work is cited with a footnote number and his name does not appear on that page.

A

Aasnaes, H. B., 55
Abragam, A., 133, 136(1)
Abramowitz, A., 58
Adam, J. P., 129
Adler, R. B., 500
Agouridis, D. C., 473
Aihara, S., 349
Aiken, W. R., 344
Alburger, D. E., 316
Alderman, D. W., 144
Alexandre, B., 124
Ammann, S. K., 55
Amsel, G., 173
Anderson, T. O., 230, 235(12, 14)
Anderson, W. A., 147
Angrist, S. W., 327
Antier, G., 124
Arque-Almaraz, H., 107
Aström B., 85

B

Bacon, K. F., 3
Baechtold, W., 473
Baelde, A., 469, 473(34)
Baicker, J. A., 313, 334
Baird, J. R., 327
Baird, P. G., 71, 73(58)
Baldinger, E., 6, 104, 110
Barber, R. C., 58
Barker, R. C., 370
Barlow, H. M., 289
Barnes, J. L., 381
Barthelémy, M., 129
Barton, G. W., 142

Beams, J. W., 305
Beck, A. H. W., 265
Becker, E. D., 144
Becking, A. G. T., 473
Bell, P. R., 316
Bell, R. E., 32, 33(7), 89, 112, 115
Belove, C., 165
Bennett, W. R., 462
Berringer, L., 6
Bertolaccini, M., 31, 33(3), 110
Best, G. C., 131
Betz, N. A., 129
Bianchetti, G., 210
Bianchi, G., 131
Biggerstaff, J. A., 113
Birks, J. B., 316
Bishop, R. L., 58
Blackwell, L. A., 285
Blalock, T. V., 110
Blankenship, J. L., 110
Bloch, F., 135, 138
Bloembergen, N., 137, 150
Blum, H., 151
Boardman, C. J., 237
Boeschoten, F., 309
Bonitz, M., 35, 36
Boothroyd, A. R., 474
Boyce, D. A., 114
Breckenridge, R. G., 323
Breune, W., 44
Bridenbough, P. M., 141
Bristow, Q., 12
Brodersen, A. J., 467, 485
Brophy, J. J., 323
Brower, R., 75
Brown, G. S., 375
Brown, H. E., 357
Brown, R. E., 372

SUBJECT INDEX

A

Ac measurements, 63–70
  average or peak, 67–70
  common-mode rejection ratio in, 74
  common-mode voltage in, 74
  current comparators in, 70
  digital value in, 76
  "guarding" in, 74
  harmonics errors in, 69–70
  Johnson noise in, 72
  noise factors in, 71–76
  operational rectifiers in, 68–69
  oscilloscopes in, 71
  phase-lock and signal averaging in, 75–76
  rectifier instruments in, 68
  shielding and guarding in, 73–74
  square-law characteristics of, 67
  thermoelements and transfer standards in, 63–65
Ac motors, for servomechanisms, 439–440
Ac voltmeter, 64
ADC, *see* Analog-digital converters
Ammeter, electrodynamic, 64
Amplidyne, 442
Amplifier(s)
  in dc measurement, 50
  difference, 449
  integrated, 485
  lock-in, 75
  magnetic, *see* Magnetic amplifiers
  noise in, 482–486
  operational, 446–452
  in servomechanisms, 442
  tuned, 482
  wideband, 482
Amplifier gain, in pulse amplitude measurement, 81
Amplitude channels, 81
Analog computers, 152–165
  applications of, 152
  computing elements in, 154

function generator in, 156, 162
operational amplifier in, 155
problem variables in, 153
quarter-square multiplier in, 159
Analog computing systems, components of, 153–160
Analog-digital converters
  digital data acquisition in, 121–122
  fast, 120–121
  vs. multidiscriminator systems, 119–120
  multi-user system in, 129
  for pulse spectrometry, 128–129
Analog indicators, in dc measurements, 49
Analog techniques, in simulation and signal processing, 160–165
Analog-to-digital converters, in pulse spectrometry, 82
Antennas
  in microwave circuits, 263
  patterns and gain in, 298–300
Arithmetic unit, in pulse amplitude measurement, 122
Associative analyzer, in pulse amplitude measurement, 128
Associative memories, in pulse spectrometer, 127–128
ATR tube, in microwave circuits, 285–286
Attenuation measurements, microwave, 295–296
Avalanche diodes, 275–277
  applications of, 280–282
  power generation capabilities of, 279
Avalanche multiplication noise, 470
Avalanche photodiode, 327–328

B

Baseline shift, in pulse amplitude measurements, 97–98
Binary counter, 3
Bi-orthogonal code, in telemetering, 224
Bipolar transistors, noise in, 471–479